Advanced Synthesis *and* Processing *of* Composites *and* Advanced Ceramics

Related titles published by the American Ceramic Society:

High-Temperature Ceramic-Matrix Composites II: Manufacturing and Materials Development (Ceramic Transactions Volume 58)
Edited by A.G. Evans and R. Naslain
© 1995, ISBN 0-944904-99-8

High-Temperature Ceramic-Matrix Composites I: Design, Durability, and Performance (Ceramic Transactions Volume 57)
Edited by A.G. Evans and R. Naslain
© 1995, ISBN 0-944904-98-X

Low-Expansion Materials (Ceramic Transactions Volume 52)
Edited by David P. Stinton and Santosh Y. Limaye
© 1995, ISBN 0-944904-92-0

Design for Manufacturability of Ceramic Components (Ceramic Transactions Volume 50)
Edited by Asish Ghosh, Basavaraj Hiremath, and John Halloran
© 1995, ISBN 0-944904-88-2

Manufacture of Ceramic Components (Ceramic Transactions Volume 49)
Edited by Basavaraj Hiremath, Allan Bruce, and Asish Ghosh
© 1995, ISBN 0-944904-87-4

Bioceramics: Materials and Applications (Ceramic Transactions Volume 48)
Edited by Gary Fischman, Alexis Clare, and Larry Hench
© 1995, ISBN 0-944904-82-3

Advances in Ceramic-Matrix Composites II (Ceramic Transactions Volume 46)
Edited by J.P. Singh and Narottam P. Bansal
© 1994, ISBN 0-944904-80-7

Advanced Synthesis *and* Processing *of* Composites *and* Advanced Ceramics

▼▲▼

Edited by
Kathryn V. Logan
Georgia Institute of Technology

▲▼▲

Ceramic Transactions
Volume 56

Published by
The American Ceramic Society
735 Ceramic Place
Westerville, Ohio 43081

Proceedings of the International Symposium on Advanced Synthesis and Processing, held January 9–11, 1995, in Cocoa Beach, FL.

Library of Congress Cataloging-in-Publication Data

Advanced synthesis and processing of composites and advanced ceramics / edited by Kathryn V. Logan.
p. cm. -- (Ceramic transactions ; v. 56)
Proceedings of the International Symposium on Advanced Synthesis and Processing, heald Jan. 9-11, 1995, Cocoa Beach, Fla.
Includes index.
ISBN 1-57498-000-9
1. Composite materials--Congresses. 2. Ceramic materials--Congresses. I. Logan, Kathryn V. II. International Symposium on Advanced Synthesis and Processing (1995 : Cocoa Beach, Fla.) III. Series.
TA418.9.C6A2832 1995
666--dc20 95-24463
CIP

For information on ordering titles published by the American Ceramic Society, or to request a publications catalog, please call 614-890-4700.

Printed in the United States of America.

1 2 3 4–99 98 97 96 95

ISSN 1042-1122
ISBN 1-57498-000-9

Contents

Advanced Processing of Composites

Novel Synthesis

Nanotechnology and Advanced Characterization

Sol-Gel Technology

Preface

The First International Symposium on Advanced Synthesis and Processing was held in conjunction with the Fall Meeting of the Engineering Ceramics Division of the American Ceramic Society, held January 8–12, 1995, in Cocoa Beach, FL.

The Fall Meeting was attended by an international audience of over 513 participants from academia, government, and industry. A total of 47 papers were presented during five sessions with eight invited speakers.

The symposium was cosponsored by the U.S. Army Research Office, Research Triangle Park, NC; the Engineering Ceramics Division of the American Ceramic Society; and the National Institute of Ceramic Engineers.

Numerous individuals contributed to the success of this year's conference and are due many thanks. The 1995 International Symposium Committee, which participated in the symposium planning and implementation, includes K.V. Logan, N. Thadhani, Z.A. Munir, R.M. Spriggs, A. Crowson, E. Chen, and I. Ahmad. The session cochairs, K.V. Logan, Z.A. Munir, I. Ahmad, T.L. Starr, H. Marcus, N. Thadhani, R.M. Spriggs, K. Niihara, E. Chen, and S. Sakka, organized and conducted sessions covering state-of-the-art technical information. Special acknowledgments are due Z.A. Munir and R.M. Spriggs, who served as coeditors of the proceedings.

The American Ceramic Society headquarters staff, including Paul Holbrook (Executive Director), Bill Douglas (Meetings Director), Jackie Davis (Technical Meetings Coordinator), and Lori Kozey (Production Coordinator, Books & Journals), ensured the success of the symposium by providing very appreciated organizational support and assistance.

The proceedings are organized by session topic as they occurred in the program and contain 38 papers from the conference.

Kathryn V. Logan

Combustion Synthesis

HISTORY OF AND NEW DEVELOPMENTS IN SHS

A.G. Merzhanov

Institute of Structural Macrokinetics,
Russian Academy of Science, Chernogolovka, Moscow oblast', 142432 Russia

ABSTRACT

The lecture focuses on the history of the discovery of the solid flame phenomenon, development of SHS, and on the arising of a new individual branch of research and development on the basis of this phenomenon. The precursors and real history of SHS are considered. The current infrastructure of SHS and the development of its individual branches are analyzed. Fundamentals of synthesis, production, advanced materials (homogeneous and inhomogeneous), and application of SHS products will be presented. Key points are illustrated by the newest accomplishments in different laboratories of the world. The organization of research in Russia, the USA, Japan, China, and other countries is analyzed in brief. Information about international links in the field (cooperation, meetings) is presented. An attempt is undertaken to outline some promising research avenues and current problems.

INTRODUCTION

In view of recent accomplishments in research and practical applications of Self-propagating High-temperature Synthesis (SHS), there is growing interest in the history of the discovery and development of these processes.

In the paper of Prof. Hlavacek [1], the rich collection of Western papers published in the nineteenth and twentieth centuries has been compiled; these papers concern the exothermic effects in the synthesis of materials. Among them, there is the famous work of Goldschmidt [2] on the out-of-furnace metallothermy. However, this collection cannot be regarded as the complete SHS history for several reasons. One is the fact that the early works cited had no influence on the subsequent ones. These studies were carried out independently and did not result in the formulation of the methodology and prompts for further research. For this reason, this review has to be called a collection but not a history. Nevertheless, the knowledge of these works is interesting and instructive. Unfortunately, the author of [1] did not mention similar works of Russian researchers, in particular, those of Prof. Beketov [3], founder of aluminothermy whose work preceded that of Goldschmidt.

It should also be noted that SHS differs from the aforementioned processes by: a distinctly defined combustion front and the conditions controlling its traveling; a deep insight into the interrelation between the processes and product characteristics, on one side, and the parameters of the "green mixture--environment", on the other; and the possibility of controlling the processes to optimize the synthesis conditions. Such an approach became possible only after the development of modern combustion theory that established the nature and mechanism of self-propagating chemical reactions. No wonder that SHS has prospered only in the 60s of the twentith century and not earlier, in spite of the existence of precursors.

In this paper, we consider the real history of the discovery and development of SHS along with the newest achievements and promising research avenues.

1. SHORT HISTORY OF SHS DISCOVERY

A real history of SHS is closely related to the discovery of the so-called "solid flame". In the 60s, the studies on the mechanism of combustion in gasifying condensed systems -- such as explosives, gun powders, and solid propellants -- were carried out in Chernogolovka (near Moscow) at the Research Center of the USSR Academy of Sciences. At that time, the role of gasification in combustion was vigorously argued. To check one of the hypotheses, the gasless iron-aluminum thermite diluted with alumina was prepared [4]. After combustion, the sample retained its initial weight because starting reactants and products remained in their condensed state during combustion. In contrast to gasifying systems, gasless combustion was consistent with the predictions of the simplest combustion theory. It was unusual and interesting.

A search for new gasless systems has lead to a remarkable discovery later called the "solid flame" phenomenon [5]. The solid flame is a solid-state combustion when reactants and products remain in their solid state even at high combustion temperatures. Poorly-reactive powders of refractory metals (niobium, tantalum, molybdenum, tungsten, etc.) and nonmetals (boron, carbon) served as reactants whereas the products were refractory compounds (borides, carbides). It soon became clear that the solid flame is an efficient method for obtaining refractory materials; the method was termed Self-propagating High-temperature Synthesis (SHS) [6].

It follows that the studies on the mechanism and theory of combustion in gasifying condensed systems as well as the experience acquired in the combustion of thermites played a key role in the build-up of SHS. The real history of SHS is schematically presented in Figure 1.

As SHS research progressed, the primary approach confined by the framework of solid flame turned out to be too limited. Further advance in research and the popularity of the term SHS has lead to a situation where the experimental work in adjacent fields began to be carried out under the "flag of SHS". The author happened to see the publications in which the classical metallothermy known since Beketov and Goldschmidt was considered as a special case of SHS. As a result, SHS unintentionally (not de jure, but de facto) left the strict frames of solid flame. To date, any combustion processes yielding valuable compounds and materials have been classified as SHS.

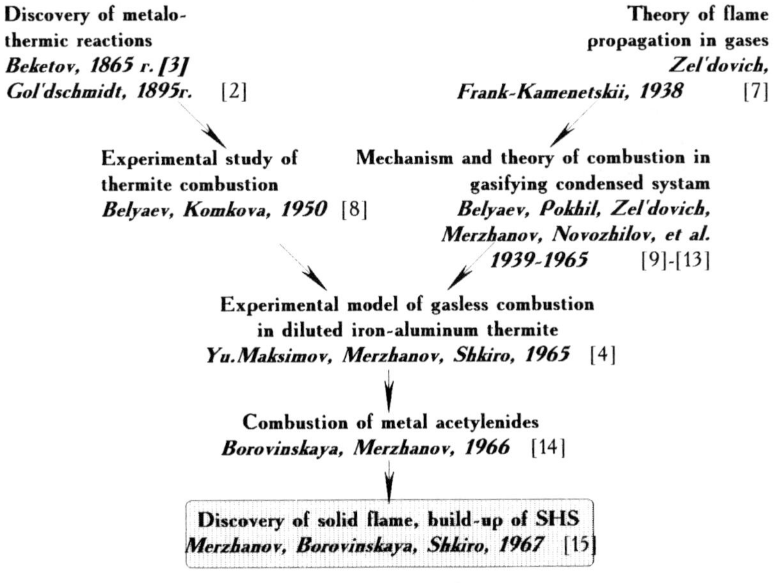

Fig. 1. Real history of SHS discovery.

2. CHRONOLOGY OF SHS DEVELOPMENT

Owing to its interesting mechanism and technical efficiency, SHS has been studied extensively. The chronology of SHS research in the former Soviet Union is schematically presented in Figure 2. Indicated are the dates of the first experimental, theoretical, and technological studies that prompted further research in various fields of SHS, as well as the stages of organizational activities. It is interesting to note that, during first 12 years, these studies were carried out in the former Soviet Union without special financial support merely on initiative of enthusiastic research workers. SHS received governmental support only in 1979. This support was used to erect new buildings, to develop a state research program, to organize a scientific council, to open a new research institute, and to launch industrial-scale production. Today, SHS entered a new stage of nongovernmental, commercial development.

SHS studies outside Russia were initiated in 1980, long after the appearance of first Soviet works. This delay might have been longer. We, Russians, consider that the brilliant review of Joe Crider [16] played the important stimulating role. In his paper, the author has given a vivid description of the birth and development of SHS in the former Soviet Union. Crider's review is another type of historical sketch.

An important role in the promotion of SHS research belongs to Dr. Holt and Dr. McCauley who performed the first SHS studies in the USA [17, 18]. Later, of significant importance were the results, lectures, and reviews by Prof. Munir [19, 20].

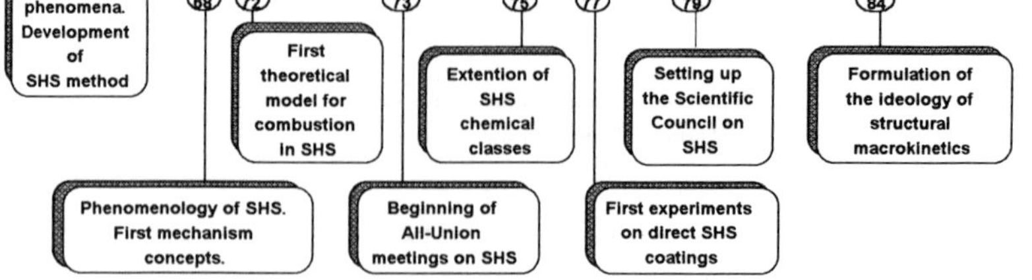

Fig. 2. Chronology of SHS: Soviet stage.

At present, when SHS studies are spread all over the world (Fig. 3), the question arises "What was the reason for the delay that allowed Soviet researchers to fix their priority and become the leaders in the field?". Many experts consider it to be a consequence of lack of belief in the potentialities of SHS, its exotic nature, and apparent impossibility to obtain high-quality materials.

1980	**U.S.A.**
1984	**Japan**
1985	**Poland**
1990-1994	**Canada, China, India, Taiwan, Spain, Brazil, Korea, France, U.K.,Finland, Italy, Holland**

Fig. 3. Chronology of SHS: International era

For this reason, it does not seem surprising that the researchers of the former Soviet Union have become the founders of not only SHS per se, but also of its individual branches. Now let us consider some of them.

3. FUNDAMENTALS OF SHS

In 1968 -1972, Merzhanov, Borovinskaya, Khaikin, et al. [6, 21 - 24] formulated the primary concepts of the combustion mechanism in SHS systems on the basis of the experimental data on

- gasless combustion in the powder mixtures of metals with carbon, boron, and silicon
- filtration combustion of porous metal samples in nitrogen gas.

These concepts have played an important role in formulating the fundamentals of SHS. In these works, the notion and the term "filtration combustion" was introduced for the first time, the notions of kinetic autoretardation of heat release and of broad reaction zones were formulated, spinning waves were discovered, etc. Later on, the following directions of SHS research were successfully developed:

- Patterns of front propagation, unsteady modes (*Borovinskaya, Shkiro, Filonenko, Holt, Munir, et al.*)
- Temperature profiles in SHS waves (*Zenin, Mal'tsev, et al.*).
- Initiation of SHS (*Barzykin, Hardt, Yuan, et al.*)
- Theory of combustion in SHS systems (*Khaikin, Aldushin, Novozhilov, et al.*)
- Mathematical modeling of SHS (*Shkadinskii, Matkowsky, Hlavacek, et al.*)
- Thermodynamic calculations (*Gold'shleger, Mamyan, Shiryaev, Sata, et al.*)

- Kinetics of high-temperature reactions (*Grigor'ev, Kharatyan, Shteinberg, et al.*)
- Mechanical stimulation of SHS (*Yukhvid, Stolin, et al.*)
- Electric and electromagnetic stimulation of SHS (*Shteinberg, Clark, Munir*)

During past years, advances have been made in studies of the processes behind the front and the structure of combustion products. These processes are unique to SHS; as a rule, they do not occur in the combustion in other systems. As a result, the methodology of a new research avenue has been formulated; within this avenue, the relationships between the rate of chemical reactions, structure, phase transformations, heat, and mass transfer are studied. This scientific discipline was termed "structural macrokinetics" [25, 26].

The advance in the studies on the mechanism of SHS is impressive. To date, three levels of diagnostics have been developed and accepted.

The first level is based on the phenomenological approach. Here one confines oneself to the observation of the pattern of front propagation and to the analysis of end products. This is a simplest and most accessible approach. Within this approach, the primary optimization of SHS can be readily achieved using the process parameters.

The second level allows us to determine the zone structure of the combustion front and to interpret some processes. The work at the 2nd level requires some experimental skill (for example, in preparing thermocouples with a junction 3 - 5 μm thick).

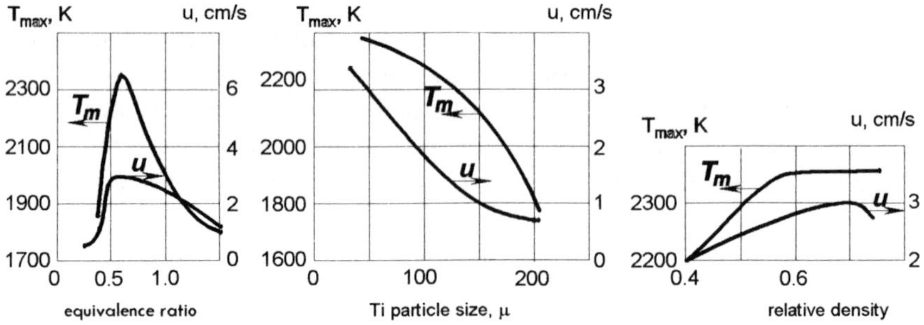

Fig. 4. Characteristics of combustion in the 5Ti + 3Si system.

The third level is the most intricate one. It requires complex and hard-to-access experimental techniques (synchrotron radiation, dynamic X-ray diffraction analysis, and/or quenching technique). As a reward, a researcher obtains direct information about the chemical and phase transformations in the combustion wave.

In Figures 4-6, the examples of the three-level approach to the study of the same system are given. Figure 4 shows the front propagation velocity and the maximum combustion temperature versus some parameters of the system [27].

The combustion thermogram, that is, the temperature profile in the wave [28], is shown in Figure 5. Arrows indicate special points and portions of the profile that admit some certain physicochemical interpretation.

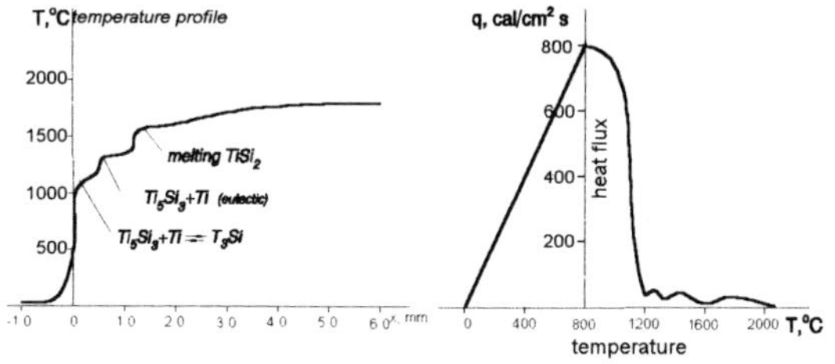

Fig. 5. Thermogram of combustion in the 5Ti + 3Si system.

Figure 6 shows the X-ray diffraction patterns recorded in the course of SHS process [29]. Using these data, we can obtain the concentration profiles for the phases formed in the wave and thus directly observe the kinetics of SHS. Detailed three-level studies give the complete information about SHS in specific systems.

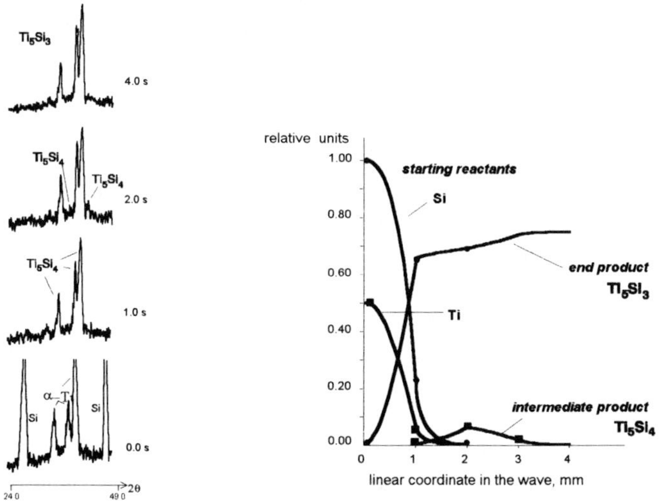

Fig. 6. Dynamic X-ray diffraction analysis of combustion in the 5Ti + 3Si system.

4. CHEMICAL SYNTHESES. TECHNOLOGY OF SHS POWDERS

The first synthesis and analysis of SHS products were carried out by Borovinskaya et al. in 1969 - 1971. These studies aimed at establishing the relationship between the chemical and phase composition of products and the composition of the green mixture and combustion conditions in order to optimize the formation of one-phase products of minimum contamination. Synthesized were nitrides, carbides, borides, and silicides of transition metals (Ti, Zr, Hf, Nb, Ta, etc.) [6, 21].

Later, the classes of SHS products were markedly extended. The following compounds were synthesized by different workers:

- nonmetal refractory compounds (*Borovinskaya, Miyamoto, Pampuch, Puszhynski, et al.*)
- intermetallics (*Itin, Nayborodenko, Kaieda, et al.*)
- hydrides (*Dolukhanyan, Nersesyan, et al.*)
- complex oxides (*Boldyrev, Vereshchagin, Nersesyan, Patil, et al.*)
- chalcogenides (*Prokudina, Kazarbina, et al.*)
- phosphides (*Chernogorenko, Muchnik*)

Since 1975, the metallothermic processes have begun to be used for synthesizing both individual compounds and composites (*Borovinskaya, Mamyan, Yukhvid, Logan, Moore, et al.*). During past years, much attention has been paid to the exploration of the mechanism of complex chemical reactions in SHS waves with the elucidation and identification of intermediate products and intermediate stages [30, 31].

The pilot-scale production of SHS powders was initiated in 1972 when an SHS reactor 20 l in volume was built in Chernogolovka (*Borovinskaya, Ratnikov, Prokudina, Merzhanov*). The green mixture preparation and processing of end products were performed by conventional powder metallurgy. The productivity of the first pilot-scale installation for manufacturing elemental powders (TiC, TiN, TiC-TiN, TiB_2, AlN, Si_3N_4, $MoSi_2$) was 10 - 15 tons per year. An important conclusion drawn was that SHS in large volumes was more efficient than in small ones: the quality of these products was higher than that of the laboratory-made ones.

Later, due to the efforts of Chernogolovka researchers, the technology of SHS powders was considerably improved: better SHS reactors were designed, modernized accessories were used, processing charts for numerous products were developed and new efficient processing methods for combustion products were found.

However, in spite of the achievements, the modern state-of-art in the technology of SHS powders should be qualified as pilot-plant scale. Work on construction of a highly automated in-line production of SHS powders is now in progress.

The important problem is manufacturing ultrafine and nanoscale powders. For a long time, it was thought that only large particles (several microns and above) could be obtained by the SHS method. And accordingly, they were thought to be unsuitable for sintering. But it was found recently that by controlling structure and improving the dispersion procedure for the primary SHS products (cakes, ingots), finely dispersed powders (0.1 - 1.0 microns) could be obtained.

A new method for processing the cakes of nonmetal nitrides -- so-called chemical dispersion -- was suggested by Prof. Borovinskaya [31]. A cake of SHS product is placed in a liquid of special composition at some specified temperature. The defect layers at the crystallite boundaries are dissolved, whereas the crystallites not. As a result, the cake disintegrates giving single-crystalline particles of the crystallite size. By chemical dispersion, different fine-grained powders were obtained: B_4C - specific surface 8.0 m^2/g, SiC - 11.0 m^2/g, Si_3N_4(α-88%) - 11.5 m^2/g, BN - 14.0 m^2/g.

Attack on the size of SHS particles is being continued. Recently, Prof. Yuan and co-workers obtained nanometric NiAl crystallites (5 - 70 nm) using mechanical alloying -- an intensive mechanical treatment of the green mixture (see [32]).

Of interest is the reverse problem -- an increase in the size of single-crystalline particles. A decrease in the cooling rate for combustion products enhances the extent of recrystallization. Performing combustion in 1 ton of green mixture, single-crystalline particles of TiC about 1mm in their size were obtained at ISMAN [31].

Obtaining ultrafine and nanometric particles as well as large single crystals illustrates the possibility of controlling SHS process in solving specific problems of materials science.

Controlling the particle size and shape in SHS products is one of the most important and constantly developing research avenues. Recently, the optimum conditions for the synthesis of single-crystalline (plate-like) particles of hexagonal BN have been found (Fig. 7). The synthesis was carried out in a 30 l SHS reactor with 100-% yield. The powders showed high quality. This was achieved by using special additives to initiate the gas-phase stages.

Composition, wt%
Nitrogen, N. 55.8
Boron total, B 43.5
Boron free, B_f. 0.04
Borous anhydride, B_2O_3. . . . . .0.04

Specific surface, m^2/g 11.9

Fig. 7. Single-crystalline SHS powders of BN.

The most important quality parameters of ultradispersed powders are their sinterability and the properties of sintered items[34].

5. DIRECT SYNTHESIS OF MATERIALS, ITEMS, AND COATINGS BY SHS

1975 turned out to be a decisive year in the development of SHS technology. The first experiments on the direct (net-shape) SHS production of consolidated (pore-free) materials and items were carried out. The idea of forced compaction (pressing) of still hot combustion product was developed. The synthesis of material components, its structuring, and, in some cases, its shaping were carried out in a single-stage process. Therefore, there was no need in comminuting an SHS product to obtain powder and then use this powder to obtain material or finished parts by powder metallurgy. For the first time, these data were published in 1981 [35].

Initially, forced compaction of SHS products was carried out by uniaxial pressing (*Merzhanov, Borovinskaya, Ratnikov*) using simple press dies. Later, the dies were modified to complicated units (*Ratnikov, Kvanin*). The revolutionary significance of the first experiments on SHS pressing was confirmed by the considerable interest shown by other workers. Through the efforts of Soviet, American, and Japanese workers, the following modifications of consolidating combustion products were elaborated:

- SHS + high gas pressure (porous items) — *Borovinskaya, Martynenko, Loryan*
- SHS + high gas pressure (pore-free items) — *Miyamoto, Koizumi, et al.*
- SHS + rolling — *Rice, Osipov, et al.*
- SHS + extrusion — *Stolin, Merzhanov, et al.*
- SHS + forging — *Meyers et al.*
- SHS + pressing (large-scale items) — *Kvanin, Balikhina, et al.*
- SHS + explosion — *Adadurov, Gordopolov, Niiler, Thadhany, Sawaoka, Rabin, et al.*

In parallel, studies were carried out in another direction but with the same objective -- the deposition of high-temperature SHS melts to obtain surface layers and coatings (*Yukhvid, Merzhanov, Odawara, et al.*). An important accomplishment here is the centrifugal casting of large-scale pipes (*Odawara et al.* [36]). To date, a wealth of important results has been obtained. Many of them have become widely known.

One of the newest achievements here is manufacturing hard-alloy items of a given surface configuration by SHS pressing [37]. An example is presented in Fig. 8. The surface acquires the shape of the press die. The processing procedure involves some special stages to provide the required quality of the surface.

For a long time, SHS technology was aimed at obtaining homogeneous materials. The homogeneity of an SHS product was the focus to prove that SHS technology produced material equivalent to that made in a furnace.

Fig. 8. SHS hard-alloy items of predetermined surface configuration.

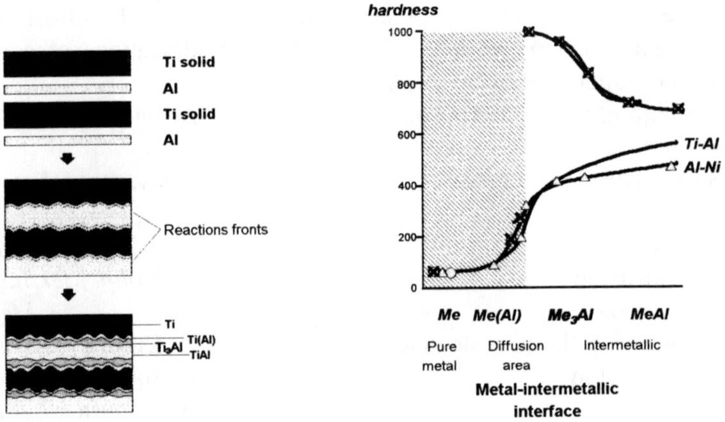

Fig. 9. Multilayered items.

But it recently became clear that the preparation of SHS materials with internationally preset inhomogeneity is a yet more important and interesting task. Of great success were the efforts of Japanese and Russian workers in preparing functionally gradient materials. A new accomplishment in the field [38] is illustrated in Figure 9. Using coils as reactants, the SHS method was applied to obtain a multilayered material.

The method of SHS pressing (initially used for the net-shape manufacturing of STIM hard alloys) is now being applied to consolidate materials of other classes and function (e.g., building and refractory ceramics, etc.). Planar heating elements made of molybdenum disilicide were prepared by this method. These elements are to be used in the heaters of unconventional configuration [39].

At the end of the seventies, work on the direct deposition of coatings by the SHS method was initiated. The following two methods were investigated:

- the formation of SHS melts on the surface of an item to be coated (SHS surfacing, *Yukhvid, Borovinskaya, Merzhanov*),
- the transfer of reactants via the gas phase to the item surface in the combustion wave (gas-transport SHS coatings, *Shtessel, Kostogorov, Merzhanov*).

Later, the following techniques were suggested:

- induction SHS surfacing *Yukhvid et al.*

- deposition of coatings in chain flames (at low temperatures) — *Azatyan, Aivazyan, Merzhanov*
- formation of coatings in chemical furnaces — *Ohianagi, Koizumi, Miyamoto, et al.*

By applying these methods for direct deposition of SHS coatings, the coating thickness can be varied within the following limits:

chain flames (<100nm)	<	*gas-transport reactions (10-100 μm)*	<	*induction SHS coating (0.1-1.0 mm)*	<	*SHS surfacing (>1 mm)*

6. INDUSTRIAL SCALE PRODUCTION AND APPLICATIONS OF SHS PRODUCTS

Attempts to find industrial application for SHS products were undertaken long ago, just after launching the pilot-scale production in Chernogolovka. Not everything was immediately OK. Initially, the unconventional structure of SHS powders posed difficulties for their application.

The first success came in 1973 when the abrasive properties of SHS-produced titanium carbide were discovered [40]. Then grinding powders and pastes based on them were prepared, and, in a short time, their industrial-scale production was developed. These abrasives were used in about 1500 different plants in the Soviet Union. This work gave a large economic benefit and for the first time demonstrated the great practical significance of SHS.

Later work proceeded in different directions. SHS products found their application in mechanical engineering, metallurgy, the chemical industry, aerospace engineering, electronics, and even in medicine. To date, applications have become numerous. And this strongly promotes further development of SHS.

Of course, important problems are: the commercialization of SHS, the development of new production lines and new marketable products. These are serious problems. To organize production and find a market for products are much more difficult to do than to carry out research and development. For this reason, this trend in SHS is developing more slowly.

The first industrial-scale production of SHS products was set up in Armenia in 1979, at the Kirovakan Plant of High-Temperature Heaters (Pogosyan, Sarkisyan, Dolukhanyan, Borovinskaya, Merzhanov). At this plant, induction furnaces producing molybdenum disilicide were replaced by SHS equipment. The effect was significant: reduction in the workspace, increase in labor productivity, improvement of the product quality. All this served as evidence for the profitability for SHS.

Later, industrial-scale production of titanium carbide powder and abrasive pastes was set up in Azerbaidzhan. After this, a governmental program of setting up other SHS production lines was adopted in the former Soviet Union. As a result, new SHS production lines for nitrided ferroalloys (*Maksimov, Lebedev*), ceramic insulators (*Kurchatkin, Borovinskaya*), and ferrite powders (*Saldugei, Avakyan, et al.*) were put into action.

To date, SHS production lines have been set up in Japan and the United States. Especially successful were the efforts of Dr. Kaieda who organized the production of SHS titanium nickelide (a material with shape memory) and a wire made of this material [41].

One recent achievement was setting up the industrial-scale production of SHS ferrites. It is based on the following reaction scheme:

$$2\alpha\mathrm{Fe} + (1-\alpha)\mathrm{Fe_2O_3} + \sum_{i=1}^{n} m_i(\mathrm{Me_2O}_R)_i + 1.5x\mathrm{O_2} \rightarrow \mathrm{Fe_2O_3} \cdot \prod_{i=1}^{n} m_i(\mathrm{Me_2O}_R)_i$$

This reaction is carried out in a self-propagating mode. The main combustion reaction is:

$$2\mathrm{Fe} + 1.5\mathrm{O_2} \rightarrow \mathrm{Fe_2O_3}.$$

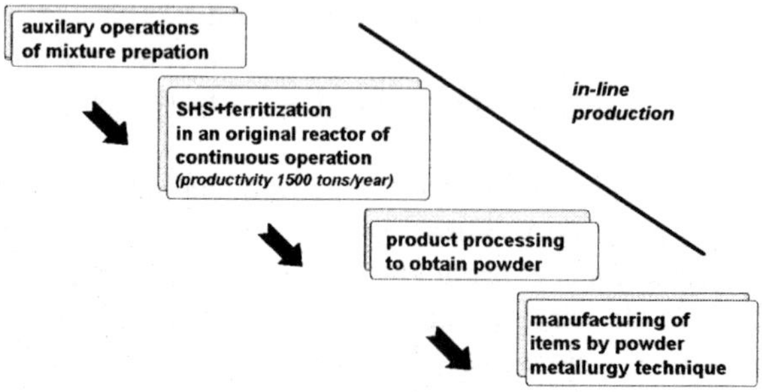

Fig. 10. Schematic diagram of the production line for manufacturing SHS ferrites.

The production line is schematically represented in Figure 10.

This production differs from others in the fact that for the first time the principle of continuous combustion was utilized. In the reactor, the green mixture is continuously supplied to the reaction zone, and product (ferrite) is continuously removed from the output. The productivity of such a unit is fifteen hundred tons per year.

The productivities of the traditional furnace and SHS technologies are not comparable. The efficiency of SHS production becomes more evident if we note that the individual stage of ferritization is absent in the SHS production method (it occurs simultaneously with SHS).

The following SHS ferrites are now being produced on an industrial scale: $Ni_{0.45}Zn_{0.42}Cu_{0.13}\cdot Fe_2O_4$, $Ni_{0.35}Zn_{0.65}\cdot Fe_2O$, $Mn_{0.72}Zn_{0.26}Ni_{0.02}\cdot Fe_2O_4$, $Mn_{0.725}Zn_{0.275}\cdot Fe_2O_4$, $BaFe_{12}O_{19}$, $SrFe_{12}O_{19}$

Table 1. Economical Efficiency of SHS powders [42]

SHS powder	Production cost with respect to the world prices				
	Russia*	U.S.A.	Korea	China	India
Elemental powders:					
aluminum nitride	0.10-0.12	0.16-0.20	0.13-0.15	0.10-0.13	0.10-0.12
silicon nitride	0.15-0.18	0.27-0.30	-	-	-
sialon	0.10-1.12	-	-	-	-
Chromium carbide, cast	0.15-0.22	-	-	0.25-0.29	-
Composite: titanium chromium carbide + nickel	0.10-0.12	-	-	-	-
Magnesiothermal powders:					
titanium diboride	0.20-0.25	-	-	0.39-0.41	-
boron nitride	0.20-0.25	0.75-0.80	-	0.28-0.31	0.31-0.33
Ferrites	0.25-0.30	-	-	0.45-0.47	-

*Production cost in Russia refers to the exchange rate on Oct. 1, 1994.

During past years, advances have been made in economical design of SHS system. The ISMAN experience in the field has been summarized [42]. Some data are presented in Table 1. The production costs are given here with respect to world prices. The calculations were carried out with due regard for the production conditions in various countries (the level of salary, the cost of electric power, raw materials, and public utilities). A simple conclusion is evident -- SHS production is profitable everywhere but especially so in the countries with inexpensive labor (Russia, India, China). The work on economical design is in progress. The author is confident that industrial-scale SHS production and SHS products will find a proper place in the worlwide market.

7. CURRENT INFRASTRUCTURE OF SHS. LEADING RESEARCH CENTERS

To date, SHS has become an individual branch of research and development. Advance in research and applications resulted in structuring of SHS, in the appearance of different research avenues. Figure 11 illustrates the current infrastructure of SHS as an individual field of research and development.

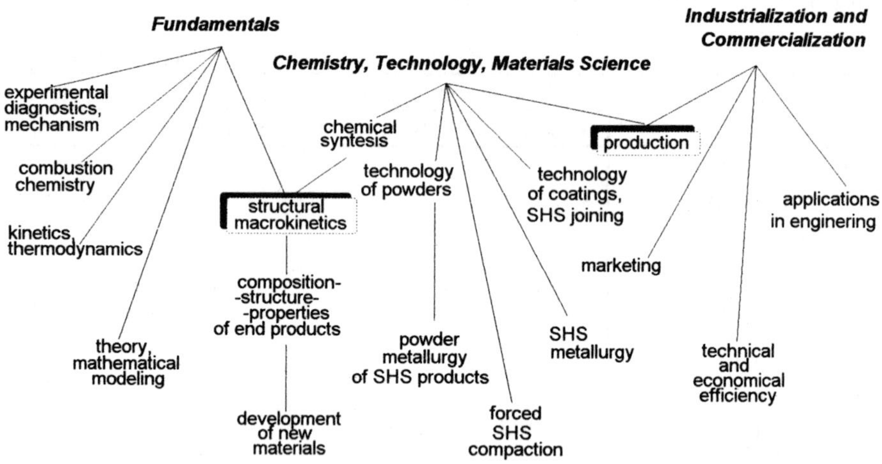

Fig. 11. Current infrastructure of SHS.

The major role in the development of SHS infrastructure belongs to the worldwide SHS research centers. Table 2 lists only the most important of them.

Good traditions in SHS research grew up in Russia. They can be characterized by a breadth of approach, by a search for new unconventional solutions, and by the eagerness to obtain the final result. Table 2 indicates those centers that continue their activities after the dissolution of the USSR.

Table 2. Worldwide leading SHS centers

Former SU

Institute of Structural Macrokinetics
Chernogolovka
A.Merzhanov

Tomsk Branch of the Institute of Structural Macrokinetics
Tomsk
Yu.Maksimov

Scientific-Education Center of SHS
MISIS-ISMAN **Moscow**
E.Levashov, I.Borovinskaya

Research Center on SHS
Samara
A.Amosov

Institute of Combustion Problems
Alma-Ata
G.Ksandopulo

Institute of Chemical Physics
Erevan
S.Kharatyan

U.S.A.

University of California
Davis *Z.Munir*

Alfred University
Alfred *R.Spriggs*

University of Notre Dame
Notre Dame
A.Varma

Northwestern University
Evanstone
B.Matkowsky

Technological Institute
Atlanta
K. Logan

University at Buffalo
Buffalo *V.Hlavacek*

Colorado School of Mines
Golden *J.Moore*

Washington State University
Pullman *S.Wojcicki*

Japan

Ryukoku University
Seta, **Otsu** *M.Koizumi*

Osaka University
Osaka *Y.Miyamoto*

National Research Institute for Metals Tokyo *Y.Kaieda*

Institute of Technology
Tokyo *O.Odawara*

China

Wuhan University of Technology
Wuhan *R.Yuan*

University of Science and Technology
Beijing *Lai-Ho-Yi*

Poland

Academy of Mining and Metallurgy
Krakow *R.Pampuch*

India

Indian Institute of Science
Bangalore *K.Patil*

Interest in SHS is continiously growing in the United States. American scientists prefer to concentrate on understanding the SHS mechanism by preforming new experiments. Such an approach has lead to repetition of previous work but, in return, had provided confidence in the correctness of approach. In the author's opinion, the fundamental studies carried out on the basis of western electronics are the best among them. The work of Dr. Holt and co-workers [43] who used the synchrotron of Stanford University in their experiments serves as an example.

The work of Japanese researchers is characterized by a focus on practical results. As a result, these studies contributed greatly to the technology of SHS (SHS + HIP, FGMs, SHS joining of diamonds, production of large-scale pipes, etc.).

One cannot help mentioning the popularity of SHS in China. The number of research groups and publications is growing rapidly.

The geographical map of SHS is presented in Fig. 12. Points indicate the places that are active in SHS research. The map was compiled by Prof. V.V. Barzykin who kindly supplied it to the author.

Concerning research centers, the leading role of the Institute of Structural Macrokinetics (ISMAN), Russian Academy of Sciences [44] has to be specially outlined. ISMAN is a large research institution working in the field of general and structural macrokinetics, theory and practice of combustion, SHS and technology of inorganic materials. The Institute was established in 1986 from the Department at the Institute of Chemical Physics in Chernogolovka. The research interests of this Department go back to the middle of fifties, when a research group at the Institute of Chemical Physics was set up to study thermal explosion in explosives. Step by step, this research avenue developed both in depth and breadth. As the research work proceeded, people grew in skill, and many of them have become well-known over the world. The author is grateful to them for their long-term cooperation.

Fig. 12. Geographic map of SHS.

8. INTERNATIONAL LINKS IN SHS

International links were established in 1988 when Dr. Holt and Prof. Munir organized the Symposium on Combustion and Plasma Synthesis of High-Temperature Materials in San Francisco. Since then, international links between the researchers active in the field of SHS have grown. To date, the number of meetings and symposia have grown significantly. Among them are:

- International Symposia on SHS
- Bilateral Seminars
- SHS Topical Sessions at the International Congresses on Combustion, Powder Metallurgy, Ceramics, etc.

- International Scientific Collaboration
- International Commercial Agreements and Contracts
- International Journal of Self-Propagating High-Temperature Synthesis

International Symposia on SHS are the most important international forums for research workers active in SHS. The first of them was held in Alma-Ata in 1991, the second one, in Honolulu in 1993. The third one will be held in Wuhan (China) this year. It promises to be interesting. In addition to regular sessions, two special advanced sessions and discussions will be organized.

The *International Journal of Self-Propagating High-Temperature Synthesis* becomes more and more important for the SHS community. Three volumes (12 issues) have already appeared. Many researchers from the former Soviet Union, which earlier had no possibility to publish in international journals, have become authors the Journal.

Of great importance for the SHS community is also the *Materials Synthesis and Processing* journal edited by Prof. Munir and Dr. Holt.

Increasingly important roles in the development of SHS are being played by international projects. One of the most important among them is the Spanish-American-Russian Project on industrial-scale production of SHS powders ("Prometheus"). The participants are: AIE Committee "SHS Espana (*M. Hernan*), United Technology Research Center (*M. Werle*), SHS Research Center at ISMAN (*I. Borovinskaya*).

Tasks and stages:

- implementation of ISMAN technology under the Western conditions
- improvement of SHS reactor
- setting up the production line with a required level of automation and electronic control
- launching the industrial-scale production of SHS powders in Spain

Powders to be produced:

AlN, α and β-Si_3N_4, SiAlON, Cr_3C_2 (*cast*), and (TiC - Cr_3C_2) + Ni

The first-year effort work yielded good results and, what is important, demonstrated the compatibility of viewpoints, approaches, and characters of individuals from different countries.

It is worth mentioning an unscheduled result of this joint work -- the synthesis of fibers from powder reactants in large reactors. The fibers of Si_3N_4 and AlN have been synthesized in the 30-l SHS reactor. They are characterized in Table 3. Impressive is a high yield of fibers -- up to 80%. It is much higher than it was found earlier in laboratory-scale and pilot-scale experiments.

Table 3. SHS whiskers, preliminary data.

	Si_3N_4	AlN
Length, mm	< 10	< 0.1
Thickness, μm (diameter)	0.5-3	0.3-2
Density, g/cm^3	3.2	3.25
Chemical composition, wt %		
N	> 39.2	> 32.5
O	0.9	< 1.0
Phase composition	α, 97 % β, 3 %	100 %
Yield (content in reactor), %	~ 80 %	~ 80 %

CONCLUSION

In conclusion, let us outline ten, in the author's opinion, of the most promising SHS research avenues. Of course, this is a subjective viewpoint of the author. But no doubt, the advance in these directions will provide the "tomorrow of SHS".

- Structural macrokinetics: the relationship between the processes of structure formation and combustion
- Multi-dimensional combustion modeling. Chaos in combustion
- Gas-phase and gas-dispersion SHS
- SHS in organic systems
- Unconventional powders
- Nonequilibrium materials
- Production of items: single-stage near-net-shape processes
- In-line SHS production. Large-scale production
- SHS mechanochemistry. Advanced SHS processes and processing of products
- Space experiments. Microgravity effects

In the author's opinion, advances in these directions will define the features of SHS in the nearest future.

ACKNOWLEDGMENTS

The author is grateful to Dr. *A.S. Mukasyan*, Dr. *A.E. Sytchev*, Dr. *Yu. B. Scheck*, Mrs. *O.B. Trushnikova*, and Mrs. *N.D. Shukhina* for their help in preparing the manuscript.

REFERENCES

[1]V. Hlavacek, Combustion Synthesis: A Historical Perspective, Ceram. Bull., v. 70, no. 2, pp. 240-243, 1991

[2]H. Gol'dschmidt, Processes for Synthesis of Metals or Metalloids, or Alloys, Ger. Pat., no. DRP 996317, 1895

[3]N.N. Beketov, A Study on Substitution of one Metals by Others, Dissertation, Khar'kov University, 1865

[4]E.I. Maksimov, A.G. Merzhanov and V.M. Shkiro, Gasless Compositions as a Simplest Model of Combustion in Nonvolative Condensed Systems, Fiz. Goreniya Vzryva, v. 4, pp. 24-30, 1965

[5]A.G. Merzhanov, Solid Flames: Discoveries, Concepts, and Horizons of Cognition., Combust. Sci. Tech., v. 98, no. 4-6, p. 307-336, 1994

[6]A.G. Merzhanov, I.P. Borovinskaya, Self-Propagating High-Temperature Synthesis of Inorganic Compounds, Dokl. Akad. Nauk SSSR, v. 204, no. 2, pp. 336-369, 1972

[7]Ya.B. Zel'dovich, D.A. Frank-Kamenetskii, The Theory of the Thermal Propagation of Flame, Zh. Fiz. Khim., v. 12, no. 1, p. 100, 1938

[8]A.F. Belyaev, L.D. Komkova, Pressure Dependence of Burning Velocity for Thermites, Zh. Fiz. Khim., vol.24, no.11, pp. 1302-1311, 1950

[9]A.F. Belayev, On Combustion of Nitroglycole, Zh. Fiz. Khim., v. 14, no. 8, pp. 1009-1025, 1940

[10]P. F. Pokhil, On the Combustion Mechanism in Smokeless Gun Powder, Fiz. Vzryva, v. 2, Moscow, Izd. AN SSSR, p. 81, 1953

[11]Ya. B. Zel'dovich, To the Theory of Combustion in Gun Powders and Explosives, Zh. Eksp. Teor. Fiz., no. 12, p. 498, 1942

[12]A.G. Merzhanov, The Theory of Stable Homogeneous Combustion of Condensed Substances, Combust. and Flame, v. 13, pp. 143-156, 1969

[13]B. V. Novozhilov, Unsteady Combustion in Propellants, p. 176, Moscow, Nauka, 1973

[14]I.P. Borovinskaya, A.G. Merzhanov, Synthesis and Combustion of Copper and Silver Acetylenides, Report Inst. Chem. Phys., Chernogolovka, 1966.

[15]A.G. Merzhanov, V.M. Shkiro, I.P. Borovinskaya, A Method for Synthesizing Refractory Inorganic Compounds, USSR Inventor's Certificate 255221, 1967

[16]J.F. Crider, Self-Propagating High-Temperature Synthesis: A Soviet Method for Producing Ceramic Materials, Ceram. Eng. Sci. Proc., v. 3, no. 9-10, pp. 519-528, 1983

[17]J.B. Holt, Combustion Synthesis of Refractory Materials, Techn. Rep. LLNL, no. YCRL-53258, Febr., 1982

[18]J.W. McCauley, N.D. Corbin, T. Resetar, and P. Wong, Simultaneous Preparation and Self-Sintering of Materials in the System Ti-B-C, Ceram. Eng. Sci. Proc., v.3, pp. 538-554, 1983

[19]Z.A. Munir, Synthesis of High-Temperature Materials by Self-Propagating Combustion Methods, Amer. Ceram. Bull., v. 67, no. 2, pp.342-349, 1988

[20]Z.A. Munir, U. Anselmi-Tamburini, Self-Propagating Exotermic Reaction: the Synthesis of High-Temperature Materials by Combustion, Mater. Sci. Repts., v.3, p. 277-365, 1989

[21]A.G. Merzhanov, I.P. Borovinskaya, Yu.E. Volodin, , Combustion of Porous Metallic Samples in Nitrogen, Dokl. Akad. Nauk SSSR, v. 206, no. 4, pp. 905-908, 1972

[22]A.G. Merzhanov, A.K. Filonenko, I.P. Borovinskaya, New Phenomena in Combustion of Condensed Systems, Dokl. Akad. Nauk SSSR, v. 208, pp. 892-894, 1973

[23]A.P. Aldushin, A.G. Merzhanov, B.I. Khaikin, On Special Features of Combustion in Condensed Systems Yielding Solid Reaction Products, Dokl. Akad. Nauk SSSR, v. 204, no. 5, pp. 1139-1142, 1972

[24]A.P. Aldushin, A.G. Merzhanov, B.I. Khaikin, The Patterns of the Layer-by-Layer Filtration Combustion in Porous Metals, Dokl. Akad Nauk SSSR, v.215, no. 12, pp. 612-615, 1974

[25]A.G. Merzhanov, Self-Propagatung High-Temperature Synthesis: Thenty Years of Search and Finding, pp. 1-53, Combustion and Plasma Synthesis of High-Temperature Materials, eds. Z.A. Munir and J.B. Holt, VCH Publ. Inc., New York, 1990

[26]A.G. Merzhanov, A.S. Rogachev, Structural Macrokinetics of SHS Processes, Pure and Appl. Chem., v. 64, no. 7, pp. 941-953, 1992

[27]T.S. Azatyan, V.M, Mal'tsev, A.G. Merzhanov, V.A. Seleznev, Combustion Patterns in the Mixtures of Titanium with Silicon, Fiz. Goreniya Vzryva, no. 1, pp. 43-49, 1979

[28]A.A. Zenin, G.A. Nersisyan, Thermal Structure of SHS Wave: Mechanism and Macrokinetics of High-Temperature Nonisothermic Interaction of Elements in the Ti--Si and Zr--Si Systems, Preprint Inst. Chem. Phys., Chernogolovka, 1980

[29]A.S. Rogachev, A. Varma, V.A.Shugaev, V.A.Khomenko, C.R. Kachelmyer, On the Mecahnism of Structure Formation During Combustion Synthesis of Titanium Silicides, Combust. Sci. and Tech., in press, 1994

[30]A.G. Merzhanov, I.P. Borovinskaya, Self-Propagating High-Temperature Synthesis in Chemistry and Technology of Refractory Compounds, Zh. Vses. Khim. O-va im. D.I. Mendeleeva, v. 24, no. 3, pp.223-227, 1979
[31]I.P. Borovinskaya, Chemical Classes of the SHS Processes and Materials, Pure and Appl. Chem., v. 64, no. 7, pp. 919-940, 1992
[32]A.G. Merzhanov, Theory and Practice of SHS: Worldwide State of Art. Newest Data, Intern. J. SHS, v. 2, no. 2, pp. 113-158, 1993
[33]I.P. Borovinskaya, N.S. Makhonin, and G.G. Khurtina, private communications, Chernogolovka, 1994
[34]R. Pampuch, L. Stobierski, J. Lis, Microstructure Development Upon Sintering of SHS-Derived and Conventional Silicon Carbide and Nitride Powders, Intern. J. SHS, v. 2, no. 2, pp. 159-164, 1993
[35]A.G. Merzhanov, I.P. Borovinskaya V.I. Yukhvid, V.I. Ratnikov, New Production Methods of High-Temperature Materials Based on Combustion, pp. 193-206, Nauchnye Osnovy Materialovedeniya, Nauka, Moscow, 1981
[36]O. Odawara, US Patent 4363832, 1982
[37]V.L. Kvanin and N.T.Balikhina, private communication, Chernogolovka, 1994
[38]J.C. Rawers, H.E. Maupin, Metal-Intermetallic Composites Formed by Reaction-Sintering Metal Foils, J. Mater. Sci. Lett., v. 12, pp.637-639, 1993
[39]S.L. Kharatyan, A.R. Sarkisyan, Manufacturing Planar Heaters by SHS, Report Armenian Inst. Chem. Phys., Yerevan, 1991.
[40]A.G. Merzhanov, G.G. Karyuk, I.P. Borovinskaya, et al., SHS Titanium Carbide as a Highly Effecient Abrasive Material, Poroshk Metall., no. 10, pp. 50-59, 1981
[41]Y. Kaieda, Properties of TiNi Intermetallic Composites Industrially Produced by Combustion Synthesis, 19th Annual Cocoa Beach Conference & Exposition, Cocoa Beach, Florida, 1995
[42]A.G. Merzhanov, I.P. Borovinskaya, V.K. Prokudina, N.A. Nikulina, Efficiency of SHS Powders and Their Production Methods, Int. J. SHS, in press
[43]J.B. Holt, J. Wong, E. Larson, B. Waide, B. Rupp. and R. Frahm, A New Experimental Approach to Study Solid Combustion Reaction Using Synchrotron Radiations, pp. 107-113, Proceedings of the First US-Japanese Workshop on Combustion Synthesis, Tokyo, Japan, 1990
[44]A.G. Merzhanov, Institute for Structural Macrokinetics of the Russian Academy of Sciences, Mendeleev Communications, no. 6, pp. 258-259, 1993

Translated by Yu. Scheck.

PROPERTIES OF TiNi INTERMETALLIC COMPOUND INDUSTRIALLY PRODUCED BY COMBUSTION SYNTHESIS

Yoshinari Kaieda
National Research Institute for Metals,
1-2-1 Sengen, Tsukuba Science City, Ibaraki 305, Japan

ABSTRACT

Most TiNi shape memory intermetallic compounds are conventionally produced by the process including high frequency induction vacuum melting and casting. A gravity segregation occurs in a cast TiNi ingot because of the big difference in the specific gravity between Ti and Ni. It is difficult to control accurately the phase transformation temperature of TiNi shape memory intermetallic compound produced by the conventional process, because the martensitic transformation temperature shifts by 10K due to the change in 0.1 % of Ni content. Homogeneous TiNi intermetallic compound is produced by the industrial process including combustion synthesis method, which is a newly developed manufacturing process. In the new process, phase transformation temperatures of TiNi can be controlled accurately by controlling the ratio of Ti and Ni elemental starting powders. The chemical component, the impurities and the phase transformation temperatures of the TiNi products industrially produced by the process are revealed. These properties are vitally important when combustion synthesis method is applied to an industrial mass production process for producing TiNi shape memory intermetallic compounds. TiNi shape memory products are industrially and commercially produced today by the industrial process including combustion synthesis. The total production weight in a year is 30 tons in 1994.

INTRODUCTION

According to the chronological accounts of Hlavacek [1], industrialization of combustion synthesis was achieved in few cases in the long history of the combustion synthesis since 1825. Thermite welding process invented by Goldschmidt and Frank-Caro process for producing $CaCN_2$ were successful applications of the combustion synthesis. Industrialization for production of advanced materials using combustion synthesis is still in developing state in U.S.A. and Japan as compared to the former U.S.S.R. [2]. However, fundamental researches on combustion synthesis are carried out by many researchers [3-7]. There are some fundamental researches on combustion synthesis of TiNi [8-16]. The author has continued the research on combustion synthesis for industrialization and commercialization of the producing process of TiNi intermetallic compound [17-23].

TiNi intermetallic compound is an important and unique material for its shape memory and superelastic properties. However, in the conventional production process, the gravity segregation of Ni content is inevitable because of the big difference in the specific gravity between Ti and Ni. It affects the homogeneity of chemical components along longitudinal direction in the big ingot of TiNi. The inhomogeneity of chemical components influences the phase transformation temperatures of TiNi intermetallic compound especially when the material is formed to a long thin wire. A 100 kg ingot of TiNi becomes 4,000 km long wire of diameter 0.07 mm. The present industrial process utilizing combustion synthesis has been developed to improve the disadvantages of the conventional production process for TiNi intermetallic compound.

Since the conventional production process of TiNi has a long history [24], it is possible to obtain access to information on the properties of TiNi. However, there is little information about the properties of TiNi produced by combustion synthesis in an industrial scale. Characterization of properties of the TiNi is important for the application of TiNi shape memory products produced by the combustion synthesis method in an industrial scale.

EXPERIMENTAL

Process

The heat from the formation of the TiNi intermetallic compound is 67.78 kJ/mol [25]. This value was enough to ignite and propagate the exothermic propagating reaction even in the production process on a large scale [26]. Since, the shape memory and super-elastic materials were used in the shapes of thin wires and plates, it was a necessary condition that not only the big ingots but also the thin and long material could be produced industrially in the new production process.

In order to solve these problems the author developed an integrated process of producing TiNi intermetallic compound from the elemental powders of Ti and Ni by using the combustion synthesis as shown in **Fig. 1**.

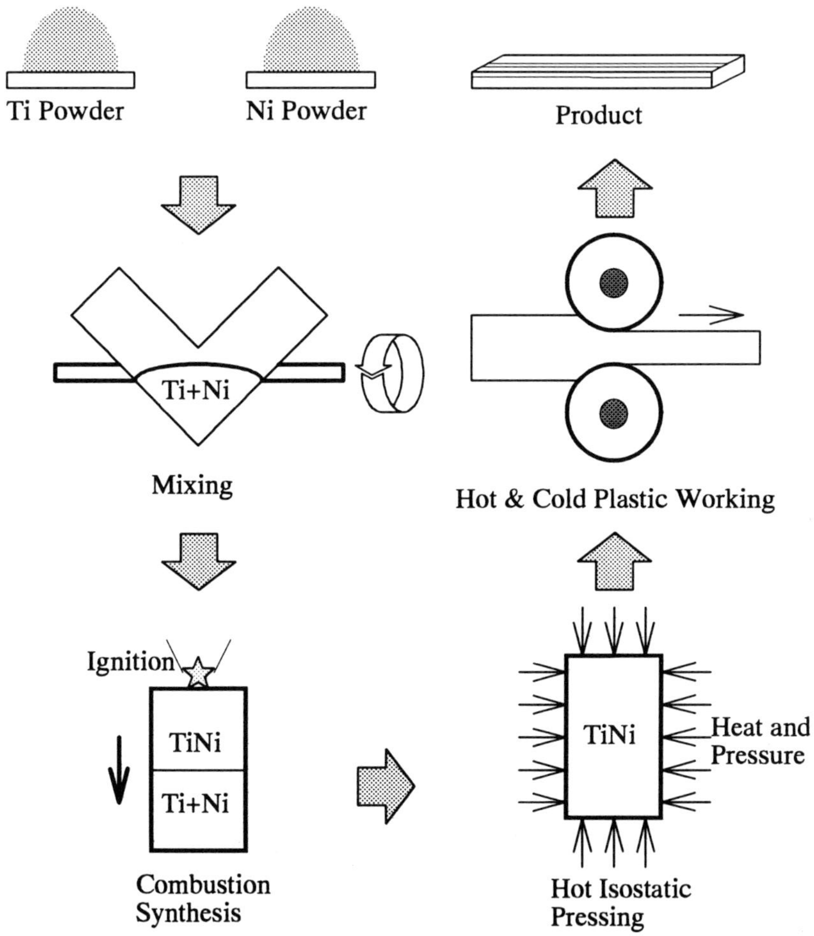

Fig. 1 New industrial production process for making TiNi intermetallic compound. The weight of a batch is 100 kg.

In the process, the weight of a batch was selected to be 100 kg in due consideration of the annual production weight and the productive facilities of the manufacturing plant. Various ratios around stoichiometry ratio (Ti:Ni=1:1) of Ti and Ni powders were weighed accurately to be 100 kg and mixed precisely by a mixing machine in a dry state of powder. The mixed powder was filled in a graphite crucible and the top of the powder was ignited by a tungsten heating wire in a vacuum (10^{-4} Pa) in a big reactor shown in **Fig.2**. There is a furnace inside the reactor to dehydrate the mixed powder.

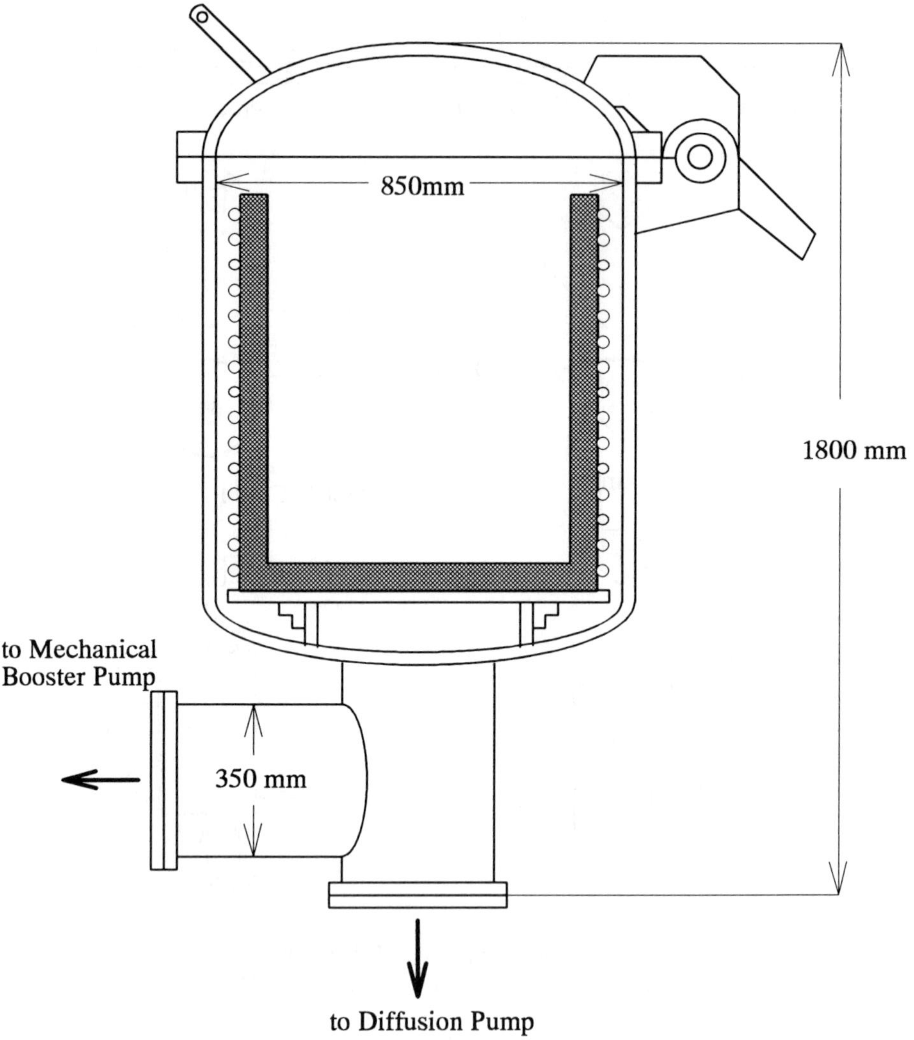

Fig. 2 Industrial Combustion Synthesis Reactor

As this stage, the degree of a vacuum and the speed of an evacuation were very important to eliminate the volatile impurities from the powder during combustion synthesis process. In order to conduct a high speed evacuation for the big reactor, those big vacuum pumps listed in **Table I** were equipped. The combustion-synthesized TiNi was encapsulated in a stainless steel can and evacuated and sealed, then sintered in a hot isostatic press (HIP). The TiNi ingot was porous and the relative density was 50 - 60 % after combustion synthesis process, but the relative density increased to 100% after a HIP process. The ingot made by the HIP process was plastically worked with hot forging, hot rolling, hot bar rolling, hot and cold rotary swaging, cold drawing and cold rolling to produce thin wires and plates. After 10 % cold plastic working, the material was heat-treated in certain conditions to get shape memory effects.

Table I. Specifications of Vacuum Pumps for the Combustion Synthesis Reactor

Type of Pump	Evacuating speed (m^3/s)	Pressure (Pa)	Power (kW)
Rotary	0.125	$7x10^{-1}$	11
Mechanical Booster	0.695	$5x10^{-2}$	5.5
Diffusion	18.5	$7x10^{-5}$	12

Starting Powders

In general, it is important to select starting powders in powder metallurgy in order to reduce the impurities in the products. The properties and deformability of TiNi intermetallic compound are very much influenced by the impurities. The excess oxygen decreases the deformability of TiNi drastically during the mechanical working process. The phase transformation temperatures of TiNi are also affected by the oxygen content. The author, for many years, investigated several kinds of starting powders for improving the quality of the TiNi products produced by the new industrial production process. As a result of the investigation, the starting powders listed in **Table II** are employed in the process at present.

Table II. Production Methods and Particle Diameters of Starting Powders

Powder	Production method	Diameter (μm)
Ti	Induction Melting Gas Atomization Process	<150
Ni	Carbonyl Process	8-15

The author selected the starting powders, which were produced by clean production methods, with relatively larger diameters than those of usual Ti and Ni powders. Chemical compositions of impurities of the starting powders are shown in **Table III**. These powders are highest grade Ti and Ni that can be obtained in a commercial market nowadays.

Table III. Chemical Compositions of Impurities of Starting Powders (mass %)

	O	N	C	Fe	Cr	Si	Mg	Mn	H	Cl
Ti powder	0.060	0.02	0.01	0.035	0.01	0.007	0.001	0.002	0.006	<0.001
Ni powder	0.06		0.1	0.01						

Characterization

The chemical analysis of Ni content, the spectrochemical analysis of impurities and gas analysis of TiNi were carried out. The phase transformation temperatures of TiNi, which are the essential properties of shape memory materials, were examined by differential scanning calorimetry method.

RESULTS

Chemical Composition

Since chemical composition plays an important role in the phase transformation temperature of TiNi, several ratios of Ti and Ni around a stoichiometric ratio are produced industrially by combustion synthesis. It is very important to control the ratio of Ti and Ni to obtain a certain phase transformation temperature for an application.

Ni content: The several TiNi intermetallic compounds with various Ni contents around a stoichiometric chemical composition were made for the experiments to investigate the dependence of the phase transformation temperatures on Ni content. The chemical analyses of Ni contents were carried out and are shown in **Table IV**.

Table IV. Ni Contents in TiNi used in the Experiment

Atomic %	48.9	49.4	49.7	49.9	50.1	50.5	50.7	51.1	51.4	51.7
Mass %	54.0	54.5	54.8	55.0	55.2	55.6	55.8	56.2	56.4	56.7

*Atomic weight: Ti=47.88, Ni=58.69.

In order to check the accuracy of Ni content control, the result of chemical analysis of Ni content in TiNi final product against Ni content in the mixed powder prepared for weighing and mixing stage is shown in **Fig. 3**. The Ni content in TiNi product is equal to the added Ni in the mixed powder. This result exhibits the advantage of the present process using powders as starting materials. It is difficult to control the Ni content so accurately in the conventional production process because the starting material is massive Ti and Ni.

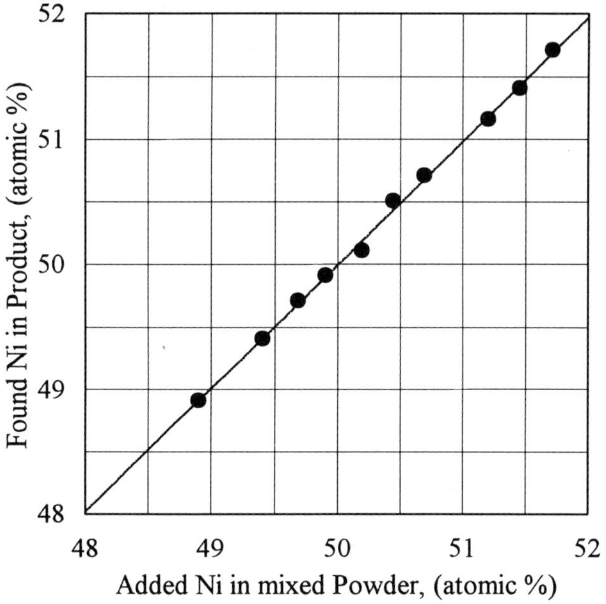

Fig. 3 Found Ni contents in TiNi product against added Ni contents in the mixed powder.

Table V. Spectrochemical Analysis of TiNi by Analytical Emission Spectroscopy

Element:	Cr	Mn	Fe	Al	Mg	Cu	Si	Co	V	Bi	B	Mo	Ta	Zr	W	Nb
Intensity:	-3	-3	-3	-3	-3	-3	-4	-4	-4	-4	-4	-4	-4	-4	-4	-4

*Rank of intensity: $-4 < -3 < -2 < \pm < +1 < +2 < +3 < .... < +10$.
*-4 means undetectable. *Light source is DC arc, 10A.

Impurities: In general, the quality of a metallic material improves with the fewer impurities. The author made efforts toward reducing impurities in the TiNi products by improving the process. The result of spectrochemical analysis of TiNi examined by analytical emission spectroscopy is shown in **Table V**. The impurities in the material are very few. Since TiNi intermetallic compounds produced by the present process are very pure, the influence of impurities on the phase transformation behavior is considered small.

Gas content: The content of oxygen in TiNi affects the phase transformation temperatures and the deformability in the producing process. The oxygen content could be reduced by using clean powder and big vacuum pumps in combustion synthesis stage in the production process. Nitrogen was also a harmful element to TiNi but could be reduced easily together with reducing oxygen. The result of the analysis of oxygen and nitrogen is shown in **Table VI.**

Table VI. Contents of Oxygen and Nitrogen

O	N (ppm)
500	200

Phase Transformation Temperature

Martensitic start temperatures (Ms) of the samples with various Ni contents listed on Table IV were measured. The dependence of the martensitic start temperatures (Ms) on the Ni content is shown in **Fig. 4**. The phase transformation temperature is influenced by the ratio of Ti and Ni, namely Ni content in TiNi, degree of cold plastic working in the process, and the conditions of heat treatment. In Fig. 4, samples were heat-treated at 783 K in 1500 s after 10 % cold working. In the range of 49.5-50.5 atomic %, Ms temperature decreases linearly by 10 K along with 0.1 % increment of Ni content. In the range of 51.4 - 51.7 atomic %, Ms temperature increases with Ni content.

DISCUSSION

Chemical Compositions

The most important property of TiNi shape memory intermetallic compound is chemical composition. The ratio of Ti and Ni influences the phase transformation temperature of the material. Impurities also influence the phase transformation temperature and affect the deformability in the production process. Oxygen is the most harmful element to the deformability and affects phase transformation temperatures significantly because oxygen reacts with Ti and makes TiO_2 and changes the ratio of Ti and Ni in the matrix of the material. Nitrogen is considered to be as harmful as oxygen. However, when oxygen content is reduced by the use of clean starting powders and high evacuation in the combustion synthesis process, nitrogen is also reduced along with the oxygen.

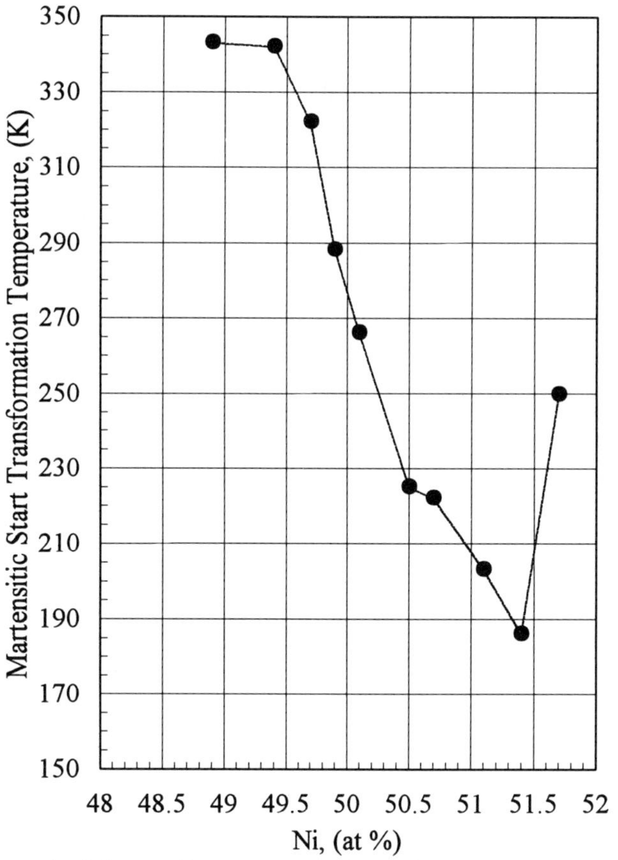

Fig. 4 Change in Martensitic start transformation temperature of TiNi wire sample heat-treated at 783K for 1500 s after 10% cold plastic working against Ni content.

In the present mass production industrial process including combustion synthesis, chemical composition of Ni, impurities, and the gas contents are examined from batch to batch. The impurities and gas contents in TiNi are on very low level and the fluctuation of the chemical content of impurities is little. The result of control of impurities makes the influence on the phase transformation temperatures negligibly small. Ni content can be controlled accurately by controlling the weighing and mixing of the starting powders. It makes the control of phase transformation temperature easier.

Production of TiNi using Combustion Synthesis

In the present industrial production process of TiNi, combustion synthesis is conducted one or two times in a day. The total production weight in a year was 30 tons in 1994. The author once used cold isostatic pressing process to form the Ti and Ni mixture of powder and to increase the relative density before the combustion synthesis process [23]. The cold isostatic pressing process is now omitted. The loose powder is settled in a graphite crucible and ignited in combustion synthesis process. The combustion synthesis of loose powder without cold isostatic pressing process makes the ignition of combustion synthesis easier. The total production cost is reduced by eliminating the cold isostatic process. Industrial production process of TiNi intermetallic compound using combustion synthesis progresses satisfactorily.

Performance of TiNi produced by Combustion Synthesis

Since there is no gravity segregation of Ni element in the ingot and content of impurities and gas elements are very low, the properties of phase transformation of TiNi are stable. There is no change in transformation temperature in each part of very long TiNi wire. The products with the desired phase transformation temperature could be obtained by accurate control of Ni content. The martensitic start temperature decreases in the range of Ni below 51.4 atomic % and increases in the range above 51.4 atomic % with the increase in Ni content in Fig.4. This tendency is similar to the other researchers' data on TiNi made by conventional method [27-29].

CONCLUSION

As compared with conventional casting process, the purity of product generally becomes worse when we use powder as starting material. However, in the present process, the purity of the product becomes better by selecting starting powder and eliminating volatile impurities with big vacuum pump during combustion synthesis process. In addition, since the starting material is powder, it becomes easier to control the ratio of Ti and Ni accurately. TiNi products produced by the combustion synthesis method are replacing TiNi products made by conventional production process from the commercial market in Japan.

REFERENCES

1. V.Hlavacek, "Combustion Synthesis: A Historical Perspective," Ceram. Bull. 70 [2], 240-243(1991).

2. Anonymous, "Soviet SHS Technology: A Potential U.S. Advantage in

Ceramics," Ceram. Bull. 68 [6], 1165-1167(1989).
3. F.Booth, "The Theory of Self-Propagating Exothermic Reactions in Solid Systems," Trans. Farad. Soc.,49,272-281 (1953).
4. J.D.Walton, Jr. and N.E.Poulos, "Cermets from Thermite Reactions," J. Am. Ceram. Soc., 42[1], 40-49 (1959).
5. Z.A.Munir, "Synthesis of High Temperature Materials by Self-Propagating Combustion Methods," Ceram. Bull., 67 [2] 342-349 (1988).
6. J.W.McCauley, N.D.Corbin, T.Resetar and P.Wong,"Simultaneous Preparation and Self-Sintering of Materials in the System Ti-B-C," 538-554, Ceramic Eng. & Sci. Proc. ed. J.D.Buckley, C.M.Packer & J.J.Gebhardt, Am. Cera. Soc., Columbus, 1982.
7. H.C.Yi and J.J.Moore, "Self-propagating high-temperature (combustion) synthesis (SHS) of powder-compacted materials" J. Mater. Sci., 25, 1159-1168 (1990).
8. V.I.Itin, V.N.Khachin, A.D.Bratchikov, V.E.Gyunter, E.F.Dudarev, T.V.Monasevich, D.B.Chernov, G.D.Timonin, and A.P.Paperskii, "Structure and Properties of Titanium Nickelide Materials Made by Self-Propagating High-Temperature Synthesis," Sov. Phys. J. 20, 1631-1633(1977).
9. A.D.Bratchikov, A.G.Merzhanov, V.I.Itin, V.N.Khachin, E.F.Dudaref, V.E.Gyunter, V.M.Maslov, and D.B.Chernov, "Self-Propagating High-Temperature Synthesis of Titanium Nickelide," Sov. Powd. Metal. & Met. Ceram. 19, 5-8 (1980).
10. V.I.Itin, V.N.Khachin, A.D.Bratchikov, V.E.Gyunter, and D.B.Chernov, "Synthesis of Cast Titanium Nickelide by Combustion Processes," Inorganic Materials, 16, 1337-1339 (1980).
11. V.I.Itin, V.N.Khachin, V.E.Gyunter, A.D.Bratchikov, and D.B.Chernov, "Production of Titanium Nickelide by Self-Propabating High-Temperature Synthesis," Sov. Powd. Metal. & Met. Ceram., 22, 156-157 (1983).
12. H.C.Yi and J.J.Moore, "SHS synthesized Ni-Ti based shape memory alloys for both low-and high temperature applications," J. Mater. Sci. Let. 8, 1182-1184 (1989).
13. H.C.Yi and J.J.Moore, "Combustion synthesis of TiNi intermetallic compounds, Part 1 Determination of heat of fusion of TiNi and heat capacity of liquid TiNi," J. Mater. Sci., 24, 3449-3455 (1989).
14. H.C.Yi and J.J.Moore, "Combustion synthesis of TiNi intermetallic compounds, Part 2 Effect of TiO2 formation," J. Mater. Sci., 24,3456-3462 (1989).
15. H.C.Yi and J.J.Moore, "The Combustion Synthesis of Ni-Ti Shape Memory Alloys," J. Min. Met. & Mater. Soc., 42, 31-35 (1990).
16. J.J.Moore and H.C.Yi, "The Effect of Addition of a Third Element on the

Combustion Synthesis of Ni-Ti Based Shape Memory Alloys," 33-47, Proc. 1st U.S.-Jpn. Workshop on Combustion Synthesis, ed. Y.Kaieda and J.B.Holt, National Research Institute for Metals, Tokyo, (1990).
17. Y.Kaieda, "Production method of shape memory alloy," Pat. No.1790370, Japan, Sept. 29, 1993, applied June 5, 1987.
18. Y.Kaieda, "Production method of chemical compounds" Pat. No.1816876, Japan, Jan. 18, 1994, applied March 2, 1987.
19. Y.Kaieda, "Production method of Ti-base shape memory alloy," Pat. No. pending, Japan, applied Sept. 8, 1987.
20. Y.Kaieda, M.Otaguchi, M.Odawara, O.Nakamura, "Self-Propagating High-Temperature Synthesis of Intermetallic Compounds," 557-562, Proc. Int. Symp. on Sintering '87 1, ed. S. Somiya, M.Shimada, M.Yoshimura & R.Watanabe, Elsevier, Tokyo, (1988).
21. Y.Kaieda, M.Otaguchi, and N. Oguro, "Self-Propagating High-Temperature Synthesis of NiTi Alloy: a New Production Process of Shape Memory Alloy from Elemental Powder," 623-628, Shape Memory Materials, Proc. of the MRS Int. Meeting on Advanced Materials, No.9, Mater. Res. Soc., Tokyo, (1989).
22. M.Otaguchi, Y.Kaieda, N.Oguro, S.Shite and T.Oie, "Possibility of Manufacturing of NiTi Intermetallic Compound by Combustion Synthesis," J. Japan Inst. Metals, 54, [2] 214-223 (1990), (in Japanese).
23. Y.Kaieda, M.Otaguchi, and N.Oguro, "Combustion Synthesis of Intermetallic Compounds," 106-113 Combustion and Plasma Synthesis of High-Temperature Materials, ed. Z.A.Munir and J.B.Holt, VCH, New York, 1990.
24. W.J.Buehler, and W.B.Cross, "55-Nitinol unique wire alloy with a memory," Wire Journal 2[6], 41-49(1969).
25. R.Hultgen, P.D.Desai, D.T.Hawkins, M.Gleiser and K.K.Kelley, 1245 "Selected Values of the Thermodynamic Properties of Binary Alloys", Am. Soc. Met., Metals Park, Ohio, 1973.
26. Y.Kaieda, "Self-Propagating Reaction Sintering of Intermetallic compounds," 395-400, Proc. 1993 Powd. Met. World Congress, ed. Y.Bando & K.Kosuge, Jpn. Soc. Powd. & Powd. Met., Kyoto, (1993).
27. J.E. Hanlon, S.R. Butler, and R.J. Wasilewski, "Effect of martensitic transformation on the electrical and magnetic properties of NiTi," Trans. Met. Soc. AIME, 239, 1323-1327(1967).
28. I.I. Kornilov, Ye. V. Kachur and O.K. Belousov, "Dilatation analysis of translation in the compound TiNi," Physics Met. Metall. [32]2, 190-193(1971).
29. T.Homma, and H.Takei, "Effect of heat treatment on the martensitic transformation in TiNi compound," J. Jpn. Inst. Met., 39, 175-182(1975)(in Japnese).

ELECTRIC FIELD-ACTIVATED COMBUSTION SYNTHESIS OF CERAMICS AND COMPOSITES

Z. A. Munir
Department of Chemical Engineering and Materials Science
University of California, Davis
Davis, CA 95616

ABSTRACT

The use of an electric field to activate self-propagating combustion synthesis is demonstrated. With the simultaneous application of a field and an ignition source, self-sustaining combustion waves can be initiated in systems which are thermodynamically excluded from such reactions at 298K. The use of the field-activated combustion synthesis (FACS) has been successfully demonstrated in a variety of single phase and composite systems. The synthesis of β-SiC is described as an example of single phase systems. A minimum (threshold) field strength of about 6 V/cm is required to initiate and propagate the reaction between silicon and graphite. At or slightly above the threshold, the wave propagates in a non-steady state (pulsating) mode. As the field is increased the velocity of propagation increases and becomes independent of time and distance, i.e. steady-state. The application of high fields (> ~ 21 V/cm) results in simultaneous combustion without the need for an ignition source.

The use of FACS in the preparation of composites was also investigated. Composites of $MoSi_2$-SiC, $MoSi_2$-Nb, $MoSi_2$-ZrO_2, and B_4C-TiB_2 were synthesized by this method. For the system (1-x) $MoSi_2$-xSiC, composites with $0 \leq x \leq 1.0$ can be synthesized by self-propagating combustion. For each value of x, a threshold field is required to initiate the wave. In the case of y B_4C-TiB_2 composites, systems with $y \leq 0.5$ can be synthesized without the field. When a field is imposed, composites with y values up to 5 have been successfully synthesized by self-propagating combustion.

1. INTRODUCTION

The imposition of electromagnetic fields during combustion synthesis has been attempted to provide activation for thermodynamically less energetic systems. Both magnetic and electric fields have been imposed; in the latter case fields have been applied in both the resistive and inductive modes. Kirdyashin, et al. [1] applied magnetic fields on metal-sulfur and metal-metal systems (with one metal being ferromagnetic) and reported an increase of the combustion wave velocity by nearly a factor of three. However, there was no correlation between the Curie temperature (1404, 1043, and 631K) of the metals used (Co, Fe, and Ni) and the observed field effect. In other words, the combustion temperatures were above or below the corresponding Curie temperature even when an effect was observed. This lack of correlation with the Curie temperature, prompted the authors to suggest that the magnetic field caused re-arrangement of the ferromagnetic particles *prior* to the arrival of the combustion wave. Such a proposed re-arrangement is believed to cause an increase in the effective thermal conductivity of the reactants by increasing contacts between particles and thus increasing the wave velocity. No other investigation on the effect of a magnetic field has been reported.

The use of electric fields in the inductive mode has been made to investigate the synthesis of titanium carbide and thermite-type reactions [2,3]. In the case of the synthesis of TiC, it was observed that the imposition of the high frequency (non-contacting) field had no effect on the velocity of the wave but a significant influence on the combustion temperature [2]. The interpretation of this observation is that the inductive field had no influence on the reaction within the combustion zone but had a substantial effect on the "after-burn" reactions. This conclusion is rather puzzling in light of the fact that reactants ahead of the wave are expected to be heated by induction. The pre-heating of the reactants give rise to higher wave velocities, in contrast to the reported observations. Furthermore, the exertion of an inductive field on thermite-type reactions did not result in ignition and wave propagation *unless* the reactant powders were placed on a conducting (metallic) substrate [3]. In this case, therefore, the effect of the field is indirect: it supplies thermal energy to the reactants by conduction from the substrate. Thus while the application of an inductive field is reported to have an effect, the exact nature of this effect is not clear and further investigations are required before a clear picture of the role of this type of field is at hand.

The effect of an electric field in a contacting mode, i.e. through resistive means has previously been examined [4,5]. In reported studies, the field is utilized as a means to impart (Joule) heat to the reactants to raise their temperature until simultaneous combustion is realized. The temperature, T_e, to which the reactants are heated by the electric field can be calculated from

$$T_e = \frac{V^2 \tau}{R_e m C_p} + T_o \quad \textbf{(1)}$$

where V is the applied voltage, τ is the duration of the application of the voltage, R_e is the electric resistance of the sample, m is its mass, C_p is its heat capacity, and T_o is the initial temperature of the reactants. According to this equation, $T_e >> T_o$ is possible when the applied voltage is high and the resistance is low. Effectively, therefore, this method is restricted to systems in which the electrical conductivity of the reactants at room temperature is high. Experimental support for this expectation is provided in a later section of this paper. In the work of Yamada, et al. [4] the powder mixture of Si and C is sufficiently conducting to provide Joule heating up to the ignition temperature. Despite this, the method is successful only when the carbon particle size is small. As stated above, even when ignition occurs, the combustion is in the simultaneous mode and no self-propagating wave is discernible.

FIELD-ACTIVATED COMBUSTION SYNTHESIS (FACS)

The primary motivation to activate combustion reactions is to extend the application of the method to energetically-inadequate systems. Low-enthalpy reaction systems cannot self-sustain and require activation, commonly by thermal (i.e. pre-heating) means. Thermal activation, however, can result in the formation of other, often undesirable, phases [6,7]. Recently, a method has been developed through which the synthesis of these systems can be successfully achieved [8]. In the new method, the reactants are subjected to a contacting field and simultaneously ignited by a heat source. Under such conditions, a self-sustaining combustion wave is initiated. This field-activated combustion synthesis (FACS) has been used to prepare a variety of ceramic and composite materials which cannot otherwise be formed by self-propagating combustion synthesis. Examples of these include SiC, B_4C, WSi_2, WC, $MoSi_2$-SiC, B_4C-TiB_2, and others. The fundamental aspects of FACS and several important experimental observations will be discussed using two systems exemplifying the synthesis of single phase and composite materials.

SYNTHESIS OF B-SIC BY FACS

The adiabatic combustion temperature of SiC (1859K) is close to the empirically determined boundary for the existence of self-propagating combustion reactions [9]. When compacts of Si and C powders at 298K are heated by an ignition source, no self-propagating reaction can be generated. However, when the same procedure is followed while the compact is subjected to a field, as shown in Figure 1, a self-propagating combustion is established. It was observed, however, that a minimum (threshold) field strength is required to initiate the wave. This is shown in Figure 2 where the dependence of the wave velocity on field strength is depicted for two different initial densities. The threshold for SiC is about 6 V/cm and at this value,

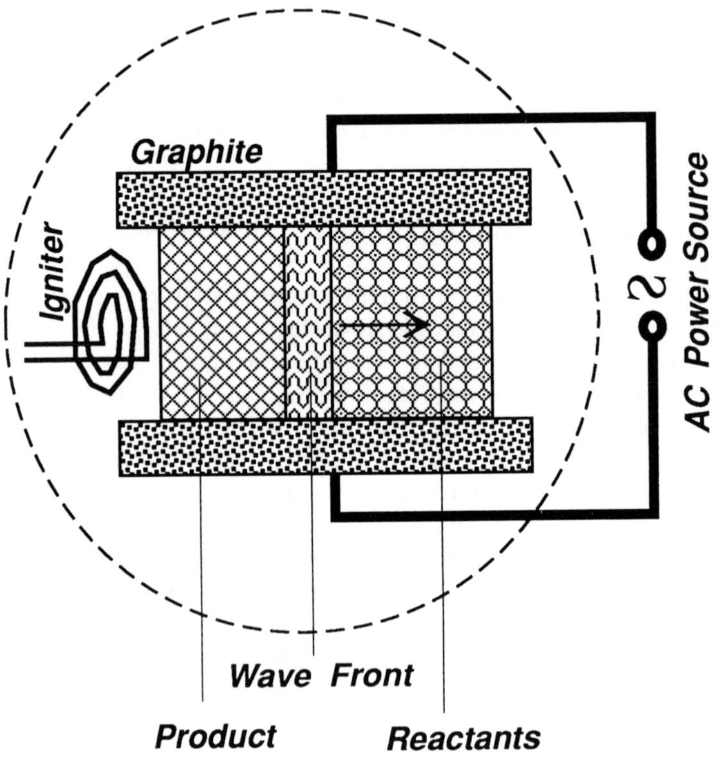

Figure 1. Schematic diagram of field-activated synthesis. (The arrow indicates the direction of wave propagation).

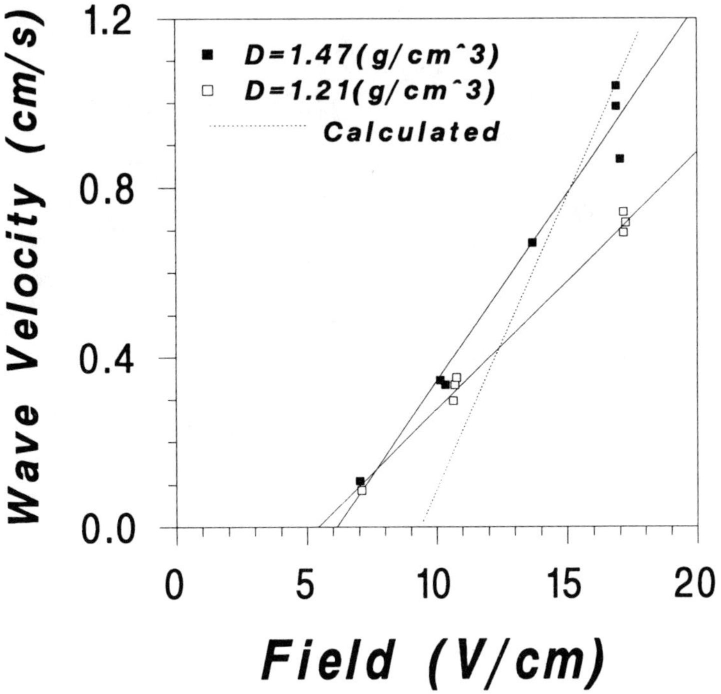

Figure 2. The dependence of the experimental and calculated wave velocity on the applied field. (The experimental results show the dependence on the initial sample density).

the wave propagates in a non-steady state (pulsating) mode. As the field is increased, the velocity increased linearly and the propagation mode became steady-state in nature. When the field is further increased to a value of about 21 V/cm, simultaneous combustion occurred without the need for an ignition source. The curve representing the calculated dependence of the velocity on field strength is obtained from modeling studies, as discussed subsequently. The product of the field-activated synthesis is β-SiC. Figure 3 shows the microstructure of the product and Figure 4 shows the x-ray diffraction peaks for the product in comparison to the reactants and a reference.

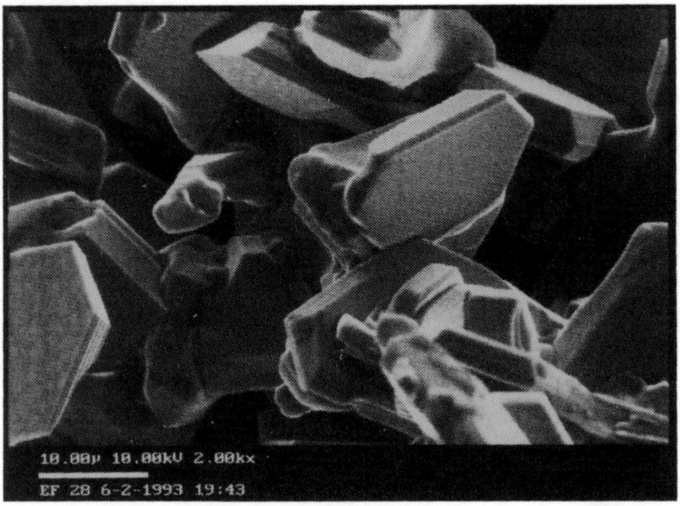

Figure 3. SEM micrograph of the β-SiC product.

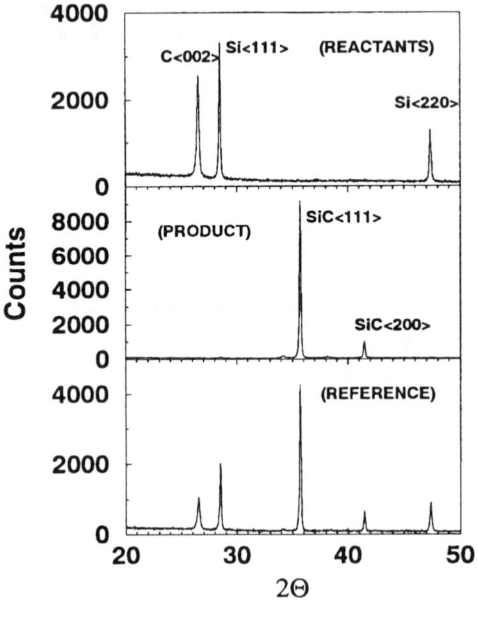

Figure 4. X-ray diffraction peaks of the reactants and product of synthesis.

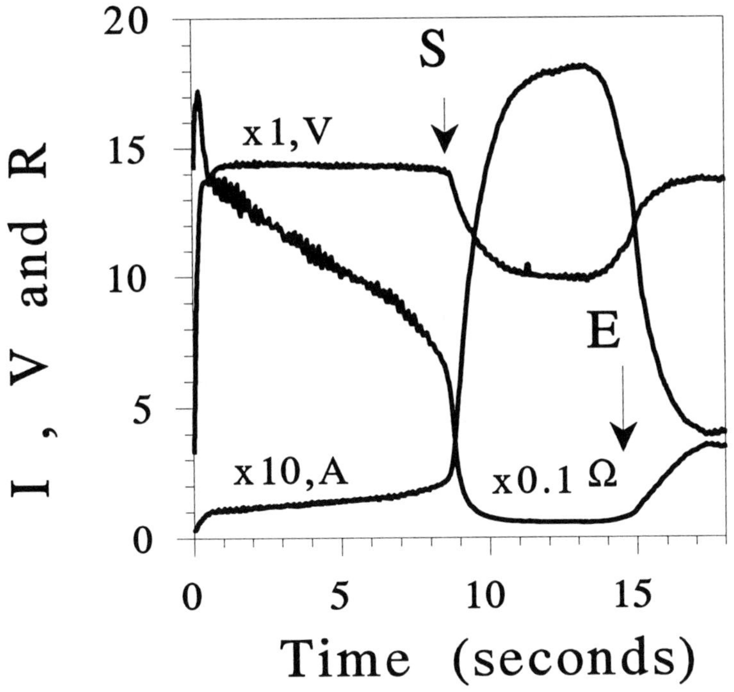

Figure 5. Current, voltage, and resistance changes during wave propagation in the synthesis of β-SiC.

The changes in the measured voltage and current during wave propagation are shown in Figure 5. Also shown in this figure is the curve for the calculated resistance. The symbols S and E refer to the start and end of wave propagation. The curves in Figure 5 show that the electric parameters (particularly the resistance) remain relatively constant during the duration of wave propagation. In view of the fact that the product (SiC) and reactants (Si+C) differ significantly in their electrical conductivity, the constant nature of the resistance during wave propagation is unexpected. To account for this observation, we proposed that the effect of the field is confined to the combustion zone [10-12]. The proposed model assumes that the current is primarily localized in the combustion zone and its contribution to the process is in the form of

Joule heating in this relatively narrow region. Accordingly, the Fourier relationship describing such reactions is now modified to include another heat source, i.e.

$$\rho C_p \frac{\partial T}{\partial t} = \frac{\partial}{\partial x}(\kappa \frac{\partial T}{\partial x}) + \rho Q \frac{\partial \eta}{\partial t} + \sigma E^2 \tag{2}$$

Eq (2), which assumes adiabatic conditions, contains the heat accumulation term (left-hand side) with ρ and C_p being the density and heat capacity, T is temperature, and t is time. The other terms (right-hand side) are the heat conduction and heat generation terms with κ being the thermal conductivity, x is the coordinate of wave travel, Q is the enthalpy of the reaction, η is the fraction converted, σ is the electrical conductivity, and E is the field. The last two terms represent heat generation rates from chemical and electrical sources, respectively.

Modeling studies were made on the basis of the modified Fourier equation (Eq 2)[11]. In doing so, the equation was generalized to account for heat loss and extended to provide a two-dimensional analysis. Temperature, current density, and conversion profiles were calculated as functions of the strength of the applied field. Figures 6(a) and (b) show , respectively, the current density and conversion profiles at E = 14.2 V/cm. Each of the curves in these figures represents the profile at a given value of time and location along the sample. Figure 6(a) shows that the current density is highly localized. At t = 2.5s the current density peak corresponds to a distance of 1.0 cm along the sample. At this same location and time, the conversion curve, Figure 6(b), shows that this condition represents the leading edge of the combustion wave. In other words, the current from the applied field is confined largely to the region where the reaction between Si and C is commencing. The high current density in this region is the consequence of higher electrical conductivity at the elevated temperature of ignition. In real systems, however, the higher conductivity may also be related to the existence of a molten phase (Si) giving rise to better particle-particle contacts. In agreement with the experimental observations, the modeling studies also showed that the velocity of the wave increased linearly with an increase in the applied field. The results of the modeling studies are shown in Figure 2 along with the experimental results. The agreement between these two sets of results is qualitatively consistent and quantitatively reasonable.

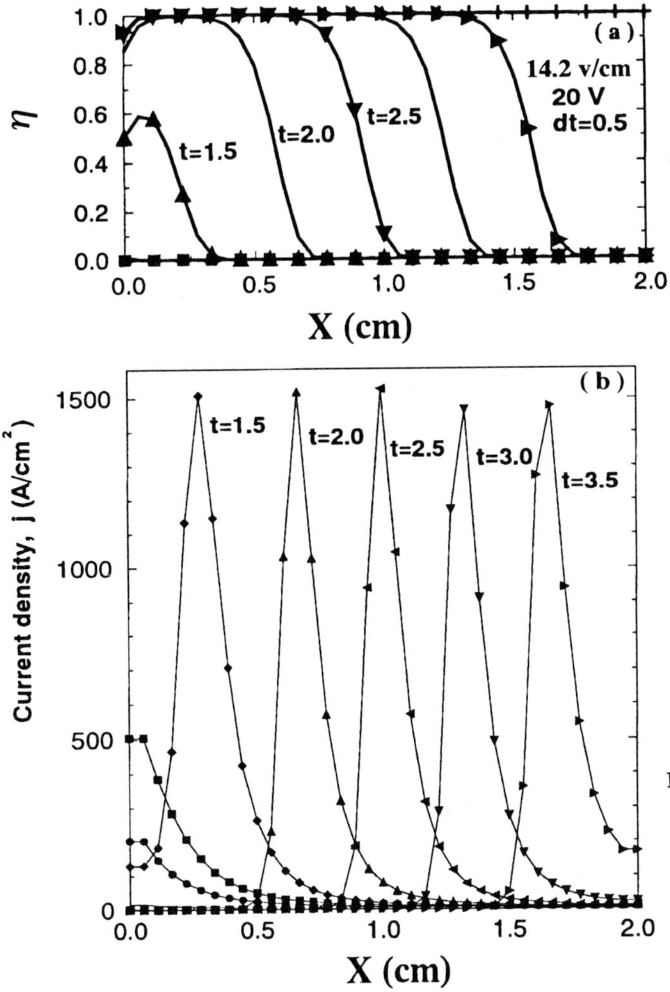

A

B

Figure 6. Calculated profiles of (a) Conversion and (b) Current density for the synthesis of β-SiC.

The synthesis of composites in one step is one of the most attractive features of the self-propagating high-temperature synthesis (SHS). Elemental reactants can be used to synthesize the final phases in one step. However, as is the case for single phases, this approach is limited by the enthalpy change associated with the synthesis reaction. Energetically less favorable composites cannot be synthesized by this method without some form of activation. For example, the composites (1-x) $MoSi_2$-xSiC can only be prepared when the SHS process is field-activated [13]. Despite its attractive oxidation resistance, $MoSi_2$ cannot be used in high temperature applications because of shortcomings in its mechanical properties [14]. The addition of SiC and other phases is made to enhance these properties. However, since the adiabatic temperatures of $MoSi_2$ (1900K) and SiC (1895K) are marginal with respect to SHS reactions, composites of these two phases cannot be synthesized directly [15,16].

The synthesis of these composites is represented by the following reaction

$$(1 - x)\,Mo + (2 - x)\,Si + x\,C = (1 - x)\,MoSi_2 + x\,SiC \qquad (3)$$

For $x = 0$, SHS reactions are possible without the application of a field. Self-propagating reactions were also observed for $0 < x < 0.18$, without the field. However, the product in such cases contains $MoSi_2$ and unreacted Si and C. In contrast, under the influence of a field, mixtures with $0 \leq x \leq 1.0$ can be reacted in a self-propagating manner and the products contain the desired phases, $MoSi_2$ and SiC as shown in Table 1. As in the case of the synthesis of SiC by FACS, combustion wave propagation is only possible when the field is above a threshold value. The voltages indicated in Table 1 are the threshold values for the indicated composite compositions.

The formation of other composites by FACS has also been investigated. Composites of $MoSi_2$ - Nb and $MoSi_2$ - ZrO_2 were synthesized by this method. In the case of the former, the reactants were Mo, Si, and Nb and in the case of the latter the reactants were Mo, Si, and ZrO_2. Composites of $MoSi_2$ - xNb, with $x < 0.45$ can be synthesized by SHS without the application of a field. At $x \geq 0.45$, however, self-propagation synthesis is only possible in the presence of a field. As indicated earlier, the presence of a field is not sufficient to initiate the wave; an ignition source is required. This is clearly demonstrated by the voltage and current curves of Figures 7(a) and (b). Figure 7(a) shows that the application of a voltage of 20V has no effect on the reactant mixture. The low current excludes the possibility of Joule heating and thus no reaction is initiated. In contrast, when an ignition source is present (Figure 7(b)), a self-propagating wave is initiated. During wave propagation, the voltage decreases and the current increases correspondingly. As in the case of the field-activated synthesis of SiC, the wave velocity increased with increasing applied voltage, as seen in Figure 8. When voltages exceeding about 25V are applied, simultaneous combustion occurs without the need for an ignition source.

Table 1. Threshold Fields and Product Composition in the Field-Activated Combustion Synthesis of (1-x) $MoSi_2$-xSiC Composites

x	Vol% SiC	Applied Voltage (V) (Threshold)	Product Phases (s)
0	0	0	$MoSi_2$
0.18	0-10	0	$MoSi_2$, Si, C
0.26	15	44	$MoSi_2$, SiC
0.32	20	33	$MoSi_2$, SiC
0.45	30	13.5	$MoSi_2$, SiC
0.72	55	11.0	$MoSi_2$, SiC
0.85	75	13.0	$MoSi_2$, SiC
1.00	100	8.5	SiC

The synthesis of composites of B_4C - TiB_2 by SHS has been previously reported [16]. However, such composites can only be synthesized at 298K when the reactants correspond to a product molar ratio, B_4C/TiB_2 (= y) of $\leq$ 0.5. Composites with higher y values could only be synthesized if the reactants are pre-heated. However, as indicated above, pre-heating resulted in the formation of other phases (TiC, TiB) in the product in addition to the desired components [7]. The goal to produce lighter and harder B_4C - TiB_2 composites is the basis for the desire to increase the molar ratio, y and hence to activate this process. Elemental mixtures of B, C, and Ti corresponding to product molar ratios of 1-5 were investigated. As before, self-propagating reactions could be generated with the application of voltages above a minimum (threshold) value. As the applied voltage increased above the threshold, the velocity of the wave also increased, as shown in Figure 9. The curves in this figure, which begin at the respective threshold values, represent composite systems with $1.0 \leq y \leq 5$. In all cases, the product contained the phases B_4C and TiB_2 only.

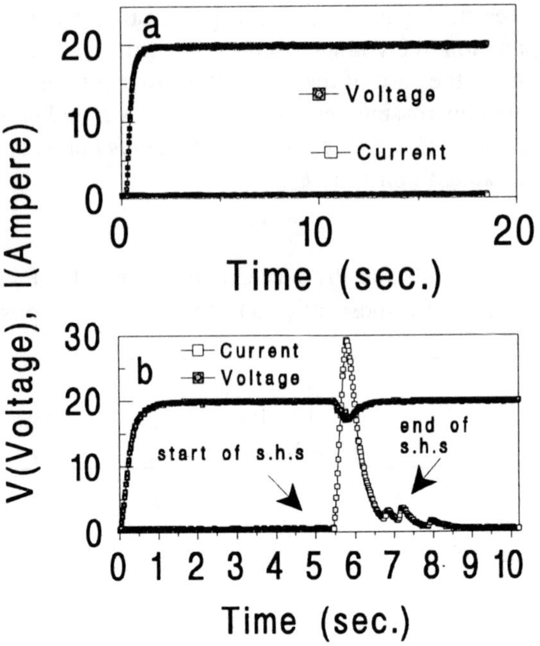

Figure 7. Voltage and current changes during the field-activated synthesis of $MoSi_2$-0.45 Nb:

(a) Without ignition (V = 20 volts)

(b) With ignition (V = 20 volts)

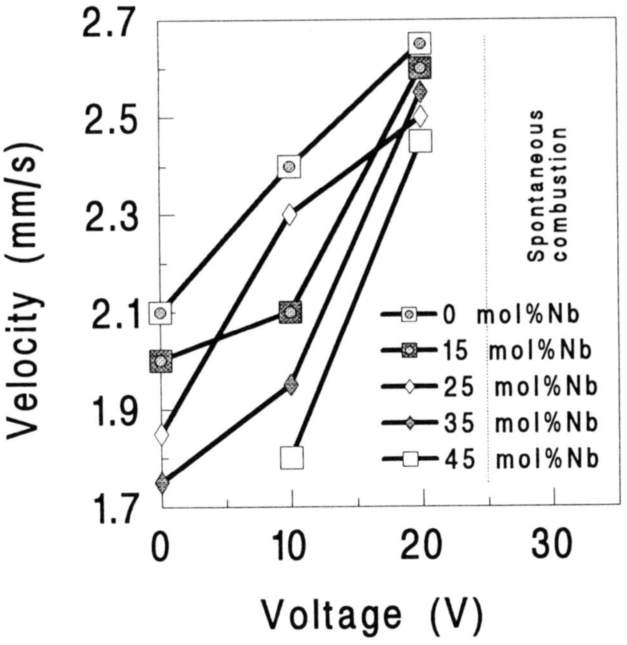

Figure 8. The dependence of wave velocity on the applied voltage in the synthesis of $MoSi_2$-xNb. (x refer to molar % of Nb in this figure)

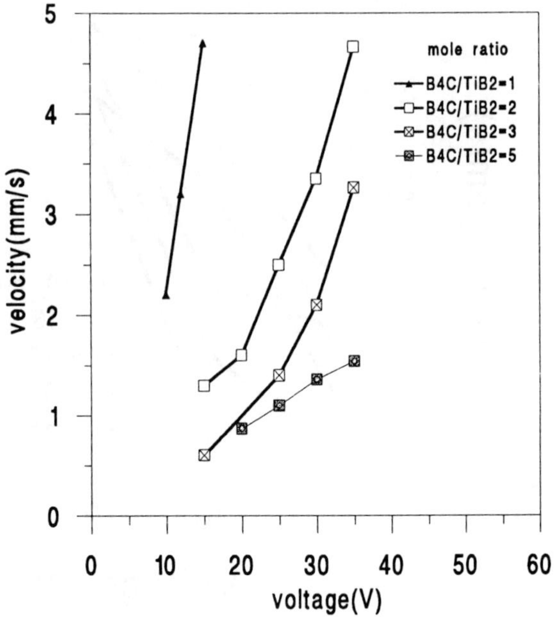

Figure 9. The dependence of wave velocity on the applied voltage in the synthesis of B_4C-TiB_2 composites.

SUMMARY

The use of an electric field to activate self-propagating combustion synthesis is demonstrated. Self-sustaining combustion waves can be established in systems which have relatively low reaction enthalpies. Wave propagation is established when the applied field is above a threshold value for any given reaction system. The velocity of wave propagation increases nearly linearly with an increase in the field. Experimental and modeling studies on the synthesis of β-SiC are described. It is proposed and subsequently demonstrated that the role of the electric field is to provide additional heat to the reaction zone at a rate of σE^2. The analysis further demonstrated that the current is localized to the leading edge of the combustion wave.

The application of the field-activated combustion synthesis (FACS) to the synthesis of composite materials is successfully demonstrated. Composites of (1-x) MoSi2-xSiC, with $0 \leq x \leq 1.0$ have been synthesized by this method. Other composites prepared by FACS include $MoSi_2$-Nb, $MoSi_2$-ZrO_2, and B_4C-TiB_2.

The above observations and analysis indicate the field can be considered as a *processing parameter* in self-propagating synthesis.

ACKNOWLEDGMENTS

The financial support of this work by the National Science Foundation is gratefully acknowledged.

REFERENCES

1. A. I. Kirdyashkin, Y. M. Maximov, and A. G. Merzhanov, "Effect of a Magnetic Field on the Combustion of Heterogeneous Systems with Condensed Reaction Products", *Combust. Explos. Shock Waves*, **22**, 700-706 (1986).

2. A. I. Trofimov, Y. I. Yukhvid, and I. P. Borovinskaya, "Combustion in Condensed Systems in External Electromagnetic Fields", *Int. J. SHS,* **1**, 67-71 (1992).

3. A. I. Trofimov and V. I. Yukhvid, "SHS Surfacing in an Electromagnetic Field", *Int. J. SHS*, **2**, 343-348 (1993).

4. O. Yamada, Y. Miyamoto, and M. Koizumi, "Self-Propagating High-Temperature Synthesis of SiC", *J. Mater. Res.*, **1**, 275-279 (1986).

5. V. A. Knyazik and A. S. Shteinberg, "High-Temperature Interaction in the TaC System Under Electrothermal Explosion Conditions", *J. Mater. Synth. Process.*, **1,** 85-92 (1993).

6. S. C. Deevi, "Diffusional Reactions in the Combustion Synthesis of $MoSi_2$", *Mater. Sci. Eng.*, **A149**, 241-251 (1992).

7. D. C. Halverson, B. Y. Lum, and Z. A. Munir, "the Combustion Synthesis of Boride Composites", *Proc. Sympos. High Temp. Mater. Chem.*, Proceedings Volume 88-5, Electrochemical Society, Pennington, NJ, 1988, p. 613.

8. Z. A. Munir, W. Lai, and K. Ewald, "Field-Assisted Combustion Synthesis", *U.S. Patent*, No. 5, 380, 409, issued January 10, 1995.

9. Z. A. Munir, "Synthesis of High-Temperature Materials by Self-Propagating Combustion Methods", *Amer. Ceram. Soc. Bull.*, **67**, 342-349 (1988).

10. A. Feng and Z. A. Munir, "Field-Assisted Self-Propagating Synthesis of β-SiC", *J. Appl. Phys.*, **76**, 1927-1928 (1994).

11. A. Feng and Z. A. Munir, "The Effect of an Electric Field on Self-Sustaining Combustion Synthesis, Part I: Modeling Studies", *Metall. Mater. Trans.*, in press, 1994.

12. A. Feng and Z. A. Munir, "The Effect of an Electric Field on Self-Sustaining Combustion Synthesis, Part II: Field-Assisted Synthesis of β-SiC", *Metall. Mater. Trans.*, in press, 1994.

13. S. Gedevanishvili and Z. A. Munir, "Field-Assisted Combustion Synthesis of $MoSi_2$-SiC Composites", *Scripta Metall. Mater.*, **31**, 741-743 (1994).

14. A. K. Vasudevan and J. J. Petrovic, "A Comparative Overview of Molybdenum Disilicide Composites" *Mater. Sci. Eng.*, **A155**, 1-17 (1992).

15. Z. A. Munir and U. Anselmi-Tamburini, "Self-Propagating Exothermic Reactions: The Synthesis of High-Temperature Materials by Combustion", *Mater. Sci. Rep.*, **3**, 277-365 (1989).

16. D. C. Halverson, Z. A. Munir, and B. Y. Lum, "Combustion Synthesis of Ceramic Preforms for Molten-Metal Infiltration", in *Combustion and Plasma Synthesis of High-Temperature Materials*, Z. A. Munir and J. B. Holt, Editors, VCH Publishers, New York, 1990, p. 262.

SYNTHESIS AND DENSIFICATION OF $MoSi_2$ BY SELF-PROPAGATING HIGH-TEMPERATURE SYNTHESIS

Gi-Wook Lee, Hyun-Woo Lee, Young-Do Kim* and Yong-Seog Kim
Department of Metallurgy and Materials Science, Hong Ik University, 72-1 Sangsu Dong Mapo Ku, Seoul, Korea.
* : Special Alloy Research Laboratory, Korea Institute of Science and Technology, P.O. Box 131, Cheongryang, Seoul, Korea.

ABSTRACT

The possibility of producing dense $MoSi_2$ by the thermal explosion mode of Self-propagating High-temperature Synthesis(SHS) was demonstrated. It was shown that the density of $MoSi_2$ produced by SHS process without applied pressure can be increased by a proper combination of processing parameters such as heating rate, addition of alloying elements, processing atmosphere, and Mo and Si powder sizes used. The heating rate and alloying elements were found to influence the density of $MoSi_2$ formed the most. Proper combinations of the processing parameter yielded $MoSi_2$ with density close to 90% theoretical density which is comparable to pressurelessly sintered $MoSi_2$.

INTRODUCTION

The morphology of $MoSi_2$ produced via SHS process is of very important concern in order for this process to be used in actual manufacturing. The morphology of the SHS products has been known to be influenced by several processing variables including combustion temperature, sizes of reacting particles, heating rate to the ignition temperature, etc.[1-3]. Especially, the combustion temperature which is affected by the preheating temperature and the diluent content was found to affect the morphology of the SHS product dramatically[4]. For instant, the size of $MoSi_2$ synthesized without preheating was found to be in a range from 2 to 3 μm in diameter. The product was found to be in a loosely agglomerated form of powder and thus could be easily milled into $MoSi_2$ powder. On the other hand, when synthesized with preheating at 500℃, the grain size increased to 20~30μm and the product had solid networks with substantial strength. This product could not be easily milled into powder. This increased agglomeration with higher preheating temperature is thought to be due to the partial melting of $MoSi_2$ during the SHS reaction. The calculation of the adiabatic temperature of the reaction indicated that the combustion temperature reaches the melting temperature of $MoSi_2$ as the preheating temperature becomes higher than 430℃[4]. Thus as the sample is preheated at is 500℃, some part of $MoSi_2$ formed should be liquid. This liquid phase may wet the particles resulting the solid network of powder upon solidification.

$MoSi_2$ has been synthesized also by uniformly heating the Mo and Si powder compact to the ignition temperature, which process is termed as the thermal explosion mode of SHS[5-6]. Since the ignition temperature of $MoSi_2$ synthesis is around 1300℃[7], a large portion of $MoSi_2$ formed during the thermal explosion mode of SHS reaction is expected to form in liquid phase. Thus proper combinations of synthesis reaction and solidification process variables might lead to simultaneous synthesis and densification of $MoSi_2$ without applied pressure. In this study, an attempt was made to investigate the effects of the processing variables on the densification of $MoSi_2$ during the thermal explosion mode of SHS.

EXPERIMENTAL PROCEDURES

Mo powder of 2~5㎛ in diameter and Si powder of -325mesh were mixed in a Spex mill for 3 hours. The powder mixture was compacted into a cylindrical shape in a steel die under a 100MPa compressive pressure. The compact with 1.3cm in diameter and 2.5cm in length was heated in a graphite heating furnace to 1450℃ to initiate the thermal explosion mode of SHS. The heating rates to the reaction temperature was varied from 20℃/min to 90℃/min. Normally the reaction was carried out in an argon gas atmosphere. Either aluminum or titanium powder was added to Mo and Si powder mixtures and its effect on the densification was investigated. The density of specimens was measured using the Archimedes' method and the phases consisting the reaction products were analyzed using XRD. The microstructures were analyzed using optical microscope and scanning electron microscope.

RESULTS AND DISCUSSIONS

Fig. 1) shows a SEM micrograph of the specimen manufactured by heating the sample at a rate of 30℃/min to 1450℃. As seen in the photograph, the sample contains many pores and its density was measured to be 60% of theoretical density(TD). XRD analysis indicated that the phase formed is single phase $MoSi_2$. This relatively poor density of the products might be attributed to several aspects of SHS process. Firstly, the large portion of the porosity could be originated from the pores of the reactant powder compact. The Mo and Si powder compact density under 100MPa pressure was measure as 63% TD. Therefore, a large portion of these pores could remain in the product unless pore filling process occurs actively during the SHS process. The pore filling and particle rearrangement should be facilitated by the addition of low melting alloying element such as aluminum to the Mo and Si powder mixture compact.

Fig. 1) SEM micrograph of the specimen produced by heating Mo + 2Si powder compact at a rate of 30℃/min to 1450℃.

Secondly, the pores could be caused by the volume shrinkage occurring during the compound formation. When elemental Mo and Si react to form $MoSi_2$, the volume is calculated to decrease by 40%. Unless the sample is allowed to shrink uniformly, the volume shrinkage should lead to the creation of pores. The volume shrinkage would be interfered by the skeleton of high melting point compounds formed at Mo/Si powder interface during heating to the ignition temperature. The degree of solid skeleton formation should be affected by the heating rate to the reaction temperature as well as by the contact area between Mo and Si powder. In addition, the smaller grain(or increased nucleation rate) may enhance the uniform shrinkage during the solidification. The smaller grain could be obtained by the addition of diluent such as presynthesized $MoSi_2$ powder. The diluents could serve as heterogeneous nucleation sites during $MoSi_2$ solidification. Thus proper heating rate and $MoSi_2$ diluent content must be determined for the densification of the reaction product. Finally, pores could be caused by the gas evolved during the reaction. In order to reduce the gas evolution, the Mo and Si powder compact was degassed in a vacuum of 10^{-3} torr at 500℃ for 3 hours and the reaction was carried in argon gas of atmospheric pressure.

A. Effects of alloying elements on the densificaiton

Fig. 2) shows SEM micrographs of the specimen synthesized with different contents of aluminum. Aluminum compositions were x = 0.1, 0.2, and 0.3 in Mo + (2-x)Si + xAl. XRD analysis of the products indicated that $MoSi_2$ as well as $Mo(SiAl)_2$ phases were formed. As seen in the figure, the size of $MoSi_2$ increased with aluminum content. The density of aluminum alloyed $MoSi_2$, however, was improved significantly compared with pure $MoSi_2$(Fig. 1).

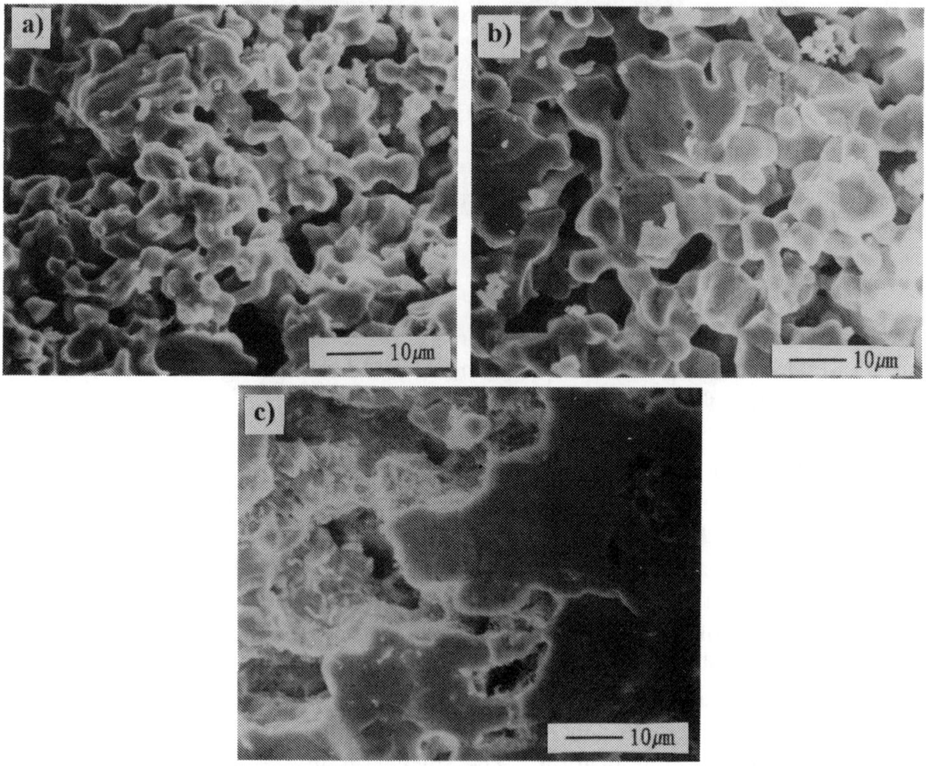

Fig. 2) SEM micrographs of the specimens with aluminum content (a) x = 0.1, (b) x = 0.2, and (c) x = 0.3 in Mo + (2-x)Si + xAl.

On the other hand, the densification of the specimens with titanium addition was improved as Ti content increased(Fig. 3). Titanium compositions were x = 0.05,

0.1, and 0.2 in Mo + 2Si + x(5Ti +3Si). When the titanium content was x = 0.1, the densification seemed to improve most. XRD analysis of the products indicated that the phases formed were $MoSi_2$ and $(MoTi)Si_2$.

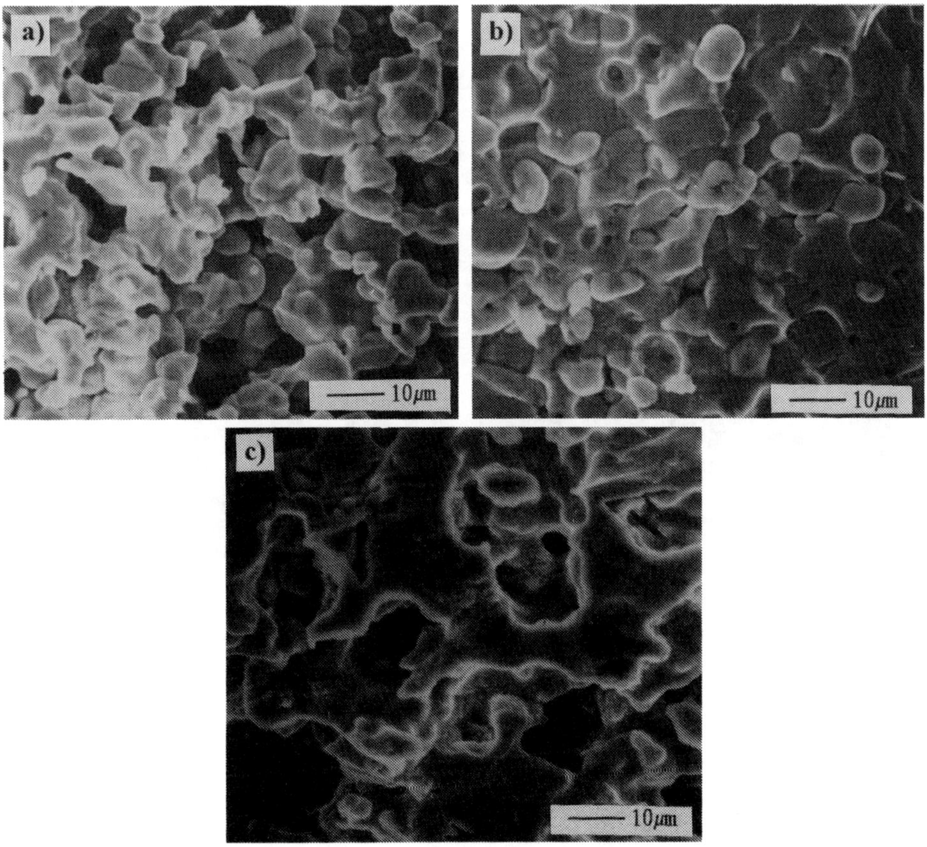

Fig. 3) SEM micrographs of the specimens with titanium content (a) x = 0.05, (b) x = 0.1, and (c) x = 0.2 in Mo + 2Si + x(5Ti+3Si).

B. Effects of the heating rate on the densification.

The microstructures of samples manufactured at several heating rates including 50, 70 and 90℃/min were investigated using SEM (Fig. 4). The composition of the powder compact was Mo + 1.8Si + 0.2Al. As the heating rate is increased, the density of the product was observed to increase. Fig. 5) shows the densities of the products produced at different heating rates. The density increased from 60% TD to 84% TD as the heating rate was increased from 20 to 90℃/min. The improved the skeleton of densification at higher heating rates might be attributed to the reduced amount of precombustion compounds as well as increased combustion temperature.

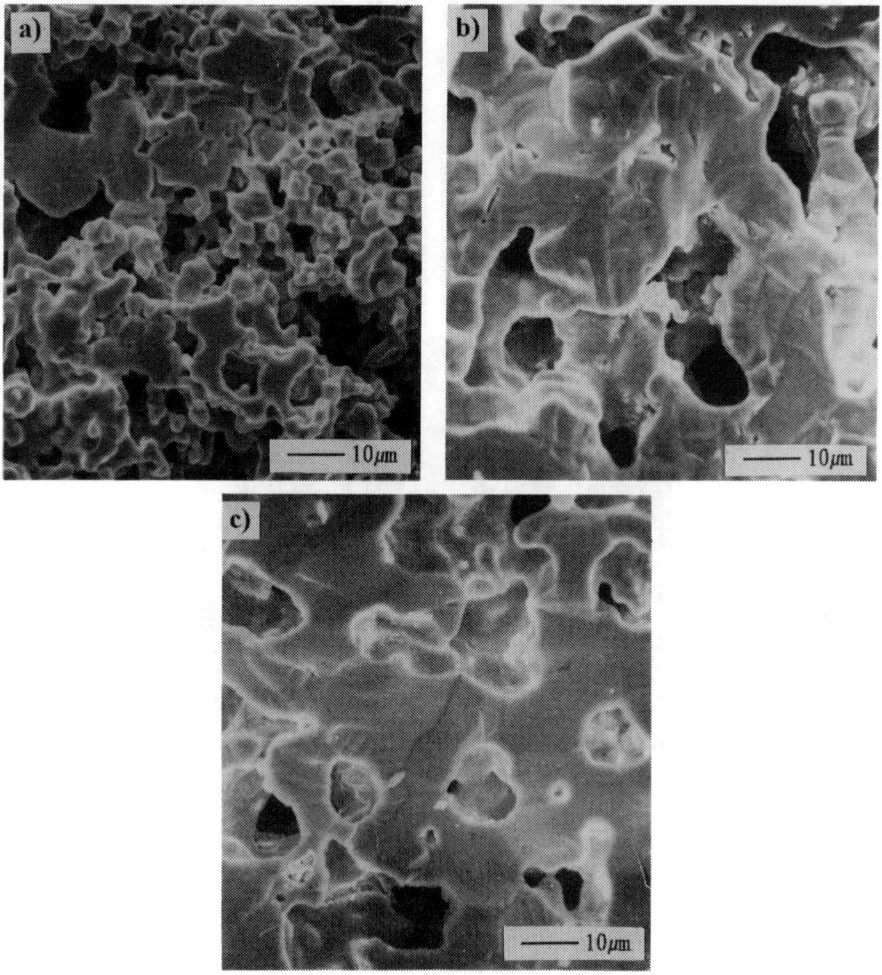

Fig. 4) SEM micrographs of the specimen with different heating rates including (a) 50℃/min, (b) 70℃/min, and (c) 90℃/min. The composition of the sample was Mo + 1.8Si + 0.2Al.

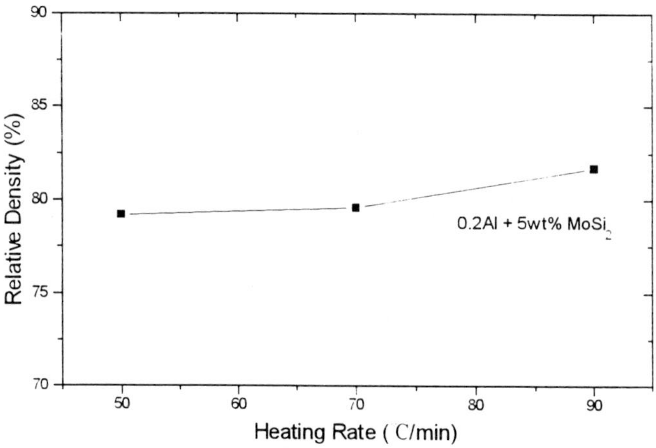

Fig. 5) Effect of the heating rate on the density of the reaction product. The composition of the sample was Mo + 1.8Si + 0.2Al with 5% $MoSi_2$ diluent addition.

C. Effects of the diluent content on the densification.

$MoSi_2$ powder was added as the diluent to Mo + 1.8Si + 0.2 Al powder mixture. The content was 2, 5, and 10% respectively. As in Fig. 6), as the diluent increases, the grain size of $MoSi_2$ formed was observed to decrease, which should be caused by the heterogeneous nucleation on the diluent and/or by the increased nucleation rate due to decreased combustion temperature. Since $MoSi_2$ formation during the SHS process is known to occur via the solution and precipitation mechanism[4], it is unlikely that the grain size refining is caused by the grain boundary pinning from the diluent. Also the pore size was observed to decrease as the diluent content increases. The density, however, peaked when the diluent content was 5%. Fig. 7) shows the SEM micrographs of the specimen with 2 and 5% $MoSi_2$ diluent additions to Mo + 2Si + 0.1(5Ti+3Si) powder compact. As seen in the figure, the sample density improved dramatically compared with unalloyed $MoSi_2$ and the large pores have been completely eliminated. The densities of the specimens with different contents of diluent are shown in Fig. 8). The maximum density of the specimen obtained was 89% TD, which is close to the pressurelessly sintered $MoSi_2$.

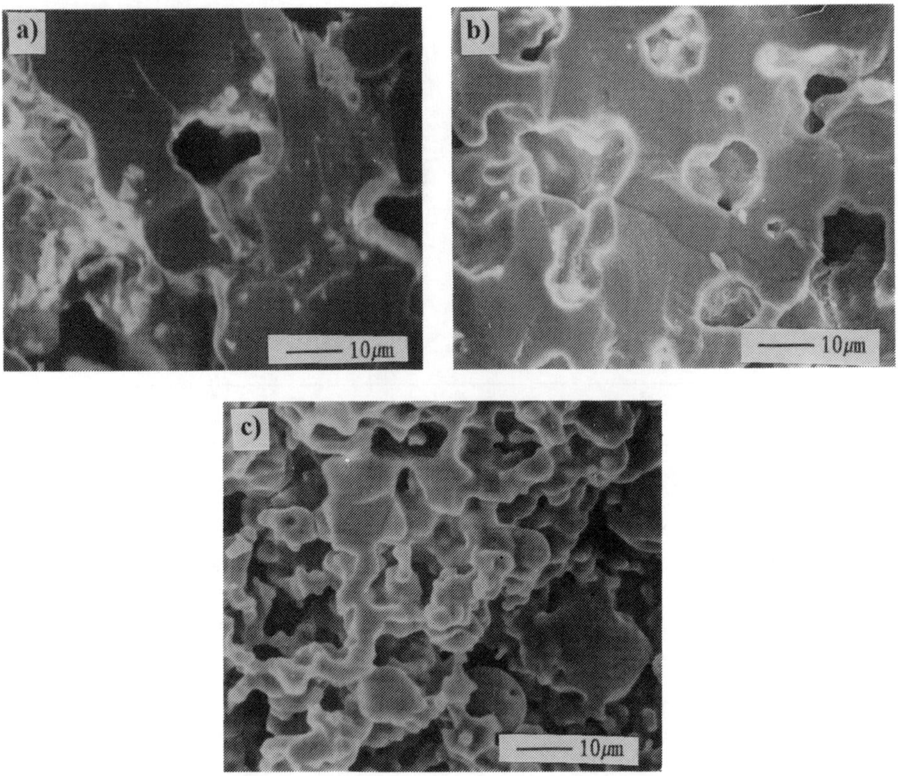

Fig. 6) SEM micrographs of the specimens with different $MoSi_2$ powder diluent contents ; (a) 2%, (b) 5%, and (c) 10 % in a Mo +1.8Si + 0.2Al powder compact.

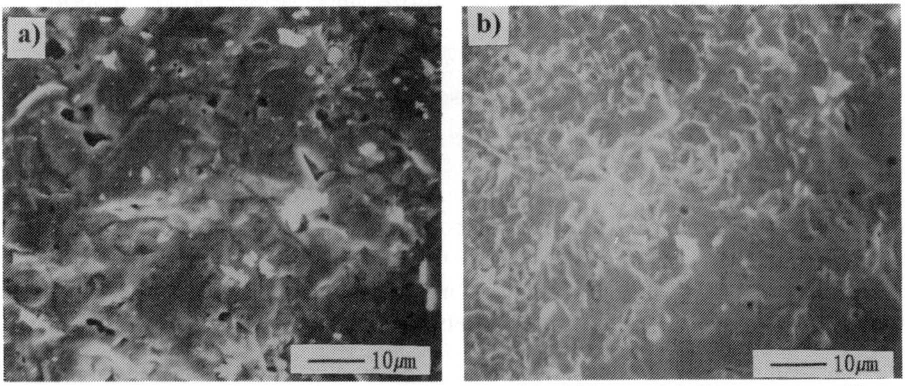

Fig. 7) SEM micrographs of the specimens with different $MoSi_2$ powder diluent contents (a) 2% and (b) 5% in a Mo +2Si + 0.1(5Ti+3Si) powder compact.

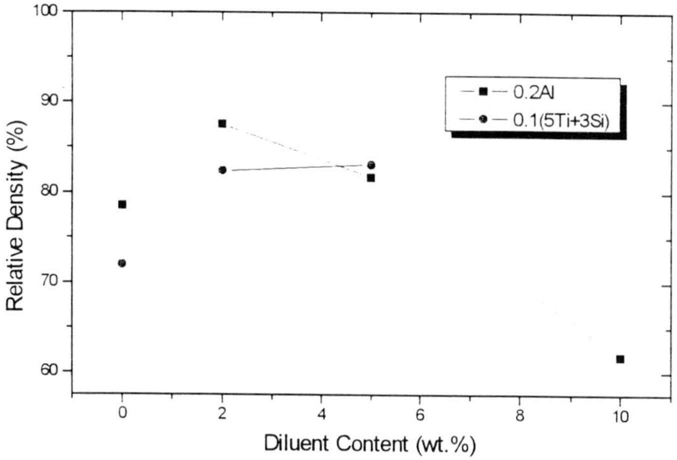

Fig. 8) The densities of the SHS reaction products with different amount of $MoSi_2$.

SUMMARY

A typical cycle of thermal explosion mode of SHS reaction yield a product with 60% TD. Processing variables including addition of alloying element such as aluminum or titanium, high heating rate and addition of diluent were found to be very effective in improving the densification of the SHS product. Combination of these processing variables yielded $MoSi_2$ with 89% TD, which is close to the density of the pressurelessly sintered $MoSi_2$.

REFERENCES

1. J. Subrahmanyam and M. Vijayakumar, "Self-propagating high-temperature synthesis", *J. Mater. Sci.* 27 (1992) 6249.
2. Y. L. Jeng and E. J. Lavernia, "Processing of molybdenum disiliside", *J. Mater. Sci.* 29 (1994) 2557.
3. S. Zhang and Z. A. Munir, "Synthesis of molybdenum silicides by the self-propagating high-temperature combustion method", *J. Mater. Sci.* 26 (1991) 3685.
4. S.W. Jo, G.W. Lee, J.T. Moon and Y.S. Kim, "On the formation of $MoSi_2$ by

self-propagating high-temperature synthesis", to be published in Acta Metallugica.
5. S. C. Deevi, "Diffusional reaction in the combustion synthesis of $MoSi_2$", *Mater. Sci. Eng.* A149 (1992) 241.
6. M. D. Ka, Y. D. Kim and Y. S. Kim, "Densification of $MoSi_2$ processed by self-propagating reaction hot press", *Proceedings of the PM'94* VOL. 2 (1994) 1477.
7. S. C. Deevi, "Differential thermal analysis of exothermic reaction in the synthesis of $MoSi_2$", *J. Mater. Sci. Lett.* 13 (1994) 868.

PROCESSING OF ALUMINUM NITRIDE WHISKERS BY COMBUSTION SYNTHESIS

Kyung-Jae LEE, Suk-Yoon KIM and Yong-Seog KIM
Department of Metallurgy and Materials Science, Hong Ik University, 72-1 Sangsu-Dong Mapo-Ku, Seoul, Korea. 121-791

ABSTRACT

AlN powder and whiskers were synthesized by direct nitridation of aluminum powder in pure nitrogen gas. The nitridation reaction of aluminum powder was initiated by heating the sample to the ignition temperature and the reaction was finished in less than 3 minutes. AlN whisker-shaped morphology was observed predominantly when the sample was heated above 900℃.

INTRODUCTION

AlN has a very high thermal conductivity (AlN single crystal : 320 watt/mK), a thermal expansion coefficient ($\alpha = 4.3 \times 10^{-6}$/K at room temperature) closer to that of silicon, relatively high electrical resistivity($\rho > 10^{13}\,\Omega$cm), and a relatively low dielectric constant ($K' = 8.9$ at 1 MHz at room temperature).[1] These properties make AlN one of the best candidate material for electronic packaging applications.[2] The high heat dissipation rate of AlN substrate could be a crucial property as the packing density of the transistor increases. Thus, a significant research effort was directed toward producing AlN powder and substrates with high thermal conductivity. On the other hand, there have been attempts to mix AlN whiskers with polymer packaging materials in order to increase the thermal conductivity of the polymer packages.[3] Various types of polymers including polyimide and epoxy are currently being used as a primary packaging material because they have low dielectric constants and are easy to fabricate, but have low thermal conductivity.[4,5] Thus, in order to dissipate the heat generated by the chip, the thermal conductivity of the polymers must be improved. It has been predicted that the thermal conductivity of the polymer matrix composites reinforced by AlN powders with acicular morphology can be improved by a factor of 70.[3] Therefore, it is expected that the thermal conductivity of polymer composites with AlN whisker reinforcement could be improved even further than composites reinforced by acicular AlN powder. However, the difficulties in manufacturing AlN powder and whiskers have to be overcome for these applications to be feasible.

AlN powder has been produced by the carbothermal reduction method, in which Al_2O_3 powder mixed with carbon is heated to 1700~1800℃ for several hours under a nitrogen atmosphere.[6] Since this method requires very high reaction temperatures and an additional nitridation step for full conversion, the production cost is extremely high, prohibiting general use of this material. On the other hand, AlN whiskers have been produced by the Vapor-Liquid-Solid(VLS) method.[7] The VLS process has been used to produce high quality SiC whiskers. This process requires several days to produce whiskers of usable size.

The Self-propagating High-temperature Synthesis (SHS) process[8] has attracted research interests in producing ceramic powders and whiskers because of its high energy efficiency, productivity, and ability to produce higher purity products compared with conventional processes. Since this process utilizes the formation energy of the compound to synthesized, it needs little external heating during the synthesis, making the process very energy efficient. The SHS process has been used in producing AlN powder and dense product.[9,10] Thus in this work, an attempt was made to study the mechanism of AlN powder and/or whisker formation during SHS reaction.

EXPERIMENTAL PROCEDURE

Al powder (<5 μm : Cerac Co. Wisconsin, USA) mixed with reaction catalysts was compacted under 14~55MPa pressure. The diameter of the cylindrical compact was 17mm and its height was 25mm. The compact was put into an alumina crucible and heated to the reaction temperature under a flowing nitrogen atmosphere. The flowing rate of the nitrogen gas was 500 ml/min and the heating rate was 5℃/min. The temperature of the sample was measured by a C-type thermocouple and acquired by a computer data acquisition system. The products of the reaction were analyzed using XRD, EDS, SEM, and TEM.

RESULTS AND DISCUSSION

A. Thermal analysis of nitridation via Combustion Synthesis

Thermal analysis of the Al compact was performed using TG-DTA (Mac Science Co. Japan) in order to determine the ignition temperature of Combustion Synthesis nitridation reaction. As shown in Figure 1 the weight of the sample continued to decrease until the aluminum melting point(~660℃) is reached. At this the temperature, a sharp exothermic peak (the first stage reaction) was detected followed by another larger peak (the second stage reaction). Also the weight of the sample increased steeply at the start of the first stage reaction and kept increasing during the second stage reaction. This indicates that nitridation reaction of Combustion Synthesis occurred via two reaction stages.

Since the compact weight of the TG-DTA sample is very small(<20mg), the temperature measured may not represent the actual reaction temperature. Thus the temperature profile of the compact during actual Combustion Synthesis reaction was measured by inserting a C-type thermocouple into a reacting compact. As seen in Figure 2, the ignition and combustion temperatures are different from the TG-DTA results, but the temperature profiles are similar. At the first reaction stage, the reaction temperature reached 1227℃ and at the second reaction stage, the reaction temperature was 1927℃. Visual observation indicated that the reaction occurs mainly at the surface of the compact during the first stage reaction and inside of the compact at the second stage reaction.

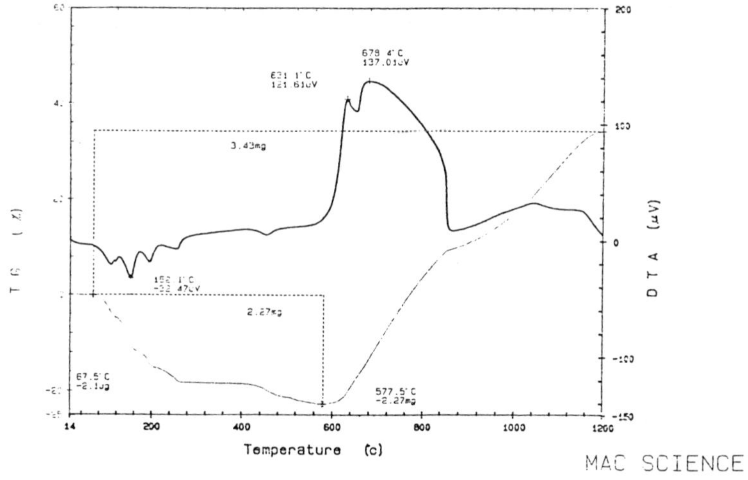

Figure1. TG-DTA result of sample reacted at a heating rate of 20℃/min in nitrogen

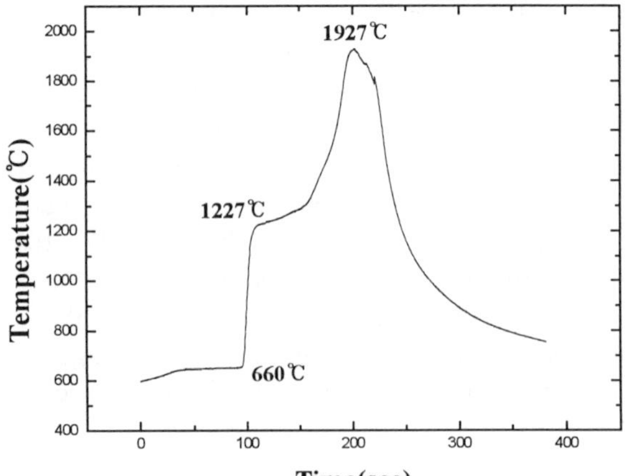

Figure 2. Temperature profile of sample undergoing Combustion Synthesis.

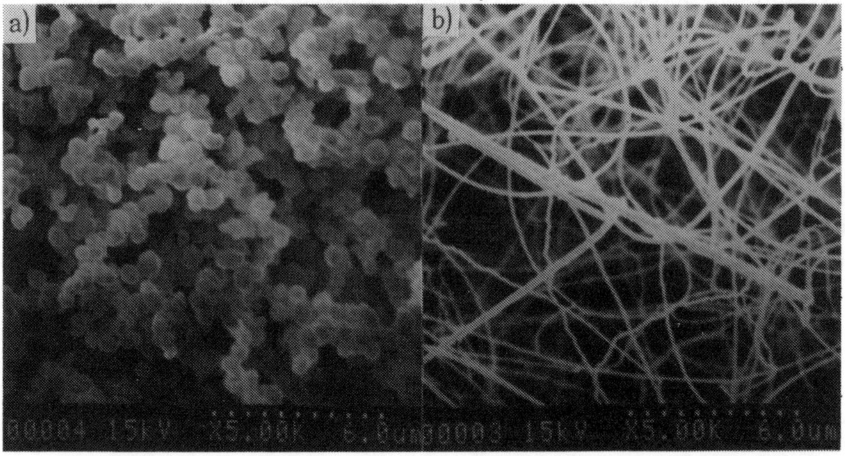

Figure 3. SEM micrograph of AlN synthesized by Combustion Synthesis reaction and kept at 900℃ for 2hr. (a) product on sample surface and (b) inside of the sample.

Figure 3 shows SEM micrographs from the various locations of the specimen after the Combustion Synthesis reaction. As seen in the figure, AlN powder was formed at the surface and whisker inside of the compact. The peaks obtained from the XRD of the sample matched with the AlN peaks of JCPDS card # 25-1133. Unreacted Al or other phases were not detected by the XRD analysis.

Figure 4 shows a transmission electron microscope(TEM) image and selected area diffraction(SAD) patterns of the whisker-shaped AlN. Although distinctive grain boundary contrasts were not observed, the SAD pattern of some AlN indicates that it is a polycrystalline fiber(Figure 4. a and c). It seems that the fiber is composed of very fine AlN crystals with slight differences in orientation. Although the dimension is very similar to the AlN fiber, the other whisker-like AlN were found to be single crystalline AlN whiskers (Figure 4. b and d).

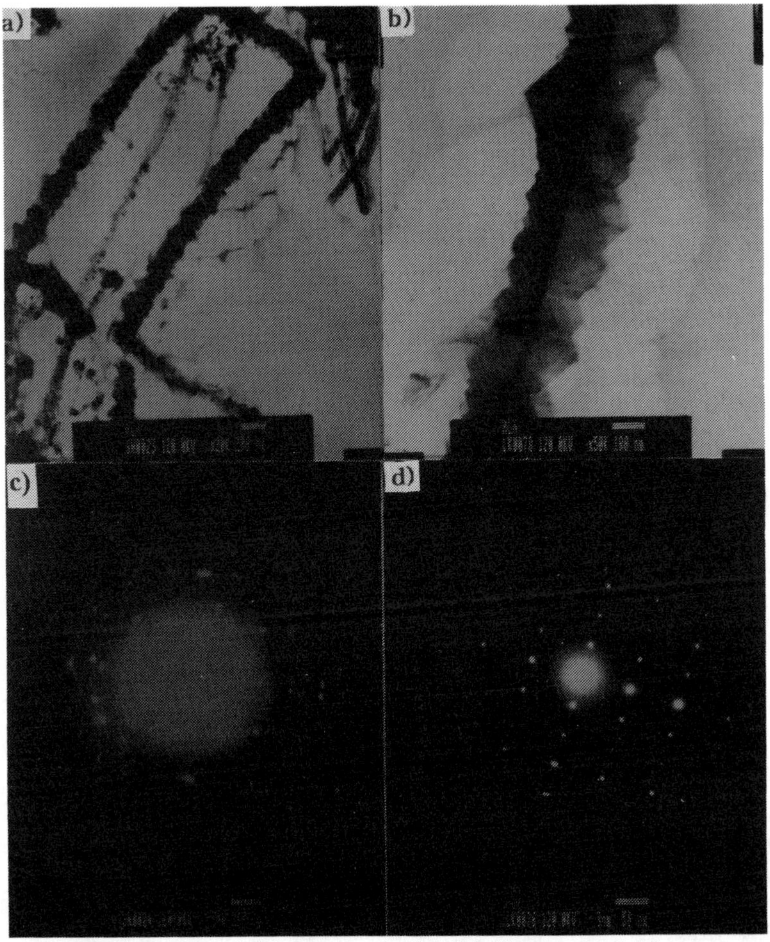

Figure 4. TEM image and SAD pattern of AlN fiber and whisker.
(a) polycrystalline fiber and (c) its SAD pattern.
(b) single crystalline whisker and (d) its SAD pattern.

B. Reaction Sequences of the AlN whisker formation

In order to study the reaction sequence of AlN whisker formation via Combustion Synthesis, the compact was heated to various temperatures and furnace cooled to room temperature. The temperatures to which the samples were heated included 400, 500, 600, 650, 670, 700, 800, and 900℃. Figure 5 shows the SEM micrographs of each specimen heated to the indicated temperatures. When the compact was heated to 400℃, Al powder remained in its spherical shape with some cracks formed on its surface. Very fine AlN particles were observed to form on the Al powder surface at 500℃ and increased its size as the temperature was raised to 600℃. At 650℃, significant amount of AlN particles were found to form on partially molten Al. Although the sample was heated to below the melting point of Al, the heat of AlN formation might have caused the melting of Al powders. When the sample was heated to 670℃, the morphology of the sample completely changed, indicating a wholescale nitridation reaction has occurred. This morphology change might have been caused by expulsion of molten Al from the liquid Al powder covered with AlN crust as proposed by Hotta.[10] XRD analysis of the specimen indicated that aluminum has completely converted to AlN. As the temperature was raised, the content of whisker-shaped AlN increased. When the specimen was heated to 900℃, the product inside the specimen was found to consist predominantly of whisker-shaped AlN.

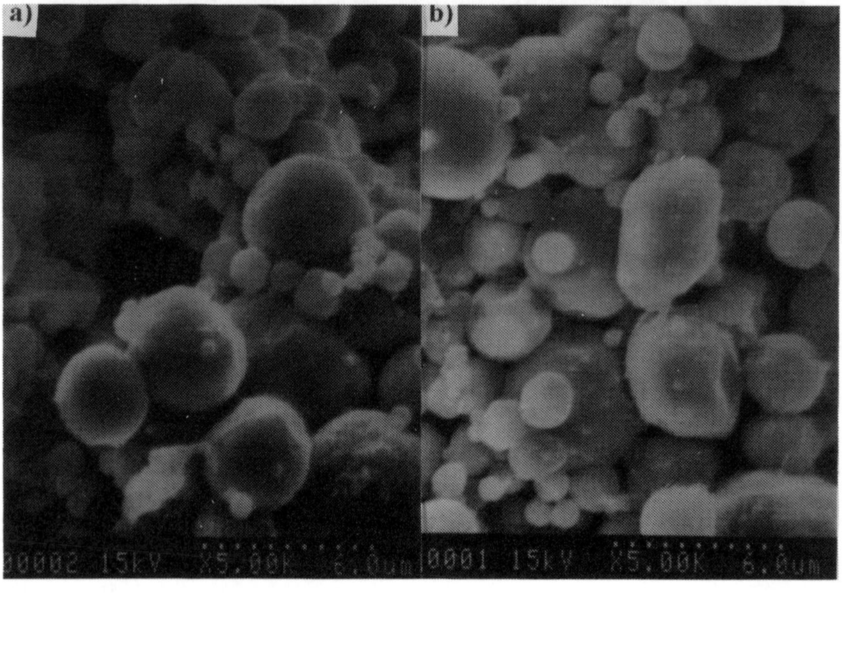

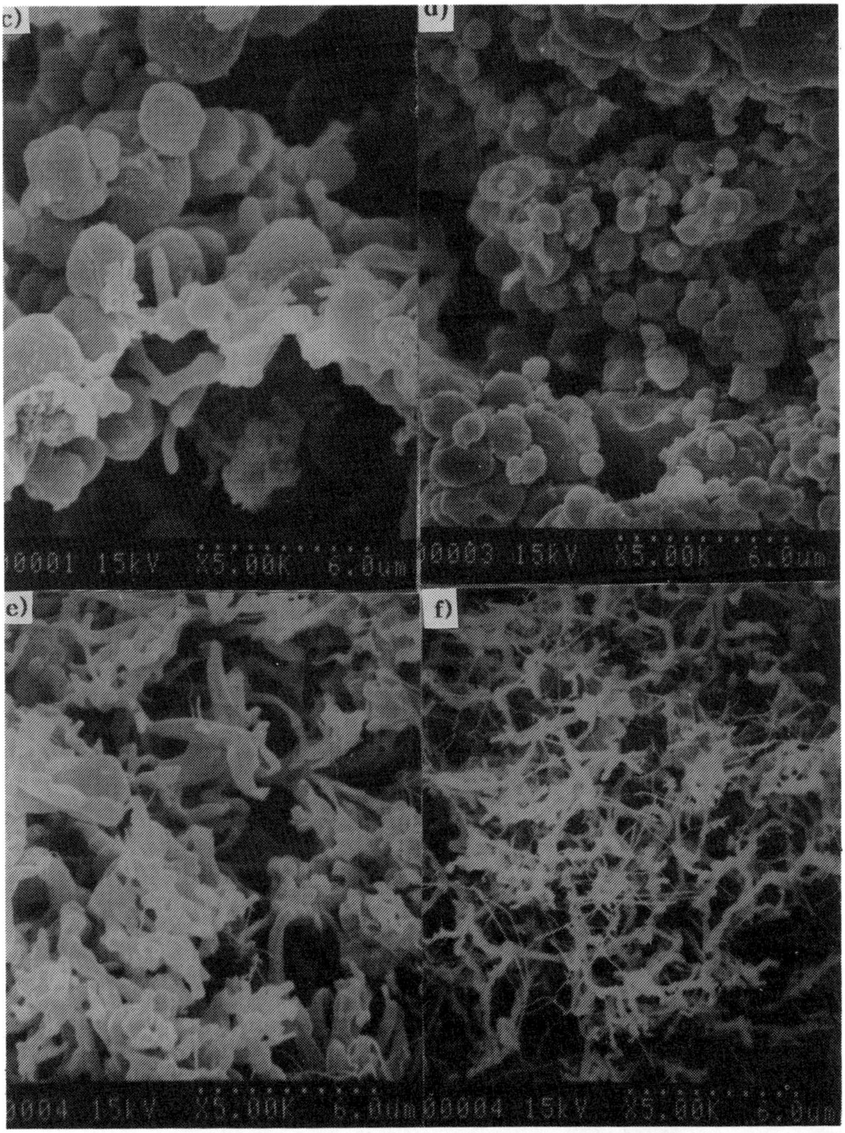

Figure 5. SEM micrographs of samples heated to (a) 400, (b) 500, (c) 600, (d)650, (e) 670, and (f) 700℃.

The compact heated to 900℃ was cut horizontally to examine the morphology change with locations. As shown in Figure 6, a thin crust of AlN powder was observed at the surface of the sample. Beneath the powder crust, the flower-like AlN which was observed from the sample heated to 670℃ was found. At the locations toward the center of the sample the AlN morphology became thinner and longer. The whisker-shaped AlN was observed at the center of the sample.

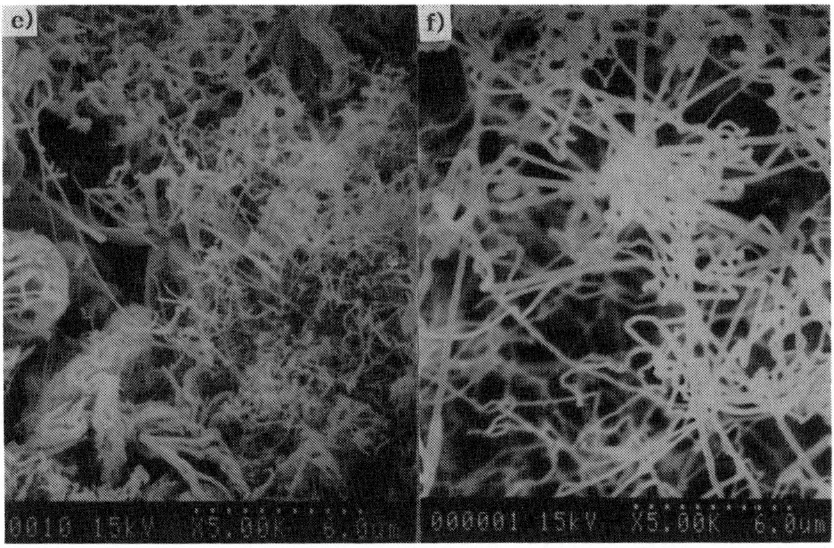

Figure 6. AlN morphology variation in the sample synthesized at 900℃. The AlN morphology changes gradually from powder shape (a) at the outer surface of the sample to whisker shape as the location move to the center of the sample from (b) to (f)

SUMMARY

AlN powder and/or whiskers was synthesized via Combustion Synthesis method. When the sample was heated to 660℃, the Combustion Synthesis reaction of AlN formation was ignited and lasted for several minutes. In the sample heated to 900℃, a complete conversion of Al to AlN was obtained. The product morphology was predominantly powder shaped at the surface and mainly whisker- shaped at the inside of the compact. Some of the whisker-shaped AlN formed were single crystalline whiskers and the others were polycrystalline fiber.

ACKNOWLEDGMENTS

This work was supported by CAAM. at POSTECH and Hong-Ik University.

REFERENCES

1. A. W. Weimer, W. G. Moore, "Rapid Process for Manufacturing Aluminum

Nitride Powder," *J. Am. Ceram. Soc.*, 77[1] 3-18 (1994)
2. "USACA: Pushing for AlN Commercialization," *Ceram. Bull.*, 69, No.11, p.1804 (1990)
3. "Annual Minerals Review : Aluminum Nitride", *Am. Ceram. Soc. Bull.*, 73, No.6, p.77, June 1994
4. "Principles of Electronic Packaging," edited by Donald P. Seraphim, Ronald C. Lasky and Che-Yu Li, McGraw-Hill, p.355
5. H. Hacker, K. R. Hauschildt, J. Huber, H. Laupenmuhlen and D. Wilhelm, "Polymeric Materials for Electronics Packaging and Interconnection," ACS Symposium Series 407, p.414, 1988, Los Angeles
6. R. Bachelard and P. Joubert, "Aluminum Nitride by Carbothermal Nitridation", *Mater. Sci. Eng.*, A109 (1989) 247-251
7. P. G. Caceres, H. K. Schmid, "Morphology and Crystallography of Aluminum Nitride Whiskers," *J. Am. Ceram. Soc.*, 77[4] (1994) 977-83
8. K. A. Golubjatinikov, G. C. Stangle and R. M. Spriggs, "The Economics of Advanced Self-Propagation High-Temperature Synthesis Materials Fabrication," *Am. Ceram. Soc. Bull.*, 72[12] (1993) p.96
9. S. D. Dunmead, J. B. Holt, and D. D. Kingman, "Simultaneous Combustion Synthesis and Densification of AlN," *Combustion and Plasma Synthesis of High temperature Materials*, edited by Z. A. Munir and J. B. Holt, p.186, 1990, New York
10. N. Hotta, I. Kimura, A. Tsukuno, N. Saito and S. Matsuo, "Synthesis of AlN by the Nitridation of the Floating Al Particles in Nitrogen gas," *Yogyo-Kyokai-Shi* 95[2] 1987, p.136

ATMOSPHERE EFFECTS ON COMBUSTION SYNTHESIS OF SiC AND TiB_2 POWDERS*

Robert L. Miller, Christopher J. Rando, John J. Moore, and Dennis W. Readey
Colorado Center for Advanced Ceramics
Colorado School of Mines
Golden, CO 80401

ABSTRACT

The effect of atmospheres on the reaction mechanism and kinetics to produce TiB_2 powders from titanium and boron and SiC from silicon and carbon is investigated. In HCl gas, high pressures of vapor species are generated which should produce significant vapor transport in both systems. With TiB_2, combustion does not occur and the product morphology consists of thin shells that are produced by gas-solid reactions in the mixed reactant powder bed. In contrast, complete reaction to SiC only occurs by combustion synthesis. The reaction to SiC occurs below the melting point of Si if the temperature is held for a period of time. Without combustion, the reaction in HCl does not go to completion and appears to hinder subsequent combustion.

INTRODUCTION

Titanium diboride, TiB_2, and silicon carbide, SiC, have been much studied in recent years due to desirable properties such as high melting temperature, high hardness, wear resistance, chemical stability, and low specific weight. These properties are attractive for cutting tools, wear-resistant components, lightweight armor, and numerous other applications. However, both TiB_2 and SiC are limited

* Sponsored by the National Science Foundation, Grant No. CTS-9202948 and ACX Technologies, Inc., Golden, CO

by poor sinterability. One way to improve sinterability is to start with submicron powders. Combustion synthesis is a process that has been used to successfully produce several ceramic and composite materials, including submicron powders.[1] Characteristics of the process include highly exothermic reactions, high reaction temperatures, and short processing times. In this method, a reactant mass is brought to its ignition temperature, at which point conversion of the reactants to products is initiated. The exothermic nature of the reaction brings the resulting products to temperatures in the 2500-3000°C range. The reactions used in this research are:

$$Ti(s) + 2\ B(s) \Rightarrow TiB_2(s)$$

$$Si(s) + C\ (s) \Rightarrow SiC(s).$$

In the presence of a reactive gas, such as HCl, vapor species are produced by reactions such as the following:

$$Ti(s) + 4\ HCl(g) \Leftrightarrow TiCl_4(g) + 2\ H_2(g).$$

These vapor species can react with the starting materials, or other gaseous species, to form the desired products, thereby facilitating the reaction. Free energy minimization calculations[2] show that significant vapor species are produced in both systems in HCl, Figures 1 and 2.

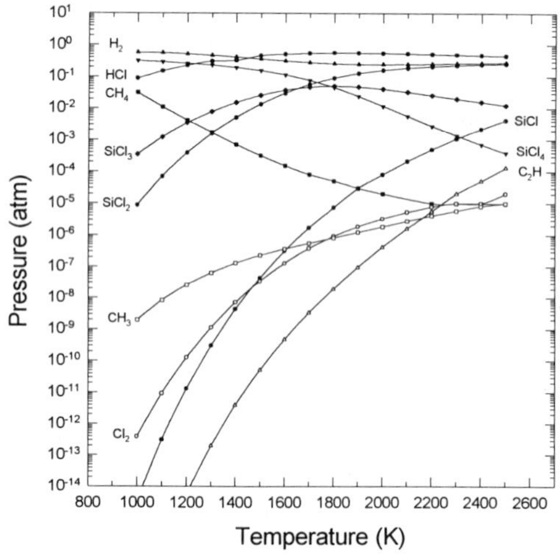

Figure 1. Calculated partial pressures over SiC in HCl.

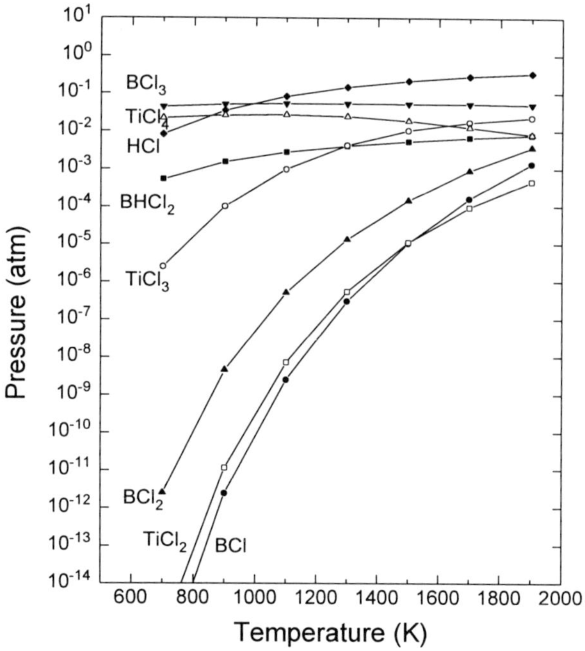

Figure 2. Calculated partial pressures over TiB_2 in HCl.

EXPERIMENTAL PROCEDURE

Titanium with boron, and silicon with graphite powders were mixed in the proper molar ratio in a polymer ball mill with alumina balls for 10-12 hours. The starting titanium and boron particles were about 20 μm in diameter; the silicon powders consisted of 10 μm diameter agglomerates of 1 μm particles; and the graphite consited of flakes about 10 μm square and 1 μm thick. The mixed powders were dry pressed into two gram pellets at 10.5 MPa. The cylindrical pellets were placed in an alumina boat in a closed-end mullite tube. This tube was sealed at the open end with a compression fitting with a valve that allowed evacuation and filling with argon, argon/100 ppm oxygen mixture, or hydrogen chloride gas. The tube was placed in a furnace and was heated to the required temperature at a slow, 20 ^{o}C/min., heating rate. A viewing port in the tube seal allowed observation of reactions in progress. Reaction products were analyzed by x-ray diffraction to determine the extent of reaction. Peak areas were compared to those of standard mixtures of products and reactants. Product powders were ball milled and SEM photomicrographs were analyzed to determine particle size distributions.

RESULTS

Titanium Diboride

At the slow heating rates used in these experiments, none of the TiB_2 samples were observed to undergo combustion. As a result, the degree of reaction was time dependent. The effect of temperature on the degree of reaction is shown in Figure 3. All samples were heated at a rate of 20°C per minute to the required temperature and held for 35 minutes. As the hold temperature was increased, the weight percent reacted to form TiB_2 increased in all three reaction atmospheres. Reactions in hydrogen chloride reached completion approximately 200°C below those conducted in argon or argon/oxygen. As shown in Figure 2, calculations predict that several gas species are available at sufficiently high partial pressures to transport both titanium and boron through the gas phase. This facilitates the reaction to TiB_2. Similar calculations for the argon/oxygen system show that some boron oxide species form, but their pressures are not high enough to cause significant vapor transport. Calculations for argon show no transport species. In

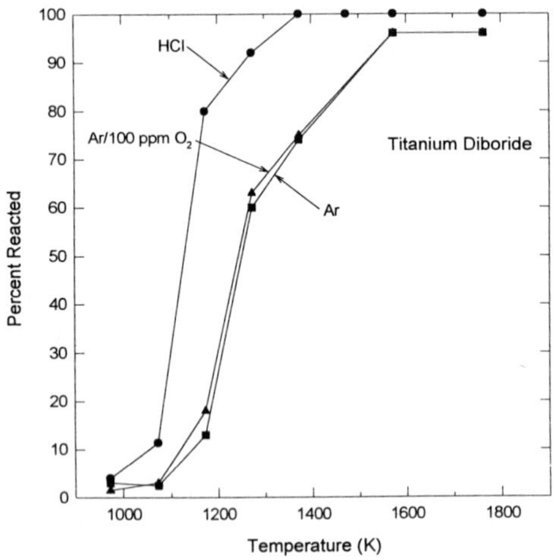

Figure 3. Fraction of Ti and B reacted in 35 minutes as a function of temperature and atmosphere.

spite of the differences in degree of reaction and presence of vapor transport species, the reacted microstructures were found to be independent of the ambient atmosphere, Figure 4. Shells of product TiB_2 formed around reactant titanium

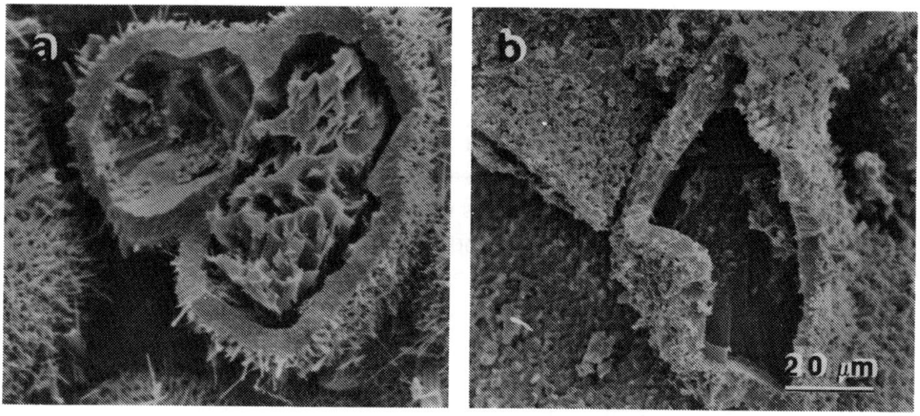

Figure 4. a. Shells of TiB_2 around partially reacted reactants. b. Hollow shells after complete reaction.

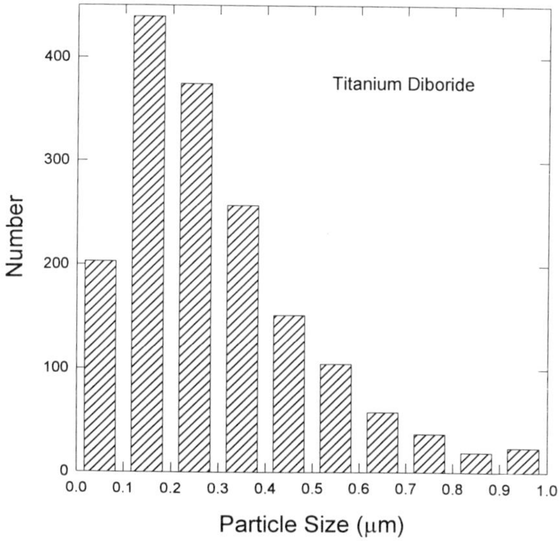

Figure 5. Particle size distribution for sample reacted in HCl at 1300 °C for 35 minutes and ball milled 15 hours.

and boron particles, Figure 4a. When the reaction had gone to completion, the reactant materials were depleted resulting in the formation of hollow shells consisting of submicron-size grains of TiB_2, Figure 4b. Ball-milling of the TiB_2 product produced small particles as shown in the particle size distribution of Figure 5.

Silicon Carbide

The effect of time and temperature on the percent of β-SiC formed in HCl is shown in Figure 6. Above 1300 °C, the samples have combusted and completely reacted. Below 1250 °C, combustion has not occurred and the samples are only partially reacted. In contrast, Figure 7 shows that no partial reaction occurs in argon, only complete reaction, the result of combustion. Note that it can take many hours before combustion occurs in argon. The combustion event was observed in video records kept of the reaction.

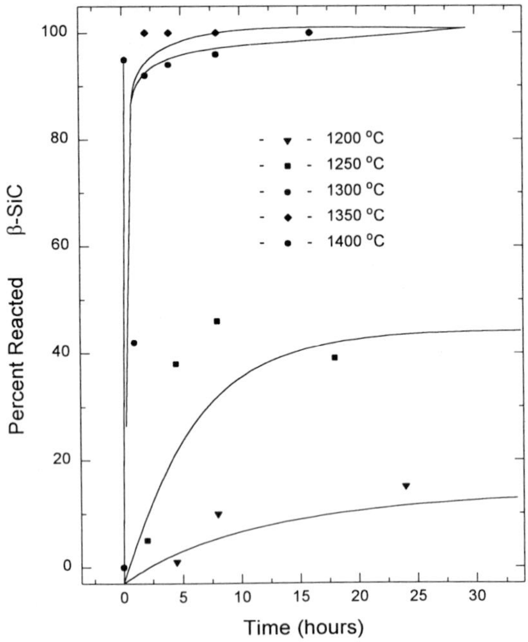

Figure 6. Percent SiC formed as a function of temperature in HCl.

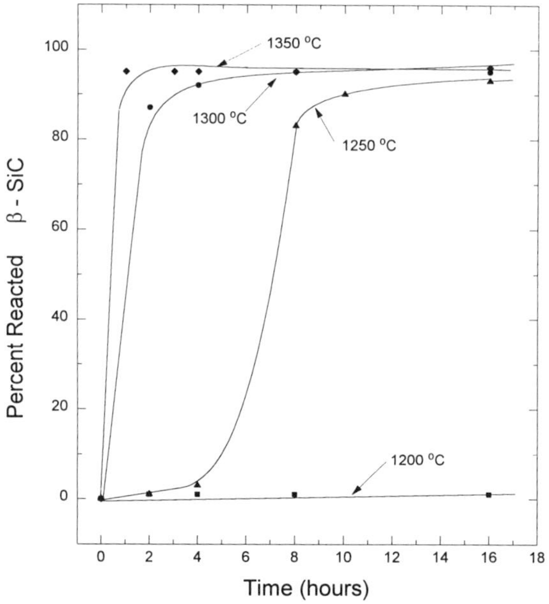

Figure 7. Percent SiC formed as a function of temperature in argon.

Figure 8. a. Partially-reacted Si and carbon in HCl. b. Fully-reacted Si and carbon in argon.

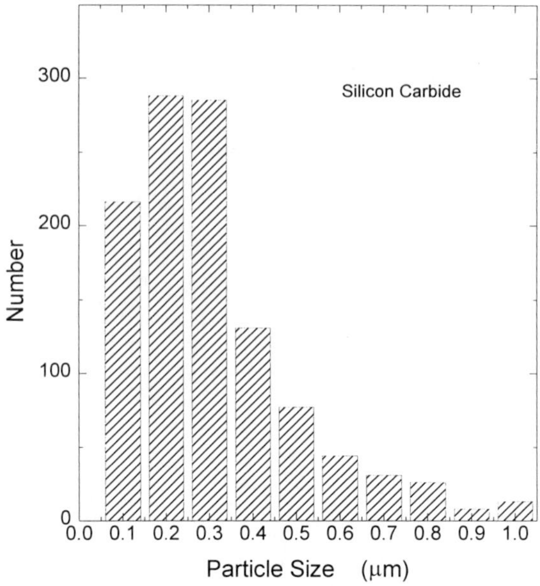

Figure 9. Particle size distribution for sample completely reacted in HCl at 1350 °C and ball milled 15 hours.

The effects of atmosphere on the resulting microstructures are shown in Figure 8. In Figure 8a, which shows a partially-reacted sample, the graphite particles show "bumps" or particles on their surfaces that are presumably SiC particles produced by reaction of $SiCl_4$ gas with the solid graphite. In contrast, Figure 8b shows a completely-reacted (as determined by x-ray diffraction), combusted, sample in argon that exhibits relics of the original graphite particles that appear to consist of small particles of SiC. Ball milling either product yields the small particle size powder desired, Figure 9.

DISCUSSION AND CONCLUSIONS

Titanium Diboride

With the slow heating rates of 20 °C/min, titanium and boron completely react to form TiB_2 without combustion within 35 minutes at temperatures above about 1200 °C. This occurs in HCl, argon, and argon/oxygen mixtures. The reaction appears to go to completion about 200 °C lower in HCl strongly suggesting that vapor species are playing a role in the reaction. Previous work has attributed the hollow shell formation to vapor phase transport[3] or to melting and flow of liquid titanium.[4] The microstructures, however, do not fully support either of these mechanisms: the hollow-shell microstructures appear in both the reactive HCl and inert argon atmospheres and there is no evidence for melting, nor should there be at these low temperatures. The product of the reaction can be easily comminuted to produce fine particle size TiB_2 powders that should be suitable for subsequent forming and sintering operations.

Silicon Carbide

Silicon and carbon partially react in HCl to form SiC without combustion. However, complete reaction only occurs with combustion in both HCl and argon atmospheres. In argon, combustion can take place many hours after the reactants have reached the desired reaction temperature. The mechanism responsible for this long incubation time is not clear. When combustion occurs in either HCl or argon, the SiC product can be easily comminuted to produce small particle size powders that should be suitable for subsequent forming and sintering operations also.

REFERENCES

1. Z. Munir and U. Anselmi-Tamburini, "Self-Propagating Exothermic Reactions: The Syntheisis of High-Temperature Materials by Combustion," Mat. Sci. Reports **3** [7-8] 277-365 (1989).

2. T. M. Besmann, "A Computer Program To Calculate Relationships in Complex Chemical Systems," Rept. ORNL/RM-5775, Oak Ridge National Laboratory, Oak Ridge, TN, 1977.

3. N. R. Perkins, J. J. Moore, and D. W. Readey, "Synergistic Effects of Coupled Combustion Synthesis and Vapor Phase Transport in the Synthesis of Advanced Materials," Proc. First Int. Conf. on Adv. Synthesis of Engineered Materials, J. J. Moore, E. J. Lavernia, and F. H. Froes, eds., (ASM International, Metals Park), 17-23 (1993).

4. H. Itoh, T. Matsudaira, S. Naka, and H. Hamamoto, "Reaction control of TiB_2 formation from titanium metal and amorphous boron," J. Mat. Sci. **24** 420-424 (1989).

CHARACTERIZATION OF SHS $MoSi_2$ AND $MoSi_2$/SiC/20w COMPOSITES

W. Chen, W. Pascheto, A. Pant, R.R. Sood
Ceramics Kingston Céramiques Inc., (CKCI),
P.O. Box 655, Kingston, Ontario, Canada, K7L 4X1

R.T. Holt and R. Berriche
Institute for Aerospace Research
National Research Council of Canada
Ontario, Canada, K1A 0R6

ABSTRACT

Molybdenum disilicide prepared by self-propagating high-temperature synthesis (SHS) using a proprietary process developed at Ceramics Kingston Céramiques Inc., has been consolidated either by hot pressing or hot pressing followed by hot isostatic pressing. In addition, the same material has been reinforced with 20 vol% silicon carbide whiskers (also manufactured by CKCI). Physical characteristics and some mechanical properties of these materials are presented.

INTRODUCTION

Molybdenum disilicide, $MoSi_2$, is an intermetallic compound with a number of properties which make it an attractive candidate for high temperature components in engines for aircraft and other aerospace vehicles. It has a high melting point, a moderate density and very good oxidation and corrosion resistances at high temperatures. Due to its tetragonal crystal structure, it has a small number of available slip systems which limits its ductility, but it does undergo a brittle to ductile transition at about 1000°C (1). Recently a number of papers have been published (2-7) which describe attempts to improve the properties through ceramic reinforcement. Examples of properties which have been achieved in various $MoSi_2$ composites are given in Table I. All this work has been presented at previous Cocoa Beach Conferences on Composites and Advanced Ceramic Materials.

One of the problems in assessing these data is that most researchers have chosen different properties to report, and, in the case of fracture toughness, different methods of performing the test. In addition, most authors have not reported the number of tests performed for each condition; in some cases neither the density nor the % theoretical density of the composite are given and in other cases samples are more finely ground or polished than the minimum requirement for the mechanical tests. Hence, even though all the various composites reported in Table I were prepared from commercially

Table I. Reinforced $MoSi_2$ - Examples of Mechanical Properties

Reference	Reinforcement vol %; form*	Compr. Y.S. (MPa)		Compr. Creep at 78 MPa, 1200 °C	K_{Ic} 20 °C+ (MPa $\sqrt{m}$)	MOR (MPa)	
		20 °C	1000 °C			20 °C	1000 °C
(2)	SiC; 20; w				8.2 (cnb)	≈ 300	≈ 360
(3)	ZrB_2; 30; pl	2300	1300				
(4)	ZrO_2;20; p				7.8 (i)		
(5)	SiC; 20; pl				7.5 (cnb)	215	≈ 330
(6)	SiC; 20; w			T_R = 195 h ε = 3%			
(7)	Al_2O_3; 25; pl				4.9 (senb)	270	490

** p = particulate, pl = platelet, w = whisker*
+ i = indentation method, cnd = chevron notched bar, senb = single edge notch beam

available $MoSi_2$ powder, it is difficult, if not impossible to compare one with another. One feature which is reasonably consistent (2,5,7) is that the modulus of rupture (MOR), obtained by 4 point bend testing, increases significantly from 20°C to 1000°C.

One potential problem associated with the use of $MoSi_2$, often called pesting, is accelerated oxidation in the temperature range 400 to 600°C. Proposed mechanisms for this behaviour have been reviewed (8) and it has been reported that pesting does not occur in fully dense (1,9) and HIPed (8) material.

The present paper describes the characteristics and some mechanical properties of a $MoSi_2$ material produced by self-propagating high-temperature synthesis (SHS). The material has been consolidated by hot pressing (HP) which in some cases was followed by hot isostatic pressing (HIP). Samples were prepared both without, and with silicon carbide whisker reinforcement.

EXPERIMENTAL

The monolithic $MoSi_2$ samples were prepared by mixing stoichiometric quantities of molybdenum powder of -500 mesh with an average particle size of about 8 μm, 99.5% purity, manufactured by Wah Chang Corporation and silicon powder of -140 mesh, average particle size about 70 μm, 99.0% purity, made by Kema Nord in Sweden, but distributed in North America by Superior Graphite. The mixed powder was then packed into a lined graphite mould, and hot pressed into discs approximately 100 mm diameter by 13 mm thick, which was within 0.25 mm (2%) of the predicted thickness based on complete combustion and zero porosity. Some of the hot pressed discs were subsequently hot isostatically pressed. For proprietary reasons, further details of the processing procedure cannot be disclosed at the present time.

For the composite pieces, 20 vol% silicon carbide whiskers, made by CKCI, were added to the ingredient powders during the mixing stage. These whiskers identified as CKC/SC-C grade whiskers are made specifically for reinforcing ceramics for tool bit and wear resistant applications. The CKC/SC-C whiskers are single crystal fcc beta (β) with a [111] longitudinal axis. They have a smooth surface, a diameter range of 0.2 to 0.9 μm (average diameter of 0.33 μm) and a length range of 8 to 25 μm (average length of 12 μm).

The discs were cut by electro-discharge machining to standard bend test bars conforming to MIL-STD 1942 type B. In general this method of machining gives a good surface finish on conducting materials; it worked very well on the monolithic material but there was increased surface roughness on the composites, which may have introduced surface defects. The nominal dimensions of a bar after wet surface grinding with 240 grit paper on an automatic grinding machine was 3 x 4 x 50 mm.

Sections for microstructural analysis were polished and viewed in the SEM using back scattered electron imaging (BEI) to reveal the grain structure. X-ray diffraction patterns were taken of the $MoSi_2$ and the $MoSi_2$/SiC/20w composites.

The following mechanical tests have been conducted on the machined samples, and the number of tests conducted for each group (monolithic, composite, hot pressed, HIPed) is also indicated.

(i) hardness (HV) - 5 indentations per group
(ii) fracture toughness by indentation method (K_{Ic}) - 5 indentations per group
(iii) fracture toughness by indentation strength method (K_{Ic}) - 2 to 4 tests per group
(iv) modulus of rupture at 20°C (MOR-20) - 3 tests per group
(v) modulus of rupture at 1000°C (MOR-1000) - 2 tests for the monolithic, 1 test for the composite.

RESULTS

The microstructures of the HP and HP/HIP conditions for the monolithic material were similar to each other, as was the case for the HP and HP/HIP composites. The HP monolithic $MoSi_2$ is shown in Figure 1; the grain size is about 37 μm and a small amount of porosity is observed. The microstructure of the HP composite is shown in Figure 2a; the $MoSi_2$ grain size is about 6μm and the black regions contain the SiC whiskers in the $MoSi_2$ grain boundaries. The whiskers are shown at higher magnification in Figure 2b, and in most cases the whiskers appeared shorter than unprocessed whiskers, indicating some breakage had occurred during processing of the composite. The observed porosity levels match the densities which are given, along with other measured properties, in Table II. Values used for calculating theoretical density (TD) are 6.24 g/cm^3 for $MoSi_2$ and 3.21 g/cm^3 for SiC, so the TD of a $MoSi_2$ /SiC/20w composite is 5.634 g/cm^3. Neither for the monolithic nor the composite did HIPing significantly alter the HP density.

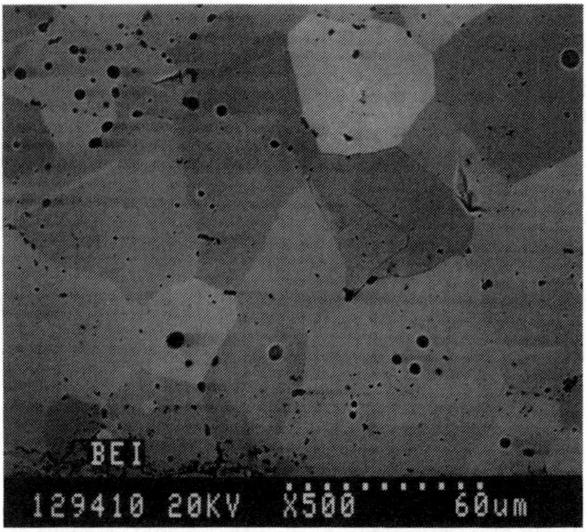

Figure 1. Microstructure of HP $MoSi_2$

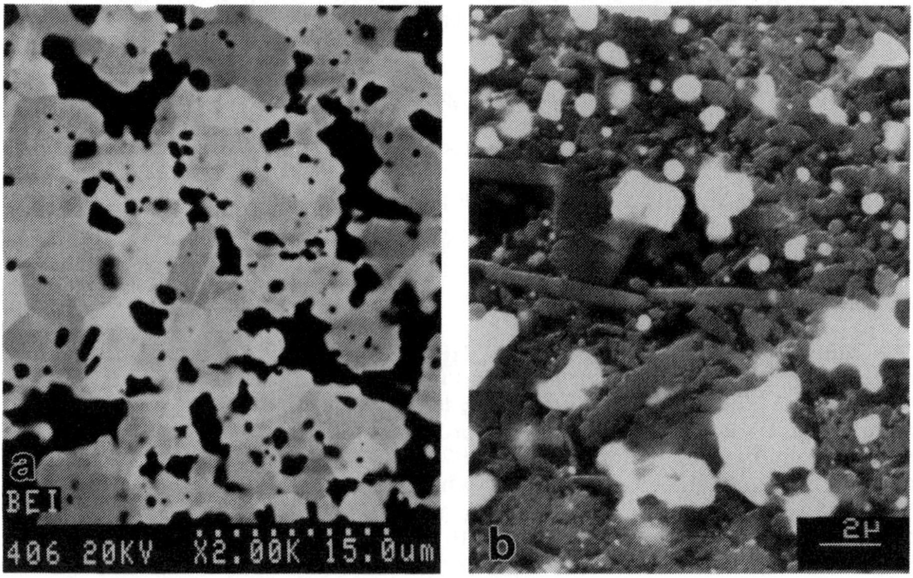

Figure 2. Microstructure of HP $MoSi_2$/SiC/20w Composite;
(a) grain structure and (b) SiC whiskers in the grain boundaries

Table II. Physical and Mechanical Properties of SHS Material

Property	$MoSi_2$		$MoSi_2$/SiC/20w	
	HP	HP/HIP	HP	HP/HIP
Density (g/cm^3)	6.19	6.17	5.33	5.40
% Theoretical Density (%TD)	99.1	98.9	94.6	95.8
Grain Size (μm)	37	43	6.2	7.2
HV (GPa)	9.9	9.4	13.5	14.7
K_{Ic} (Indentation Method) (MPa $\sqrt{m}$)	2.4	2.7	3.7	3.5
K_{Ic} (Indentation Strength) (MPa $\sqrt{m}$)	3.82	4.18	4.84	5.35
MOR at 20 °C (MPa)	170.7	148.7	171.8	174.2
MOR at 1000 °C (MPa)	174.4	204.3	126.0	154.7

The x-ray diffraction patterns (Figure 3) show that the SHS process is complete to the limit of detection by this method. In the monolithic material only peaks for the tetragonal $MoSi_2$ phase were observed (Figure 3a) and these same peaks were dominant in the composite material (Figure 3b). However, in the composite, additional peaks were observed which were identified as hexagonal Mo_5Si_3, identified by the prefix M (and the reflecting planes) in Figure 3b and beta silicon carbide, identified by the prefix S. The Mo_5Si_3lines matched those for the hexagonal allotrope with 2.5% carbon stabilization (JCPDS-PDF card file 8-429) and from line intensities, it is estimated that the Mo_5Si_3, content in the composite is less than 5%. In support of Figure 2b, the presence of the SiC lines in Figure 3b indicates that the whiskers were not transformed during processing. Neither in Figures 3a nor 3b were peaks for Mo or Si found, indicating that SHS combustion was complete.

The hardness tests were conducted on a standard Vickers machine, according to ASTM E 384 using a 10 kg (98N) load. Hardness values are given in Table II where it can be seen that the composite material was significantly harder than the monolithic material despite having a lower %TD. Indentations at 10 kg load were used for calculating indentation fracture toughness (K_{Ic}) according to the formula developed by Anstis et al. (10), equation (1),

$$K_{Ic} = \zeta_V^R (E/HV)^{1/2} (P/c^{3/2}) \qquad (1)$$

where ζ_V^R is a material-independent constant (0.016) for Vickers-produced radial cracks, E is the modulus of elasticity, HV the Vickers Hardness, P is the indentation load (N) and 2c is the total length of the radial crack developed from opposite corners of the indentation. The value of E for the HP monolithic

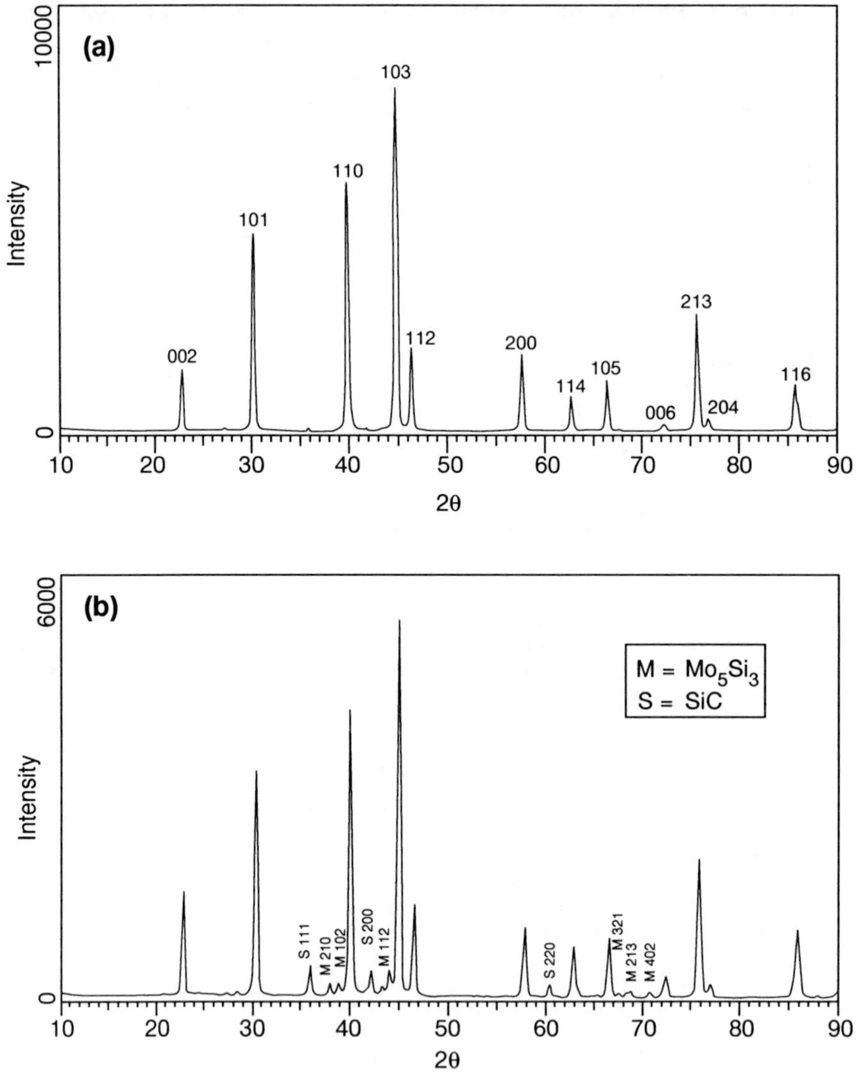

Figure 3. X-Ray diffraction patterns for hot pressd SHS material, (a) monolithic $MoSi_2$ and (b) $MoSi_2$/SiC/20w composite. The $MoSi_2$ peaks in (a) also occur in (b) and are unmarked. The three digit numbers identify the crystallographic plane indices.

$MoSi_2$ was 410 GPa, as determined by the resonant frequency method using a GrindoSonic instrument.

The calculated K_{Ic} values for the monolithic $MoSi_2$ were about 2.5 MPa√m and for the composite about 3.6 MPa√m. Although these results are lower than those shown in Table I, variations in testing and interpretation make comparison difficult but the composite does appear to have better fracture toughness than the monolithic $MoSi_2$.

According to Quinn et al. (11), a more reliable method to measure fracture toughness in brittle materials is the indentation strength method where a standard Vickers indentation is used to start a crack in a MOR specimen and the bar is then loaded to fracture. The fracture toughness is then calculated from equation (2).

$$K_{Ic} = 0.59\,(E/HV)^{1/8}\,(\sigma_c P^{1/3})^{3/4} \qquad (2)$$

where σ_c is the flexure strength (MPa) and P is the indentation load (N).

Using this method, the fracture toughness of the composite (about 5 MPa√m) was again better than that of the monolithic material, (about 4 MPa√m) as shown in Table II. These results are not directly comparable to the values obtained by different methods as given in Table I.

The MOR results on $MoSi_2$ at room temperature are comparable with those found for other $MoSi_2$ samples and are given in Table II. The MOR results for the $MoSi_2$ at 1000°C and the MOR results for the $MoSi_2$ /SiC/20w composite, both at 20°C and 1000°C are slightly lower than expected. However, these results are preliminary; only one sample at each condition was tested for the composite at 1000°C. As mentioned earlier, some surface damage might have been introduced into the composites by the electro-discharge machining. For future testing, all samples will be prepared by diamond cutting. We are encouraged by the results to date and, feel that with improved processing techniques and the optimizing of ingredients, the resulting densities should increase. This, together with better specimen preparation, should improve the mechanical properties. Since it is a relatively inexpensive process we feel that this new method offers good potential for the fabrication of net shape parts for elevated temperature applications.

CONCLUSIONS

1. Monolithic $MoSi_2$ and $MoSi_2$/SiC/20w composites have been prepared by self-propagating high-temperature synthesis using a low cost proprietary process which can be used for net shape production of components.

2. The density of the monolithic material was 6.19 g/cm^3 (about 99%TD), and the properties compared favourably to those published elsewhere.

3. The grain size of the $MoSi_2$ in the composite was about 7 μm and the whiskers were uniformly distributed in the grain boundaries.

4. Despite a somewhat lower density (about 95%TD), both the hardness and fracture toughness properties of the composite were significantly higher than those of the monolithic material. However, the MOR and fracture toughness properties of the composite were slightly lower than published elsewhere.

ACKNOWLEDGEMENTS

This work was conducted under the terms of an Agreement for Collaborative Work between NRC/IAR and Ceramics Kingston Céramiques Inc., under project code JHR07. Partial funding for this work was covered by two contracts from the Canadian Department of National Defence, identified as FE144692NRC and 05SV.W.2207-2-AF15. Mr. Andre Van Leuven of J.W. Lemmens Inc., kindly performed the GrindoSonic resonant frequency tests.

REFERENCES

1. A.K. Vasudévan and J.J. Petrovic, "A Comparative Overview of Molybdenum Disilicide Composites", Mat. Sci and Eng., A155, pp. 1-17, 1992.
2. D.H. Carter, J.J. Petrovic, R.E. Honnell and W.S. Gibbs, "SiC-$MoSi_2$ Composites", Ceram. Eng. Sci. Proc. 10, pp. 1121-1129, 1989.
3. R.M. Aiken, Jr., "Structure and Properties of insitu Reinforced $MoSi_2$", Cer. Eng. Sci. Proc., 12, pp. 1643-1655, 1991.
4. J.J. Petrovic, R.E. Honnell and T.E. Mitchell, "ZrO_2 - reinforced $MoSi_2$ Matrix Composites", ibid, 12, pp. 1633-1642, 1991.
5. K.K. Richardson and D.W. Freitag, "Mechanical Properties of Hot Pressed SiC Platelet-reinforced $MoSi_2$", ibid, 12, pp. 1679-1689, 1991.
6. K. Sadananda, H. Jones and J. Feng, "Creep of Monolithic and SiC Whisker-reinforced $MoSi_2$", ibid, 12, pp. 1671-1678, 1991.
7. S.C. Tuffé, K.P. Plucknett and D.S. Wilkinson, "Processing and Mechanical Properties of Alumina Platelet Reinforced Molybdenum Disilicide Laminates", ibid, 14, pp. 1199-1208, 1993.
8. D.A. Berztiss, R.R. Cerchiara, E.A. Gulbransen, F.S. Pettit and G.H. Meier, "Oxidation of $MoSi_2$ and Comparison with Other Silicide Materials", Mat. Sci. Eng. A155, pp. 165-181, 1992.
9. P.J. Meschter, "Oxidation Kinetics of $MoSi_2$ and $MoSi_2$/Reinforcement Mixtures", Mat. Res. Soc. Symp. Proc. Vol. 213, pp. 1027-1032, 1991.
10. G.R. Anstis, P. Chantikul, B.R. Lawn and D.B. Marshall, "A Critical Evaluation of Indentation Techniques for Measuring Fracture Toughness: 1, Direct Crack Measurements", J. Amer. Cer. Soc., 64, No. 9, pp. 533-542, 1981.
11. G.D. Quinn, J. Salem, I.B. Cho, K. Cho, M. Foley and H. Fang, "Fracture Toughness of Advanced Ceramics at Room Temperature", J. Res. Natl. Inst. Stand. Technol., 97, pp. 579-607, 1992.

FABRICATION OF TiC REINFORCED METAL MATRIX COMPOSITES FROM PLASMA DISSOCIATED ILMENITE

S.B. Bhaduri, P.R. Taylor, G.L. Watt
Department of Metallurgical & Mining Engineering
University of Idaho, Moscow ID 83844-3024
M. Manrique
Simon Bolivar University, Caracas, Venezuela

ABSTRACT

In-situ mixed TiC-Fe powders were produced by feeding Ilmenite concentrate into a non-transferred arc plasma reactor. The powders were subjected to sintering runs up to 1500°C for one hour. It is believed that the densification takes place in these samples by a liquid phase sintering route due to the presence of an eutectic at 1475°C in the Ti-Fe-C ternary system. The resultant microstructures were examined using a scanning electron microscope and image processing software. The average hardness is calculated to be 17GPa.

INTRODUCTION

It is well known that brittle materials can be considerably toughened by the addition of a ductile phase. Examples of such cermets are Lanxide™ [1], WC-Co [2], B_4C-Al [3] etc. It is believed that the primary mechanism for the increase in toughness is due to the bridging of intact ligaments of ductile phase or crack deflection by the matrix/reinforcement interface. Bridging shields the crack tip from the applied stresses thereby enhancing the reliability of the composites.

The above examples are also representatives of the different processing routes for fabricating them. Lanxide™ composites are typically produced by the directional oxidation of aluminum alloys [1]. Cemented carbides are conventionally produced by powder metallurgy routes i.e., mixing powders of the carbide (e.g., WC) and binding phase (Co) and sintering the pressed compacts. Very similar procedure is also used in making ductile phase toughened intermetallics [4]. Alternatively, porous preforms can be infiltrated with molten metals, typically aluminum. The examples of such composites include B_4C-Al, Al-AlN etc. [5]. Yet another approach is to use the combustion synthesis process whereby the heat generated by the highly exothermic reactions melt the ductile phase to fabricate ductile phase toughened brittle composites [6,7].

The focus of the present research is to produce TiC-Fe composites using a modified powder metallurgy process, the modification being the introduction of a plasma processing step. Such composites have been produced in the past by mixing TiC grains and Fe (or steel) and sintering them appropriately. Choe et al. [8] and Chun et al. [9] reported fabricating TiC reinforced composites with various binding phases including Fe. Kakeshita and Wayman [10] processed TiC reinforced composites with tailored binder phase of austenite steel, which is heat-treated to tailor the properties. These materials are now commercially available [11]. Typical applications of such composites include pelletizer knives, punches, extrusion dies etc. In spite of the high ceramic content, these materials are reported to be machinable, wear and fracture resistant. By choosing a right binder phase, they can be further heat-treated to achieve the desired properties. However, the main problem about the processing these material is that the TiC particles must be carefully dispersed within the binding phase such that both the phases are distributed homogeneously. If there is an uneven distribution of the two phases, the composites are going to have inhomogeneous properties. Furthermore, high TiC content products may be expensive due to the high cost of TiC. The plasma process modification on the other hand, produces homogeneously mixed powders in-situ, easily consolidated later on. The process can be further economized by using cheap raw materials and recirculating the plasma gases. In fact, Terry and Chinyamakobvu [12,13] reported producing TiC reinforced metal matrix composites by reducing Ilmenite with coal in pelletized forms. They reported the viability of such a procedure but did not carry out the process further in terms of fabricating the composite.

This paper reports the initial results of formation of TiC-Fe composites produced in a non-transferred are plasma from Ilmenite ($FeTiO_3$). This work is a part of larger work on plasma synthesis of value added ceramics at the University of Idaho [14].

EXPERIMENTAL

A schematic of the plasma reactor used for this work is shown in Figure 1. The system consists of a tubular reactor, non-transferred arc thermal torch, rapid quench section, filter, powder feeder and data acquisition system. A mixture of argon and nitrogen was used as plasma gas. The solid particles of Ilmenite concentrate (RGC Mineral Sands Inc. USA) was fed using a powder feeder and the gases were fed through a water cooled injection probe, inserted at angle near the plasma jet. The torch operating conditions are given in Table I. These conditions as well as the optimum condition for maximizing the yield of TiC are obtained from thermodynamic calculations with the overall reaction given by-

$$FeTiO_3(s) + 4\,CH_4(g) = TiC(s) + Fe(l) + 3CO + 8H_2O(g)$$

The as produced powders were subjected to X-Ray diffraction. The powders were collected and directly compacted at 76 Ksi using a 1/2 " diameter die. Such large pressures of compaction at the green state are useful in subsequent sintering operations. As compacted pellets were sintered in graphite heating element furnace between 1200°C and 1500°C, the highest temperature yielded the best material. The as-sintered samples were polished with diamond paste were examined using a scanning electron microscope. The samples did not require any coating. The microstructures were further analyzed by the

Table I. Operating Conditions for the Plasma Torch

PARAMETER	RANGE
Voltage, V	165-185
Current, A	175-190
Power, KW	28.9-35.2
Argon Flow, SCFM	3.5-4.0
Nitrogen Flow, SCFM	0.5-1.0
Carrier Gas, SCFM	0.25

Image^{TM} software. Hardness values were obtained by indenting the polished surfaces with a Vicker's indenter at a load of 1.5Kgs.

RESULTS AND DISCUSSIONS

Figure 2 shows a X-Ray diffraction pattern of the as-synthesized TiC-Fe in-situ mixed powders. Figure 3 shows a low magnification microstructure of the composite sintered at 1500°C for 1 hour. The interesting features of the microstructure are discussed in the following.

First, the lighter grey phase is the binding phase while the particulates are the TiC reinforcements. The sample is well sintered at 1500°C. The pores present are few and about 2 to 5μm in size. It is noted here that a combination of high pressure consolidation and a sintering temperature 1500°C is useful for sintering such composites. However it is difficult to calculate the relative density.

Second, the shape of the particulates are of specific shape (e.g. triangular, square or rectangle) with edges rounded off.as shown in Figure 4. The average particle size is 3μm as per our image analyses. Such microstructures have not been commonly reported in powder metallurgically produced composites with the exception of the work of Panchal and Trela [15]. In most of the previous cases, the particles are much larger with sharp angles, a signature of the original TiC particle used in blending the powders. Panchal and Trela argue that the edge rounding is a sought after feature of the microstructure because it provides lubricity during machining.

Third, it is obvious that there is some kind of liquid phase present in sintering of these composites, which not only results in liquid phase assisted sintering process but also helps in dissolving the edges of the TiC particulates. Panchal and Trela enhanced the the edge rounding by extending the sintering time. In the present case the rounding happens during the one hour soaking period. The presence of the liquid phase can be understood by examining Fe-Ti-C ternary system [16]. According to the pseudo-binary diagram of this system, there is significant dissolution of TiC in Fe with an eutectic at 1475°C. Terry et al. argue [13] that TiC is reasonably wetted by Fe. Because of these two reasons, the liquid phase sintering is so effective in this composites.

Fourth, an impression of the Vicker's indenter is shown in Figure

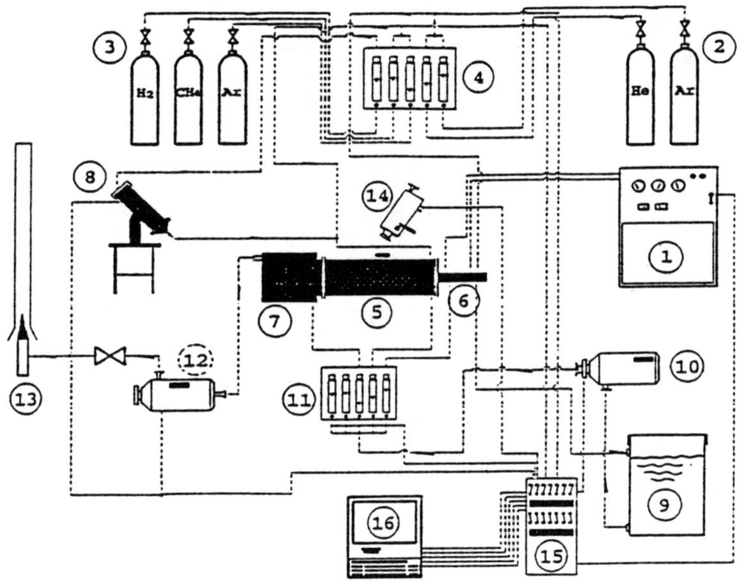

Figure 1. Schematic of the experimental system. 1- power supply, 2-plasma gases, 3-carrier gases, 4-flow meters, 5-reactor, 6-torch, 7-quench chamber, 8-powder feeder, 9-water tank, 10-water pump, 11-water flow meter, 12-filter, 13-burner, 14-pyrometer,15-data acquisition system.

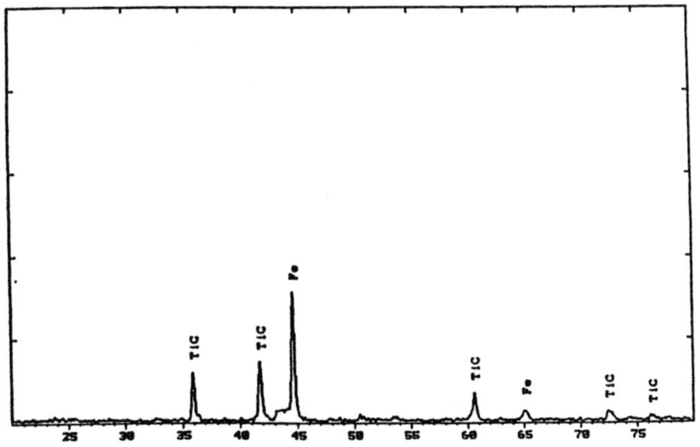

Figure 2. X-Ray diffraction pattern of the as-synthesized in-situ mixed powder

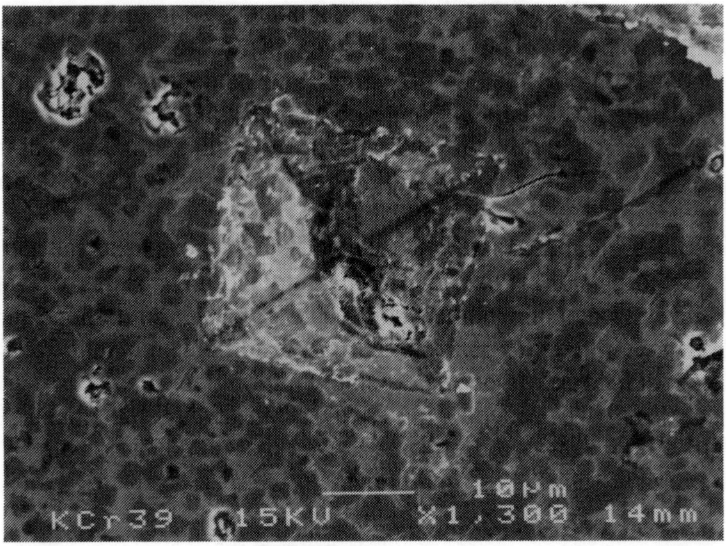

Figure 3. Low magnification microstructure of sintered TiC-Fe composite.

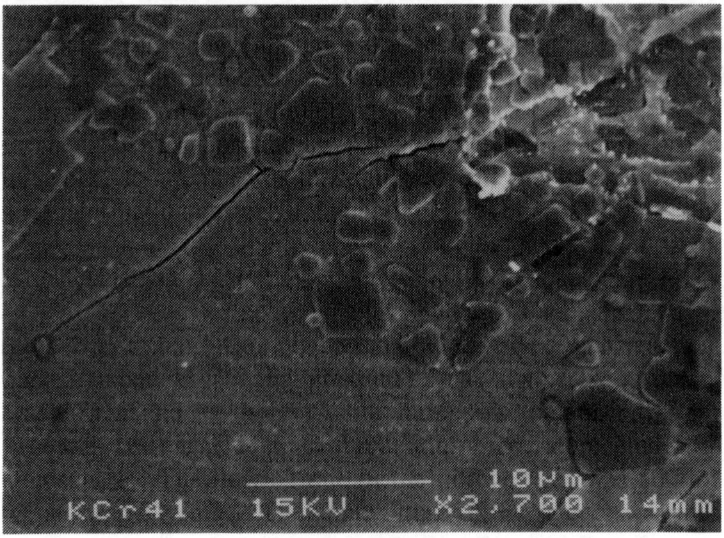

Figure 4. High magnification microstructure of sintered TiC-Fe composite.

4. The average value of the hardness is obtained to be 17GPa from ten such measurements. Cracks are also seem to be emanating from the corners of the indentation. It is possible to obtain a value of toughness from such measurements. However, we have not estimated the Young's modulus of this composite yet. Such results will be reported later on. A direct comparison of the hardness value can not be made with respect to the previous papers because either it has not been reported or the value is expressed in Rockwell scale. But considering the hardness value of TiC, the number obtained here seems to be plausible.

Finally, Figures 3 and 4 show that there is both crack bridging by the ductile phase as well as deflection of the crack front around the TiC particulates. Therefore two types of toughening mechanisms are occurring in this composite, which should lead to enhanced toughness.

CONCLUSIONS

The results presented here show that by dissociating Ilmenite in a non-transferred arc plasma reactor it is possible to produce in-situ mixed powders of TiC and Fe. The as-produced powders can be densified via liquid phase sintering process to a high density after cold isostatically pressing followed by sintering at 1500°C for one hour. The microstructure consists of homogeneous distribution of TiC grains with rounded off edges in Fe matrix. The average hardness was 17GPa. There is evidence of crack bridging as well as deflection.

ACKNOWLEDGEMENT

The authors acknowledge the support of NSF grant no. CTS-9220252.

REFERENCES

1. E. Breval, M.K. Aghajanian and S.J. Luszcz, " Microstructure and Composition of Alumina /Alumina Composites Made by Directed Oxidation of Al" J. Am. Ceram. Soc., 73 2610-2614 (1990).

2. L.S. Sigl and H.E. Exner, "Experimental Study of the Mechanics of Fracture in WC-Co Alloys" Met. Trans., 18A 1299-1308 (1987).

3. A.K. Bhattacharya and J.J. Petrovic, "Ductile Phase Toughening and R-Curve Behavior in B_4C-Al Cermet" J. Mater. Sci., 27 2205-2210 (1992).

4. H.E. Deve and M.J. Maloney, " On the Toughening of Intermetallics

with Ductile Fibers: Role of Interfaces' Acta Met. et Mater., 39 2275-2284 (1991).
5. C. Toy and W.D. Scott, "Ceramic Metal Composite Produced by Melt Infiltration" J. Am. Ceram. Soc., 73 97-102 (1990).
6. S.D. Dunmead, Z.A. Munir and J.B. Holt, "Simultaneous Synthesis and Densification of TiC/Ni-Al Composites" J. Mater. Sci., 26 2410-2416 (1991).
7. H.P. Li, S.B. Bhaduri and J.A. Sekhar, "Metal-Ceramic Composites Based on Ti-B-Cu Porosity System" Met. Trans., 23A 251-261 (1992).
8. K.W. Chae, D.I. Chun, D.Y. Kim, Y.J. Baik and K.Y. Eun, "Microstructural Evolution During the Infiltration Treatment of Titanium Carbide-Iron Composite", J. Am. Ceram. Soc., 73 1979-82 (1990).
9. D.I. Chun and D.Y. Kim, "Microstructural Evolution During the Sintering of TiC-Mo-Ni Cermets", J. Am. Ceram. Soc., 76 2049-52 (1993).
10. T. Kakeshita, C.M. Wayman, "Martensitic Transformation in Cermets with Metastable Austenitic Binder: II. TiC-Fe-Ni-C", Materials Science and Engineering, A147 (1991) 85-92.
11. J.M. Panchal, T. Vela,"New HIP Process for the Production of Carbide Wear Parts" pp. 95-114 in Advances in Powder Metallurgy, edited by E.R. Andreotti and P.J. McGeehan, APMI, Princeton, NJ (1990).
12. B.S. Terry and O. Chinyamakobvu, "Dispersion and Reaction of TiC in Liquid Iron Alloys", Mater. Sci. and Tech., 8 399-405 (1992).
13. B.S. Terry and O. Chinyamakobvu, "Carbothermic Reduction of Ilmenite and Rutile as Means of Production of Iron based Ti(O,C) Metal Matrix Composites", Mater. Sci. and Tech., 7 842-848 (1991).
14. P.R. Taylor, M. Manrique, S.A. Pirzada and M. Abdel-Latif, "Formation of Titanium Carbide from Ilmenite Concentrates in a Thermal Plasma Reactor" Submitted to Plasma Chemistry and Plasma Processing (1994).
15. J.M. Panchal, T. Vela and T. Robisch,"Ferro-TiC Cermets for Erosion Resistant Applications", pp. 125-140 in Advances in Powder Metallurgy and Particulate Materials, edited by J.M. Capus and R.M. German, APMI, Princeton, NJ (1992).
16. S. Mazumdar and V. Ramaswamy, "The Carbon-Iron-Titanium System", J. Alloy Phase Diagrams, 1 27-38 (1985).

CONSOLIDATION OF TiB_2 POWDERS PRODUCED BY PLASMA SYNTHESIS OF CHEAP RAW MATERIALS

G.L. WATT, S.B. BHADURI, R. RADHAKRISHNAN and P.R. TAYLOR

Department of Metallurgy & Mining Engineering
University of Idaho, Moscow, ID 83844-3024

Abstract

TiB_2 powders were produced in a non-transferred arc thermal plasma reactor. The starting raw materials were rutile and boron oxide, which reacted in the plasma chamber using methane to form TiB_2. As received powders were treated to remove any adherent oxide. The powders were consolidated using two different approaches. In the first case the powders were CIPed at 30 ksi, sintered at 1825°C, followed by HIPing at 1825°C and 28 ksi. In the other approach, the samples were CIPed followed by direct HIPing. The as compacted samples were characterized for microstructure and mechanical properties. The results will be reported in this paper.

Introduction

Titanium diboride exhibits several physical properties which make it a desirable engineering material. Principal among these are it's high hardness, low density, and high melting point. The refractory nature of the material allows it to be used in many high temperature applications. Dense bodies are particularly interesting due to their ability to resist oxidation and chemical attack at high temperatures [1]. Another area of application is the use of titanium diboride for cathodes for aluminum extraction [2]. The high hardness value of titanium diboride makes it a good candidate as a lightweight armor [3].

The main drawback of titanium diboride is its inherent brittleness, which is due to its covalent bonding. In order to have reasonable strength of the sintered material, it is important to produce compacts with as small a critical flaw size as possible. Fine reactive powders are therefore essential in achieving that goal resulting in smaller pores and higher density values. As a result, components with high strengths can be produced.

The use of a plasma process for fine TiB_2 powder production has already been reported in the literature. Baumgartner and Steiger synthesized submicron TiB_2 powders in a plasma arc reactor [4]. The process used $TiCl_4$ and BCl_3 as the reactants. These raw materials are expensive and HCl is produced as a byproduct, which leads to corrosion of the reactor and is an environmental hazard. The product is tabular hexagonal platelets of submicron size. At the University of Idaho, TiB_2 powders have been produced from cheap oxide raw materials in a plasma reactor, thus eliminating the production of HCl [5,6]. The main problem here is the feeding of the powders into the plasma flame as opposed to the liquids. However, this problem has been solved and submicron TiB_2 has been produced. This paper will report on the sintering aspects of the plasma synthesized TiB_2 powders. Such studies are essential in order to prove the viability of the process.

Consolidation of Titanium Diboride

The sintering of titanium diboride is difficult due to its high melting point and covalent bonding. Several studies have been reported that deal with the addition of compounds to improve the densification of the diboride during sintering as well as hot isostatic pressing. The additives are typically low temperature phases such as nickel and chromium boride [7-11]. The literature shows that the additives result in enhanced shrinkage of the compacts, and also inhibition of grain growth which can lead to microcracking. Kang et al. sintered TiB_2 at 2000°C in an argon atmosphere using a graphite element furnace. They reported that the addition of iron caused abnormal grain growth and resulted in a lower sintered density [12]. They also found that the addition of boron carbide acted as a grain growth inhibitor as long as the iron content of the mixture was less than 5 percent. Ananthapadmanabhan et al. studied the effect of oxygen pickup on the properties of titanium diboride [13]. The main effects were a reduction in the electrical conductivity values as well as a tendency for the material to absorb water due to the presence of boron oxide.

Recently, a great deal of interest has been shown in the use of microwave sintering. In this type of processing, a microwave furnace adapted to an inert gas atmosphere has been used to sinter TiB_2 to greater than 90 percent of theoretical. While the sintering temperature is the same as before, in the range of 1900-2100°C, the holding time at the sintering temperature is reduced to 30 minutes [14].

Hot isostatic pressing is also supposed to be effective in the production of components from titanium diboride powders. Besson et al., studied the densification of titanium diboride using hot isostatic pressing to produce near net shape components [15]. Green preforms were produced using cold isostatic pressing. The initial green density was shown to be 56 to 58 percent of theoretical density. The samples were then encapsulated under vacuum in titanium cans, hot isostatically pressed at 1873°C and 200 MPa of Ar pressure for 1 to 2 hours under a rapid heating rate to quickly reduce the porosity of the compacts and ensure a high degree of densification. Under these conditions it was shown that a relative density of 99% could be achieved, leading to an increase in materials properties [15].

Table 1. Consolidation schedule for various powders

Sample #	Powder Type	Sinter/ Time	HIP/ Time
1	Plasma	Yes/2hrs	Yes/2hrs
2	Comm.	Yes/2hrs	Yes/2hrs
3	Comm.	No	Yes/1hr
4	Plasma	No	Yes/1hr
5	Comm.	No	Yes/1hr

Experimental

In the plasma processing portion of this work, the effects of different variables on the conversion of the oxide precursors to titanium diboride are studied. The effects of the precursor ratio, plasma gas type and input power were all studied. Since the yield was the highest under the operating conditions of 30 kW input power, stoichiometric reactant ratios, and nitrogen-argon plasma, large quantities of powders were produced under such conditions and were treated to remove any residual oxide impurity. The powders obtained were then characterized using x-ray diffraction (XRD) operating at 40 kV and 30 mA, chemical analysis, transmission electron microscopy (TEM) operating at 80 kV, and scanning electron microscopy (SEM).

Nickel (1 wt %) and chromium diboride (2 wt %) with -325 mesh particle size, procured from Johnson Matthey were used as sintering additives. Titanium diboride powders produced by Advanced Ceramics Corporation were also obtained in -325 mesh particle size and subjected to the same consolidation processes as the plasma powders. This was carried out to allow for a comparison of the microstructure and materials properties.

Following the addition of the sintering agents, 2 weight percent polyvinyl alcohol was added to act as a binder. The powder mixture was then CIPed at 30 ksi to form green compacts.

Initially, some of the pellets were sintered at 1825°C in an inert atmosphere furnace (Centorr, NH), being careful to first allow time for binder removal. Due to the fact that very little shrinkage occurred, this approach was deemed unsuccessful and it was decided that the samples should be hot isostatically pressed.

In order for the HIPing to be successful, the samples had to be encapsulated. Due to the high processing temperatures required for the densification of titanium diboride, quartz tubing was chosen for the encapsulation. The tubing was first sealed at one end. The sample was then placed in the tube and surrounded by boron nitride powder to prevent reaction with the quartz, which was sealed under vacuum (better than 10^{-3} torr). The encapsulated samples were HIPed at 1825°C and 28 ksi for 1 hour. One sample of each material was CIPed, followed by HIPing. To the best of our knowledge, the successful use of quartz as container for HIPing has not been reported in the literature, presumably due to the difficulties associated with working with quartz. The details of the processing conditions are given in Table 1.

The density of the consolidated samples was then determined. Following this determination, the samples were cut, polished, and etched. The microstructure of the samples was then studied using SEM. Finally, the hardness values of the samples were determined using a LECO microhardness tester, fitted with a Vicker's indenter and 1kg of load.

Results and Discussion

Powders from experiments under optimized conditions were analyzed and oxide impurities were removed to allow for their consolidation. An XRD pattern of powders produced after this beneficiation experiment is shown in Figure 1, where all of the peaks are assigned to TiB_2.

Table 2. Properties of TiB_2 Powder Consolidated in Different Ways

Sample #	Final Density	Hardness (GPa)	Grain Size (Microns)
1	93	22.89	10-12
2	91	20.34	2-5
3	88	20.81	8-10
4	80	20.76	4-5
5	89	25.53	4-5

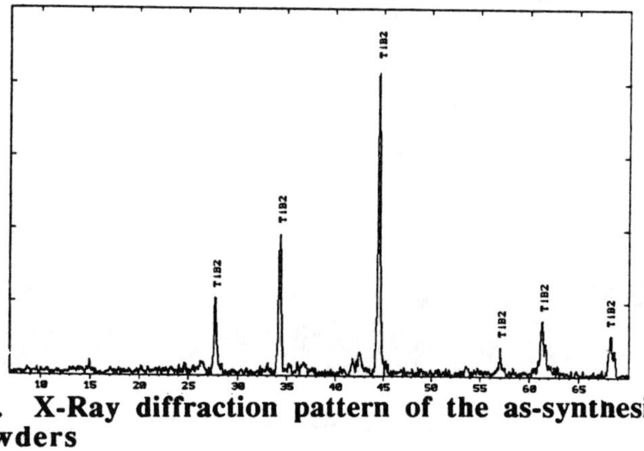

Figure 1. X-Ray diffraction pattern of the as-synthesized TiB_2 powders

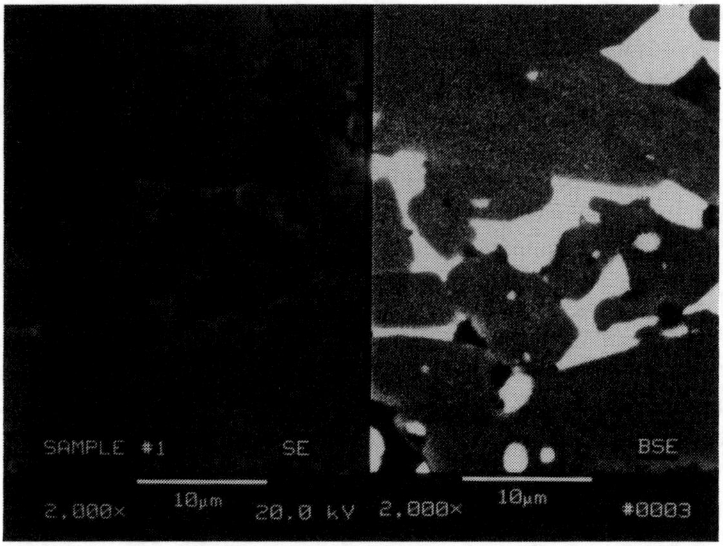

Figure 2. A representative microstructure of the as-consolidated plasma synthesized sample

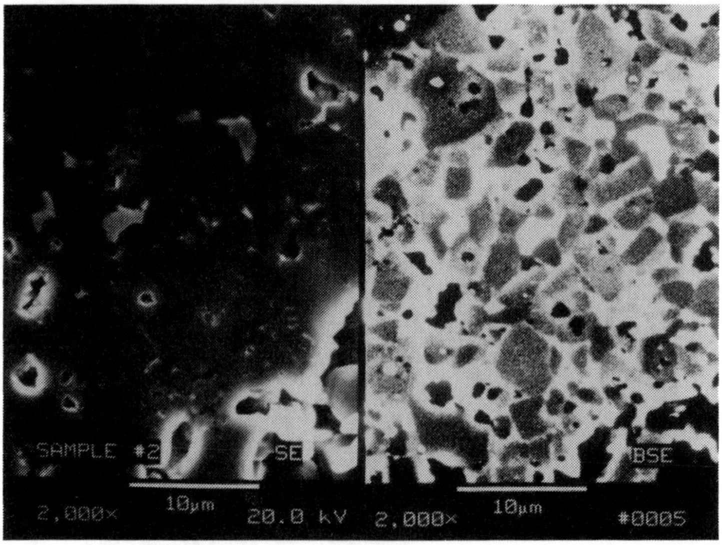

Figure 3. A representative microstructure of the as-consolidated commercial sample

The final density of the consolidated samples (sintered plus HIP) was found to be 93 percent for the plasma powder and 91 percent for the commercial powder. The samples which were HIPed after CIPing showed a density of 88 percent for the commercial powder and 80 percent for the plasma powder. The values are shown in Table 2.

The microstructure of the samples was similar with the only difference being in the grain size and porosity. All of the samples showed elongated grains of titanium diboride with nickel lining the grain boundaries and filling pores. An example of the microstructure for each type of material may be found in Figures 2 and 3. The results were comparable also to those found in the work by Ouabdesselam and Munir, who used the same sintering additives [7].

Finally, the hardness measurements showed that the plasma powder compact had an average hardness of 22.89 GPa. This was comparable to the commercially available powder which was shown to have an average hardness value of 20.34 GPa. Furthermore, the data compare well with the reported values of TiB_2 between 18 to 27 GPa [16].

Conclusions

Titanium diboride powders were produced in a non-transferred arc thermal plasma reactor. The results showed that the conversion of the oxide precursors was maximized at stoichiometric reactant ratios in an argon-nitrogen plasma operated at 30 kW. The samples were then consolidated by CIPing, sintering, and HIPing. The consolidated samples were found to be comparable with the commercially available powders with regard to density, microstructure, and hardness.

Acknowledgement

The authors acknowledge the support of NSF grant no. CTS-9220252.

References

1. J. R. Ramberg and W. S. Williams, "High Temperature Deformation of Titanium Diboride", J. Mater Sci., 22, 1815-26, 1987.
2. C. E. Ransley, "Refractory Carbides and Borides for Aluminum Reduction Cell", J. Metals, 14, 129-35, 1962.

3. D. J. Viechnicki, M. J. Slavin, and M. I. Kliman, "Development and Current Status of Armor Ceramics", Bull. Am. Ceram. Soc., 70, 1035-1039, 1991.
4. H. R. Baumgartner and R. A. Steiger, "Sintering and Properties of Titanium Diboride made from Powder Synthesized in a Plasma-Arc Heater", J. Am. Ceram. Soc., 67, 207-11, 1984.
5. P. R. Taylor and S. A. Pirzada, "Synthesis of Ceramic Carbides in a Non-transferred Arc Thermal Plasma Reactor", Thermal Plasma Applications in Materials and Metallurgical Processing, TMS, Warrendale, PA, pp. 249-265, 1992.
6. G.L. Watt, " Synthesis of Ultrafine TiB_2 in a Non-transferred Arc Thermal Plasma Reactor ", M.S. thesis, University of Idaho, 1994
7. H. R. Baumgartner, "Mechanical Properties of Densely Sintered High Purity Titanium Diboride in Molten Aluminum Environments", J. Am. Ceram. Soc., 67, 490-497, 1984.
8. M. Ouabdesselam and Z. A. Munir, "The Sintering of Combustion-Synthesized Titanium Diboride", J. Mater. Sci., 22, 1799-1807, 1987.
9. W. A. Zdaniewski, "Solid Solubility Effect on the Properties of Titanium Diboride", J. Am. Ceram. Soc., 70, 793-797, 1987.
10. D. Moskowitz and C. W. Phillips, "Methods of Making and Using a Titanium Diboride Comprising Body", U.S. Patent 4,880,600, November 14, 1989.
11. T Wantanabe, et al., "High Density Metal Boride Based Ceramic Sintered Body", U.S. Patent 5,036,028, July 30, 1991.
12. E. S. Kang, et al., "Effect of Iron and Boron Carbide on the Densification and Mechanical Properties of Titanium Diboride Ceramics", J. Am. Ceram. Soc., 72, 1868-72, 1989.
13. P. V. Ananthapadmanabhan, et al., "Effect of Oxygen Pickup on the Properties of Plasma Sprayed Titanium Diboride Coatings", Thin Solid Films, 224, 148-152, 1993.
14. C. E. Holcombe and N. L. Dykes, "Microwave Sintering of Titanium Diboride", Cer. Eng. Sci. Proc., 10, 857-867, 1989.
15. J. Besson, et al., "Densification of Titanium Diboride by Hot Isostatic Pressing and Production of Near Net Shape Components", J. Mat. Eng. and Performance, 1, 637-649, 1992.
16. C. Mroz, "Minerals Review", Bull. Am. Ceram. Soc., 73, 136 - 137, 1994.

Reaction Path in the Formation of Titanium Diboride by a Magnesium Thermite Process

V. Sundaram, K. V. Logan and R. F. Speyer
School of Materials Science and Engineering, Georgia Institute of Technology, Atlanta, GA 30332-0245.

ABSTRACT

TiB_2 was formed by a thermite reaction amongst Mg, amorphous B_2O_3 and TiO_2 powders. Mixtures consisting of 2Mg-TiO_2, 3Mg-B_2O_3 and 5Mg-TiO_2-B_2O_3 were heat treated using DTA and separately via ignition with a nichrome wire; product phases were identified using XRD. MgO and Ti were products from the first mixture reacted in argon, whereas an incomplete transformation forming Mg_2TiO_4 occurred in air. For the second mixture, a reaction forming $Mg_3B_2O_6$ occurred in air, but no reaction occurred in argon due to deficiency of oxygen. Minor amounts of $Mg_3B_2O_6$ and Mg_2TiO_4 were detected in addition to the predicted product phases of MgO and TiB_2 for the third mixture ignited both in air and argon. Based on available evidence, a path for this reaction was deduced; Mg particles in contact with TiO_2 reacted to form Ti, which in turn reacted with B_2O_3 to form TiB_2. TiB_2 product particles from the reaction in argon had a more faceted appearance than those formed during the reaction in air. This was interpreted to be the result of glassy B_2O_3 surface layer formation on TiB_2 particles in air.

INTRODUCTION

Titanium diboride (TiB_2) is a refractory material of extreme hardness (3400 kg/mm^2 Knoop hardness with 500 g load[1]) with potential applications in cutting tool inserts, wear resistant coatings and electrode materials for electrolysis of corrosive media. It has demonstrated good thermal shock and wear resistance as well as chemical stability. Processing methods currently being used to produce TiB_2 include reduction of a mixture of boron anhydride and titanium oxide with carbon[2], borothermic reduction of TiO_2, $TiO_x(OH)_y$, or Ti(O-n-Bu)$_4$-derived polymers[3] and electrolysis of fused mixtures of TiO_2, B_2O_3 and the oxides and fluorides of calcium, magnesium or sodium[2]. Recently, laser induced chemical vapor deposition has been utilized to deposit TiB_2 films on ceramic substrates[4].

Thermite reactions utilize the thermal energy released in the reaction to become self-sustaining. The reaction is initiated at one end of a powder compact and propagates through its length to completely convert the reactants to products. Use of thermite reactions in formation of advanced ceramics have shown a number of advantages. Low cost raw materials, e.g. ceramic oxides, can be used. High reaction temperatures permit rapid firing without the use of furnaces. A variety of advanced ceramics and composites such as SiC, B_4C, Al_2O_3-SiC and MgO-B_4C have been fabricated via thermite reactions[5]. In previous work at Georgia Tech, thermite synthesis involving oxidation-reduction reactions have been used to produce pure TiB_2 and an Al_2O_3-TiB_2 composite[6, 7]. The objective of this work was to characterize the reaction path in the formation of TiB_2 by a magnesium-thermite process and study the effect of process variables on the reaction behavior.

EXPERIMENTAL PROCEDURE

The starting powders were magnesium (Reade Manufacturing Co.), amorphous boron oxide (Fisher Scientific) and titanium oxide (Fisher Scientific). Average particle sizes (Microtrac II laser particle size analyzer) were measured to be 15 μm for amorphous B_2O_3 and 1.6 μm for TiO_2. The particle size of Mg was not measured, but manufacturers data showed that it was less than 45 μm. The powders were weighed according to the stoichiometry of reactants in equations 1-3, and then mixed through extensive gentle stirring to ensure homogeneity. The reactant mixtures were then ignited in bulk, or studied using smaller masses in a differential thermal analyzer.

For bulk ignition studies, starting powder mixtures of 25 g were sifted into cylindrical slip cast fused silica crucibles (4 cm dia., 8 cm height, 0.4 cm wall thickness). For each crucible, a resistance heated nichrome wire was inserted ~1 cm into the sample powder from above. The power input to the heating element was manually controlled using a variac. The crucible was covered with a slip cast fused silica lid. A 450 g weight was placed on the lid to maintain its position during reaction. The sample was removed from the crucible and crushed in an alumina mortar for characterization. These reactions were also performed in a similar fashion in an argon atmosphere. A moderate vacuum was drawn on a sealed chamber, and backfilled with argon, repeated 3 to 5 times, to create the 1 atm argon atmosphere.

For differential thermal analysis (DTA) studies, 15 mg samples of reactant mixture were placed in an alumina crucible. Alumina powder was used as the reference standard. The sample and reference crucibles were introduced into a Pt-wound DTA (Perkin-Elmer DTA1700 with an Innovative Thermal Systems control and data acquisition interface). Scans were run at heating rates of 10°C/minute and 50°C/minute, under flowing air (from a cylinder of compressed dry air) and flowing argon, up to a temperature of 1200°C. All DTA traces were repeated to ensure reproducibility. Argon atmospheres were created similarly as described for bulk ignition. The ordinate of the DTA traces represents true sample temperature. Temperatures recorded from the DTA sample thermocouple were corrected using linear correlations, based on the onset locations of the Mg melting endotherm (melting point: 650°C[9]) and the eutectic melting temperature between $Mg_3B_2O_6$ and MgO (1358°C[10]).

The crushed samples from bulk ignition were analyzed using X-ray diffraction (XRD, Philips PW1800) with a step size of 0.015° and a sampling time of one second. Reaction products from DTA studies were not analyzed with XRD since the quantity of powder was too low. Scanning electron microscopy (SEM, Hitachi S-800) was utilized in the secondary electron mode to view typical morphologies of as-reacted products. Energy dispersive spectrometry (EDS, Kevex Systems) was used in conjunction with SEM to provide information about local elemental compositions .

RESULTS

Reactions based on three stoichiometric mixtures were considered:

$$2Mg + TiO_2 \longrightarrow 2MgO + Ti \qquad \Delta H_{298K} = -263.8 \text{ kJ/mol} \qquad (1)$$

$$3Mg + B_2O_3 \longrightarrow 3MgO + 2B \qquad \Delta H_{298K} = -550.4 \text{ kJ/mol} \qquad (2)$$

$$5Mg + TiO_2 + B_2O_3 \longrightarrow 5MgO + TiB_2 \quad \Delta H_{298K} = -1093.6 \text{ kJ/mol} \quad (3)$$

The standard enthalpies[9] of formation of the various reactants and products at 298 K as well as sensible heats of product phases as a function of temperature were used to determine the adiabatic reaction temperatures. Latent heats of fusion for products phases were taken into account. For the above reactions, these temperatures were 2110, 2460, and 3350 K, respectively. Based on calculations from standard state Gibbs energy of formation for the above three reactions, conversions of reactants to products are favorable above room temperature up to 2230, 2380, 2660 K, respectively.

DTA traces of powdered samples of stoichiometries of reactants in equations 1-3, as well as Mg alone, are shown in Fig. 1.

In bulk ignition experiments, a reaction in the $3Mg$-B_2O_3-argon mixture did not initiate up to an applied filament voltage of 45V—at which point the ignition wire fused. For the $3Mg$-B_2O_3-air mixture, the reaction initiated at 40V. This reaction was not self-propagating and occurred in spurts. After the reaction, regions of reactant powders were visibly present. The $2Mg$-TiO_2 reaction took place with spectacular fury, both in air and argon. The reaction initiated when the voltage was increased to 25V. The silica lid and the 450 g weight were blown off and ~1/3 of the powder in the crucible

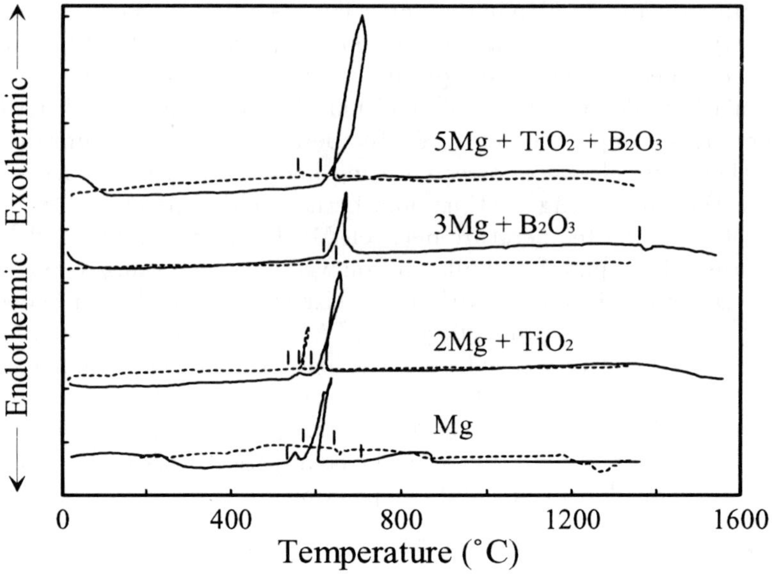

Figure 1: DTA traces for various compositions heated in air (solid lines) and argon (dashed lines) environments at a heating rate of 10°C/min.

was spilled onto the chamber floor. The $5Mg-B_2O_3-TiO_2$ reaction in air and argon initiated at about 30V in air and the reaction front propagated toward the crucible base. Gas evolution was observed during all the above reactions. The reaction products in all the systems appeared as a porous solid.

XRD results in Fig. 2 show that the reaction products from the $2Mg-TiO_2$ powder in argon were MgO and Ti. However, reaction products in air were MgO and Mg_2TiO_4. Fig. 3 shows that the reaction products for the $3Mg-B_2O_3$ powder in air were MgO and $Mg_3B_2O_6$. The XRD trace (not shown) for this mixture treated similarly in an argon atmosphere showed no reaction had taken place. Fig. 4 shows that in both air and argon, the $5Mg-B_2O_3-TiO_2$ powder reacted to form MgO and TiB_2 as major phases, with Mg_2TiO_4 and $Mg_3B_2O_6$ as minor phases. Based on relative peak heights, no difference in the relative amounts of these products can be assessed after ignition in air or argon. Fig. 5 shows the products from the $5Mg-TiO_2-B_2O_3$ powders after ignition in air and argon. In the micrographs, the larger grains are MgO, while the smaller particles are TiB_2, as determined based on EDS results.

DISCUSSION

The exotherms for $5Mg-TiO_2-B_2O_3$-air and $2Mg-TiO_2$-air traces in Fig. 1 show a crossing feature resulting from sample temperature being the ordinate,

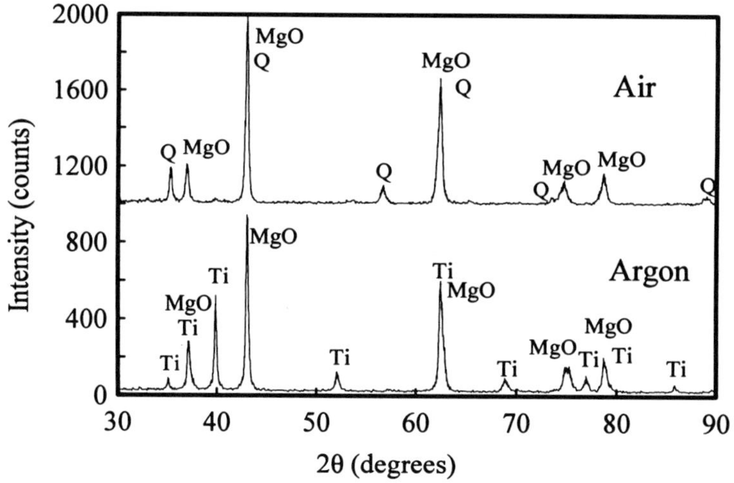

Figure 2: XRD traces of products from ignition of a $2Mg\text{-}TiO_2$ powder using a electrically heated nichrome wire. Q: Qandlite (Mg_2TiO_4).

and the reaction being highly exothermic causing rapid temperature rise in the sample. When the transformation was complete, the sample temperature cooled (negative direction on the ordinate) to match that of the surroundings, which in turn were increasing in temperature at 10°C/min. The advantage of using sample temperature as the ordinate (as opposed to setpoint, reference, or furnace temperatures) is that onsets of endotherms/exotherms closely represent the true transformation temperatures of the sample.

From Fig. 1, minor endotherms corresponding to the fusion of Mg are apparent at 648°C for the Mg-argon, $3Mg\text{-}B_2O_3$-argon, and $5Mg\text{-}TiO_2\text{-}B_2O_3$-argon traces. The absence of this endotherm in the $2Mg\text{-}TiO_2$-argon trace implies that Mg was extinguished by reaction with TiO_2 prior to its fusion temperature. This endotherm was also absent from all DTA traces in air, implying oxidation of Mg particles by the atmosphere, or reaction with other batch constituents, prior to reaching its melting temperature.

The Mg-air DTA trace in Fig. 1 shows three exotherms, with onsets at 551, 572, and 737°C. Only one exotherm corresponding to simple oxidation was expected. If $MgCO_3$ formed from reaction with atmospheric O_2 and CO_2, then a later endotherm associated with decomposition ($MgCO_3 \longrightarrow MgO + CO_2$) would be expected. All three peaks were also observed when using platinum crucibles for sample and reference, although they shifted to lower temperatures owing to higher thermal conductivity of Pt. Thus, oxidation as a result of eutectic melting with the Al_2O_3 crucible is also ruled out. Thus,

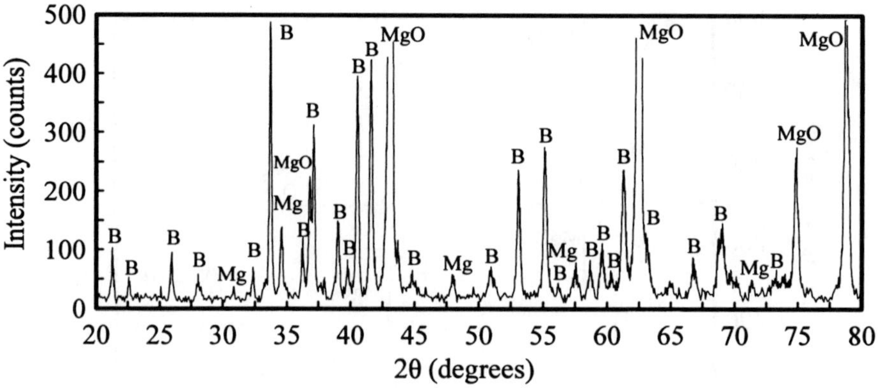

Figure 3: XRD trace of $3Mg\text{-}B_2O_3$ powder ignited in air. B: $Mg_3B_2O_6$.

based on available evidence, it is interpreted that all of these exotherms represent oxidation of Mg. It is speculated that the exotherms represent MgO coating formation on Mg, which in turn fragmented due to lattice mismatch. These fragments then formed oxide coatings, and so on.

The low intensity exotherm onset at 551°C in the $2Mg\text{-}TiO_2$-air trace matches well with the first oxidation exotherm of Mg in the Mg trace, and is hence interpreted as the formation of oxide coatings on Mg particles. The exotherm is broader and less intense due to the TiO_2 diluent in the sample. The onset of the exotherm at 592°C for the $2Mg\text{-}TiO_2$-air trace corresponds to reaction between the batch constituents and atmospheric oxygen to form Mg_2TiO_4. This product would be thermodynamically favorable regardless of whether Mg-containing reactant was in the elemental ($2Mg + O_2 + TiO_2 \rightarrow Mg_2TiO_4$) or oxidized ($2MgO + TiO_2 \rightarrow Mg_2TiO_4$) form. The MgO in the bulk ignition product detected by XRD indicates incomplete reaction. It is interpreted that the fervent combustion reaction caused appreciable gaseous convection, preferentially blowing the lighter TiO_2 reactant particles out of the container. This deviation from reaction stoichiometry resulted in the detection of MgO in the product powder.

XRD results from bulk ignition indicate that the products in the $2Mg\text{-}TiO_2$-argon system were MgO and Ti metal. Since only one peak is present in the $2Mg\text{-}TiO_2$-argon trace at 560°C, this peak corresponds to the reaction $2Mg + TiO_2 \longrightarrow 2MgO + Ti$.

The presence of an exotherm for the $3Mg\text{-}B_2O_3$-air trace, and its complete absence for the $3Mg\text{-}B_2O_3$-argon trace agrees with results of XRD for bulk ignition. In the $3Mg\text{-}B_2O_3$-argon trace, no exotherm is observed since elemental magnesium and B_2O_3 could not react to form $Mg_3B_2O_6$ due to a lack

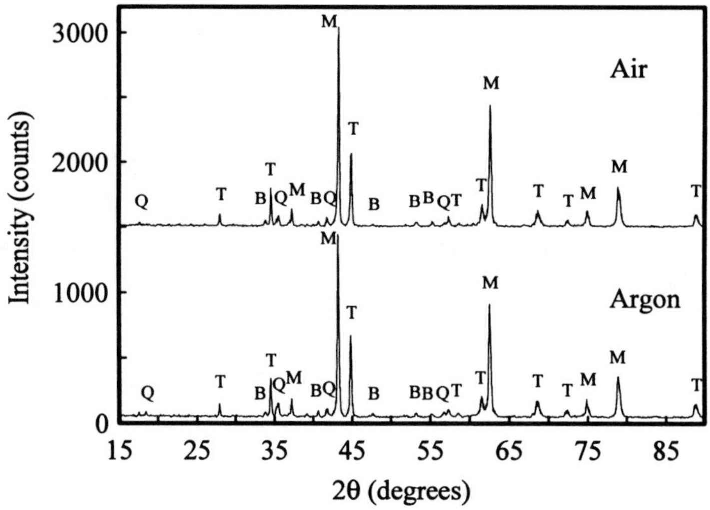

Figure 4: XRD traces of products from ignition of a $5Mg$-TiO_2-B_2O_3 powder using an electrically heated nichrome wire. M: MgO, T: TiB_2, B: $Mg_3B_2O_6$, Q: Mg_2TiO_4.

of oxygen. While the formation of MgO and elemental boron by equation 2 is thermodynamically favorable, this reaction is not evident from DTA, or XRD of bulk reaction powders. These two components were in intimate contact, so this result was not anticipated. In air, no exotherm is apparent corresponding to the oxidation of Mg at 551°C. It is interpreted that B_2O_3 fluid[1] was drawn between Mg particles by capillary action, prohibiting or slowing oxidation of the metal. From the shape of the baseline of this trace between 551 and 626°C, a broad low intensity exotherm may be inferred which corresponds to slow oxidation of Mg via diffusion of atmospheric O_2 through the liquid phase. The exotherm onset at 626°C for the $3Mg$-B_2O_3-air trace corresponds to the formation of $Mg_3B_2O_6$ via $3Mg + B_2O_3 + \frac{3}{2}O_2 \longrightarrow Mg_3B_2O_6$, or $3MgO + B_2O_3 \longrightarrow Mg_3B_2O_6$. The presence of this product is further confirmed by the endotherm onset at 1358°C, corresponding to congruent melting of $Mg_3B_2O_6$ (1366°C), or eutectic melting between MgO and $Mg_3B_2O_6$ (1358°C)[10]. Eutectic melting is implied by the broad width of the endotherm[8]. MgO detected by XRD in the powder ignited in air implies incomplete reaction. Based on the lack of a crossing feature on the exotherm for this DTA trace, the above reaction occurred comparatively more slowly. This may be due to the requirement of diffusion of atmospheric oxygen through the fluid B_2O_3.

[1] $B_2O_{3(l)}$ is quite fluid at these temperatures—it has a viscosity of ~300 Pa·s at 600°C[11].

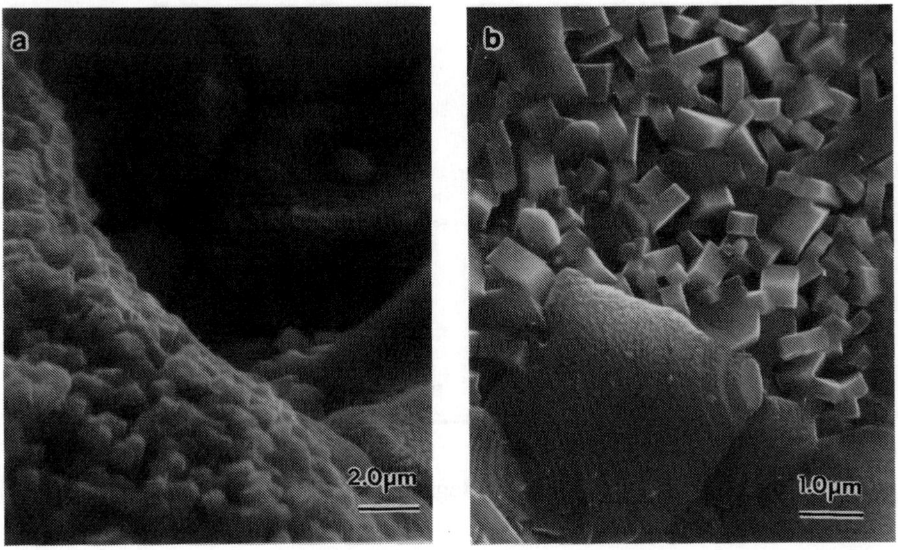

Figure 5: Products from the $5Mg$-TiO_2-B_2O_3 powders ignited in (a) air and (b) argon.

There is no exotherm onset at 551°C in the $5Mg$-TiO_2-B_2O_3-air trace as was observed in the Mg-air or $2Mg$-TiO_2-air traces, since oxidation was hampered by capillary flow of B_2O_3. The onset of the intense exotherm in the $5Mg$-TiO_2-B_2O_3-air trace matches well with the exotherm at 626°C in the $3Mg$-B_2O_3-air trace. This exotherm corresponds to $Mg_3B_2O_6$ product formation from MgO and B_2O_3. No exotherm onset is apparent at 592°C in the $5Mg$-TiO_2-B_2O_3-air trace, implying suppression of the reaction between Mg and TiO_2. It is therefore interpreted that the fluid B_2O_3 phase initially reacted with Mg particles and O_2 from the atmosphere, while keeping MgO and TiO_2 particles from mutual contact. After initial heat generation from this transformation, varying reactions in different regions of the batch fervently occurred. If B_2O_3 liquid had formed a seal around the Mg particles so that they did not oxidize prior to the exotherm onset at 616°C, then TiB_2 could reasonably be a product of a reaction where elemental boron and titanium (formed via $2Mg + TiO_2 \longrightarrow 2MgO + Ti$ and $3Mg + B_2O_3 \longrightarrow 3MgO + 2B$) combine. However, this is unlikely since Mg particles in contact with B_2O_3 fluid in air ($3Mg$-B_2O_3-air trace) showed a $Mg_3B_2O_6$ product, with no evidence of solid boron formation. More likely, Mg_2TiO_4, and $Mg_3B_2O_6$ product formation would be represented by this exotherm. The process of bulk ignition in air did not allow time for Mg oxidation or B_2O_3 flow. Thus, the TiB_2 product (in addition to Mg_2TiO_4 and $Mg_3B_2O_6$) detected by XRD of these powders indicates that the reaction was too rapid for equilibrium to be reached, since TiB_2 should not form in an oxidizing atmosphere.

The lack of intense exotherms for the $5Mg$-TiO_2-B_2O_3-argon trace contrasts with the results from XRD for this composition powder ignited with the nichrome wire. The crystalline phases resulting from ignition of these powders in argon were the same as those ignited in air. It is interpreted that the rapid ignition of the bulk powders did not allow time for capillary flow of B_2O_3 to separate Mg and TiO_2 particles. Hence Mg and TiO_2 particles in contact reacted to form Ti and MgO. This Ti then reacted with the B_2O_3 fluid to form TiB_2 ($5Ti + 2B_2O_3 \rightarrow 2TiB_2 + 3TiO_2$). This appears to be a more favorable reaction path than the reduction of B_2O_3 by Mg ($3Mg + B_2O_3 \longrightarrow 3MgO + 2B$) followed by reaction with Ti product ($2Mg + TiO_2 \longrightarrow 2MgO + Ti$) to form TiB_2[2]. This is based on the demonstrated unwillingness of Mg and B_2O_3 to react in argon, even when heated up to 1200°C ($3Mg$-B_2O_3-argon trace in Fig. 1). For the DTA specimen of this composition in argon, where no significant exotherms were observed, adequate time was allowed for B_2O_3 to become fluid, wherein it was drawn between, and separated, Mg and TiO_2 grains. Under such conditions, very limited Ti intermediate products would form (a minute exotherm in the $5Mg$-TiO_2-B_2O_3 trace at 560°C can be seen), in turn limiting the formation of TiB_2.

The TiB_2 product particles from the $5Mg$-TiO_2-B_2O_3 mixture ignited in air shows a smoother shape than those ignited in argon (Fig. 5). It is suggested that the smoother shape is a result of the TiB_2 surface having oxidized to form an amorphous B_2O_3 surface layer during cooling of the products.

CONCLUSIONS

DTA traces of the $5Mg$-TiO_2-B_2O_3 mixture represent different reactions than those which occurred during ignition of bulk powders. Under slow heating in the DTA in argon, amorphous B_2O_3 was permitted time to become fluid and flow between Mg and TiO_2 particles by capillary action, prohibiting their reaction. Hence, an essential feature of the thermite reaction is that time is not allowed for this to occur. Based on available evidence, it was interpreted that the thermite reaction path involved reaction of Mg and TiO_2 particles in contact to form elemental titanium, which in turn reacted with amorphous B_2O_3 to form TiB_2.

FUTURE WORK

XRD data of larger masses of powders heated at 10°C/min in argon and air are required to confirm interpretations of DTA traces. Simultaneous

[2]The reaction of B with TiO_2 ($2B + TiO_2 = TiB_2 + O_2$, $p_{O_2} = 0.2$ atm) is not thermodynamically favorable in the temperature range studied.

thermogravimetric analysis will show the temperature ranges and extent of magnesium oxidation from the atmosphere. Rapid quenching in the midst of the thermite reaction via placement of the lower portion of the reactant powder in a wedge-shaped liquid N_2-cooled enclosure may show intermediate reaction products. This will shed further light on the reaction path.

ACKNOWLEDGEMENTS

We would like to express our appreciation for the assistance provided by Guillermo Villalobos, Gautam Patel and Charles Huthmaker in this work.

References

[1] K. V. Logan, "Elastic-Plastic Behavior of Hot-Pressed Composite Titanium Diboride/Alumina Powders Produced Using Self-Propagating High-Temperature Synthesis", Ph. D. Thesis, Georgia Institute of Technology, 1992.

[2] G. V. Samsonov, *et al.*, "Boron, Its Compounds and Alloys", Publishing House of the Academy of Sciences Ukrainian SSR, Kiev, pp. 388-393, 1960.

[3] Z. Jiang and W. E. Rhine, "Preparation of Titanium Diboride from the Borothermic Reduction of TiO_2, $TiO_x(OH)_y$, or $Ti(O\text{-}n\text{-}Bu)_4$-derived Polymers", *J. European Ceramic Society*, **12**, 5, pp. 403-411, 1993.

[4] J. Elders and J. D. W. van Voorst, "Laser-induced CVD of Titanium Diboride and the Influence of Atomic Hydrogen", *Applied Surface Science*, **69**, 1-4, pp. 267-271, 1993.

[5] W. L. Wang, Z. A. Munir and Y. M. Maximov, "Thermite Reactions: Their Utilization in the Synthesis and Processing of Materials", *J. Materials Science*, **28**, 14, pp. 3693-3708, 1993.

[6] K. V. Logan and J. D. Walton, "TiB_2 Formation Using Thermite Ignition", *Ceram. Eng. Sci. Proc.*, **5**, pp. 712-738, 1984.

[7] K. V. Logan, J. T. Sparrow and W. J. S. McLemore, in *Combustion and Plasma Synthesis of High-Temperature Materials*, edited by Z. A. Munir and J. B. Holt, VCH, New York, pp. 219-228, 1990.

[8] R. F. Speyer, *Thermal Analysis of Materials*, Marcel Dekker, Inc., New York, 1994.

[9] "JANAF Thermochemical Tables", *J. Physical and Chemical Reference Data*, **14**, Supplement No. 1, Amer. Chem. Soc. and Amer. Inst. Physics for National Bureau of Standards, Washington DC, 1985.

[10] E. M. Levin, C. R. Robbins and H. F. McMurdie, "$MgO\text{-}B_2O_3$ Phase Diagram", p. 111 in *Phase Diagrams for Ceramists*, M. K. Reser editor, American Ceramic Society, Columbus, Ohio, 1964.

[11] R. H. Doremus, in *Glass Science*, John Wiley and Sons Inc., New York, pp. 101-114, 1973.

Advanced Processing of Composites

PROPOSED NEW NONDESTRUCTIVE TESTS FOR Si_3N_4 BY CRYOGENIC HEAT CAPACITY MEASUREMENTS

Toyohiro Hamasaki, Tomohiro Yanai, Yoshihiro Okumoto and Kozo Ishizaki
Nagaoka Gijutsu-Kagaku Daigaku (Nagaoka University of Technology)
Nagaoka, Niigata 940-21, Japan

ABSTRACT

The quantity and quality of grain boundary phases of Si_3N_4 can be evaluated by measuring cryogenic specific heats. The specific heats of Si_3N_4 ceramics have been measured at different temperatures between 10 and 40 K. The temperature dependency of measured cryogenic specific heats provides quantitative and qualitative (amorphous or crystalline) information of grain boundary phases. This method is useful for evaluating small changes in the amount of grain boundary phases. The sensitivity of this method is very high, and allows to clarify different heat histories of sintering process for ceramics. This proposed method is nondestructive, and eventually leads to a new quality control method of ceramics to clarify relative difference of heat histories of ceramic processing.

INTRODUCTION

Si_3N_4 ceramics have become increasingly a promising material to use as heat engine components because of their excellent performance in high-temperature environments. Fully densified Si_3N_4 ceramics are obtained through liquid phase sintering to overcome the difficulty of sintering due to the strong covalent bonding. Metal oxides are normally used to facilitate the sintering of Si_3N_4 ceramics by liquid-phase formation. During densification, these additives react with Si_3N_4 surface oxides to form a liquid-phase. [1] When the liquid phase is cooled, it remains at the grain boundaries and forms glassy or crystalline phases.

The presence of a grain boundary glassy phase in sintered Si_3N_4 influences mechanical properties, particularly at high temperatures. [2, 3] Therefore, the quantitative evaluation of the grain boundary phase is of great importance.

Quantitative evaluation of the grain boundary has been determined usually by a destructive method, using metallographic techniques by chemically or thermally etching polished sample surfaces. [4-6] The authors reported a quantitative evaluation method of the grain boundary glassy and crystalline phases by cryogenic specific heat measurements.[7-9]

In this study, measured specific heats have been compared with the calculated specific heat from the Debye theory and measured specific heats of standard samples. The temperature dependence of the specific heat gives information about not only the glassy phase but also the crystalline phase at the grain boundary. [7-9] The proposed method is nondestructive, and the sensitivity is extremely high. This method can be very useful for evaluating small changes in the amount of grain boundary crystalline and glassy phases, which can reveal different heat histories of sintered bodies, [7,8] and eventually leads to a nondestructive quality control method of ceramics.

EXPERIMENTAL PROCEDURE

Four types of sintered Si_3N_4 ceramics were investigated in this study. The first type (sample A) was sintered from a commercial α-Si_3N_4 powder (UBE-SN-E10, UBE Industries, specific surface area of 11.1 m^2/g and average particle size of 330 nm). The second type (sample B) was made from the same powder with surface modification by aqueous Soxhlet washing for 70 hours. The modified powder had less surface oxides. [10-12] The third type (sample C) was sintered from the same powder with oxidation treatment in air at 1000℃ for 12 hours. The fourth type (sample D) was made from the same powder having a carbon coating surface modification by thermal decomposition of CH_4 gas at 800℃. [13, 14] Powders of Y_2O_3 (Shin-Etsu Chemical Co.; Ltd.; 1.3 m^2/g, 900 nm) and Al_2O_3 (Onoda Cement Co., Ltd.; 1.7 m^2/g, 900 nm) were added as sintering aids (3 mol% each) to all samples.

The powders were mixed for 24 hours in ethanol using a ball milling machine with Si_3N_4 balls in a plastic vessel. The powder slurries were dried in a vacuum furnace at 50℃ for 3 hours, and crushed. The resultant powders were hot pressed at 1800℃ under 30 MPa for one hour in nitrogen atmosphere. Density measurements by the Archimedes method showed full densification (above

99.9%). The sintered bodies were machined to rectangular bars of 3×4×36 mm. Three-point bending strengths were measured from room temperature to 1500℃. The crosshead speed was 0.5 mm/min, and the span was 30 mm.

Simulated grain-boundary glassy and crystalline phases were synthesized. The composition of the glass was fixed by determining the maximum nitrogen solubility by X-ray diffraction determination, under the same heat treatment, assuming that the surface oxygen of Si_3N_4 must react with sintering additives to form a grain-boundary glassy phase. [1, 8, 15-17] The composition of the crystallized grain-boundary phase was $Y_5N(SiO_4)_3$ ($4.8Si_3N_4$ $18.4SiO_2$ $76.8Y_2O_3$, in wt%), which was determined from the X-ray diffraction analysis of heat treated sample A after crystallizing treatment.[8]

The cryogenic specific heat was measured at temperatures from 10 to 40 K using an adiabatic calorimeter designed by this laboratory. [7] The temperature was measured by using a carbon glass resistor. [7]

RESULTS AND DISCUSSION

At cryogenic temperatures the difference between an isochoric specific heat of a substance and its isobaric specific heat becomes negligibly small (less than 0.03% at 50 K for Si_3N_4). [8] In the present work, the difference between the isochoric specific heat (i.e. the theoretical specific heat from Debye theory) and the isobaric specific heat (i.e. the empirically obtained specific heat) is ignored.

The isochoric specific heat of an electrically insulating material with a perfect crystalline phase at low temperatures depends on temperature in accordance with the Debye T^3 law. The transition to noncrystalline state is associated with a departure from the Debye T^3 law. [18] The difference between the specific heat of noncrystalline and crystalline phase has been referred to as an abnormal specific heat. It is caused by extra phonon modes characteristic to glassy phases. [19] In the case of Si_3N_4, the difference which exists between the measured cryogenic specific heats of the samples and the calculated value gives evidence of the existence of a grain boundary glassy phase in sintered Si_3N_4.

Figure 1 shows the results of the specific heat measurements for sintered Si_3N_4 ceramics. In Figure 1, the solid line indicates the Si_3N_4 lattice specific heat calculated from the Debye theory using a Debye temperature of 1100 K. [20] The solid marks correspond experimental values, which lie on the solid line at the higher temperatures (>100 K). Some of the higher temperature data were reported previously.[20]

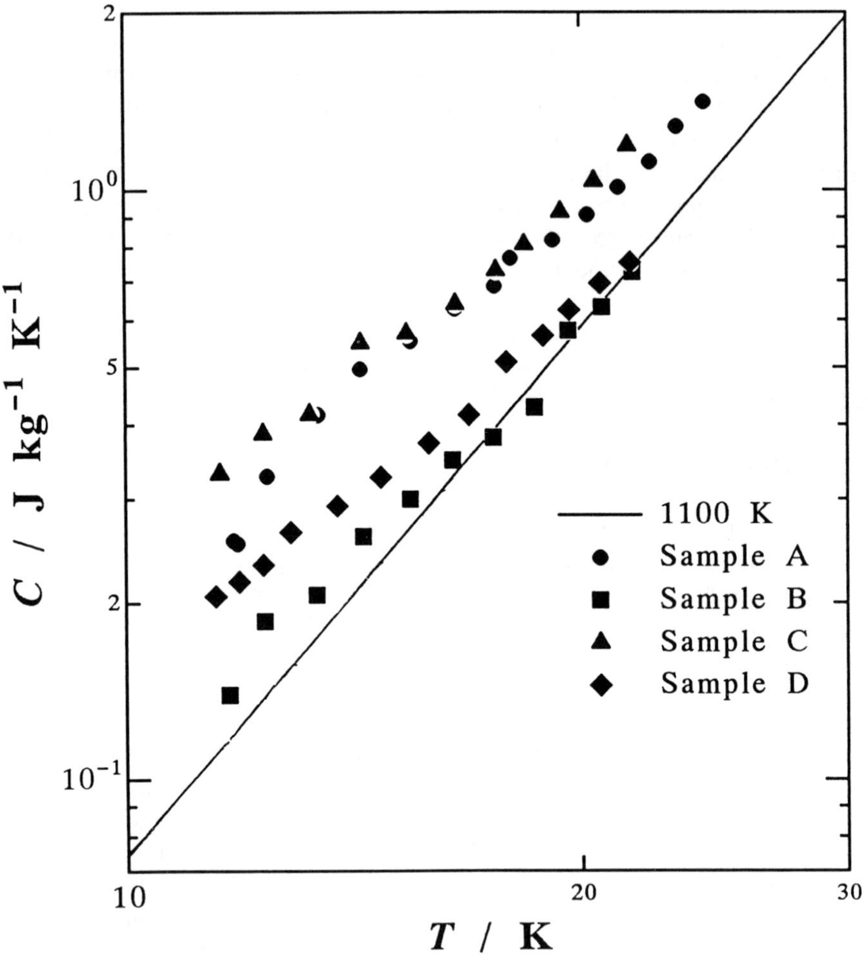

Figure 1. Specific heats, C, of four types of Si_3N_4 versus temperature, T, ((●) (sample A) (made from commercial powder), (■) (sample B) (made from surface modified powder by aqueous washing), (▲) (sample C) (made from surface modified powder by oxidized), and (◆) (sample D) (made from carbon coated powder), and (—) calculated specific heat of Si_3N_4 from Debye temperature of 1100 K).

The specific heat of a sintered body can be represented by

$$C = xC_1 + yC_2 + zC_3 \qquad (1)$$

where x, y, and z are, respectively, the amounts of grain lattice, grain boundary crystalline phase, and grain boundary glassy phase, and C, C_1, C_2, and C_3 are, respectively, the total specific heat of the sample, the lattice specific heat of Si_3N_4, the specific heat of the grain boundary crystallized phase, and the specific heat of the grain boundary glassy phase. C_1 and C_2 can be estimated by the Debye theory using the Debye temperatures of 1100 K [20] and 350 K that was calculated by the cryogenic specific heats of a simulated partial crystalline grain boundary phase,[8] respectively. C_3 is calculated by using the measured cryogenic specific heats of simulated glassy grain boundary phases. [21] The amount of each phase can be evaluated by solving Eq. (1).

Table I and Figure 2 show the results of the quantitative evaluation of the grain boundary phases. The grain boundary phases of sample D decreased with respect to the sample A, because SiC particles are precipitated by in-situ reaction between surface silicon oxide phase and coated carbon during sintering.

Table I. The Amount of Lattice Si_3N_4 Phase, x, Grain Boundary Recrystallized Phase, y, Grain Boundary Glassy Phase, z, and Precipitated SiC Phase, SiC. All of them in weight fraction.

	Sample A	Sample B	Sample C	Sample D
x	0.936±0.001	0.974±0.004	0.916±0.001	0.936±0.002
y	0.010±0.001	0.020±0.002	0.016±0.001	0.032±0.001
z	0.054±0.001	0.006±0.003	0.068±0.001	0.012±0.002
SiC	-	-	-	0.020

The Debye temperature of SiC lattice is about 1090 K [22] or 1150 K [23] which is close to Si_3N_4. The amount of the lattice Si_3N_4 calculated by this method must include the precipitated SiC content. The precipitated SiC content is estimated by the difference between Si_3N_4 contents of sample A and sample D. The amount of the precipitated SiC was estimated to be about 2 wt% by this method.

Figure 3 shows the relationship between the surface oxygen content of starting powder and the amount of grain boundary phases in sintered Si_3N_4. The surface oxygen content of the starting powder influences the amount of grain boundary phases in sintered Si_3N_4 ceramics.

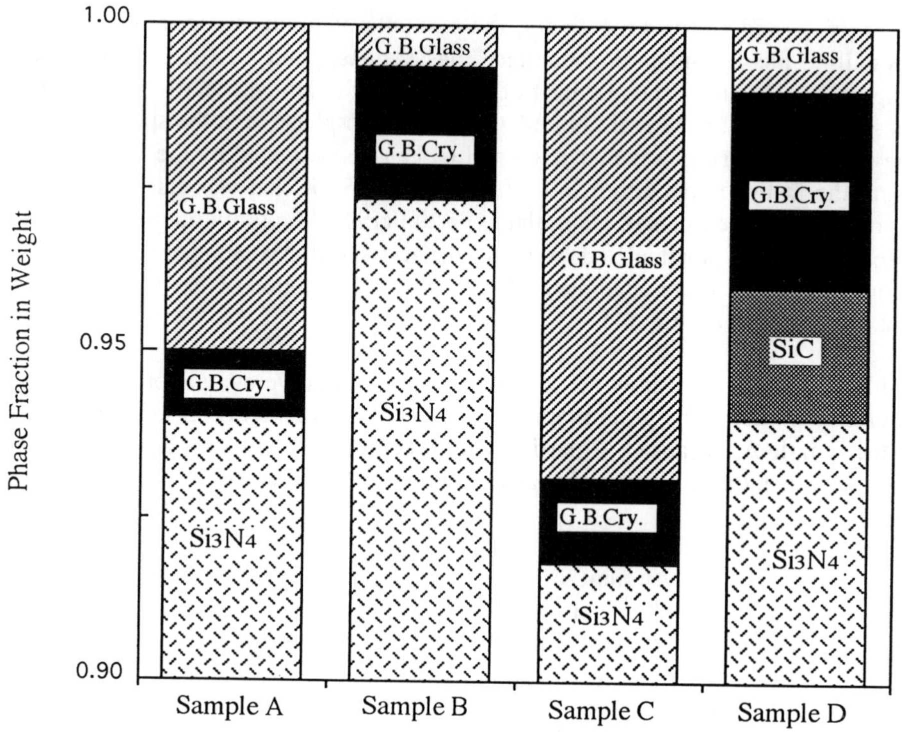

Figure 2. The results of the quantitative evaluation of constituent phases for hot pressed Si_3N_4.

Sample B has the most lattice Si_3N_4 phase. For the surface modified powder that has the least surface oxides, the least liquid phase was formed during the sintering process, and consequently, the least grain boundary phase is found in sample B.

Sample C has the most grain boundary phases. The raw powder of sample C had the highest amount of surface oxides. The highest amount of grain boundary phases is found in sample C.

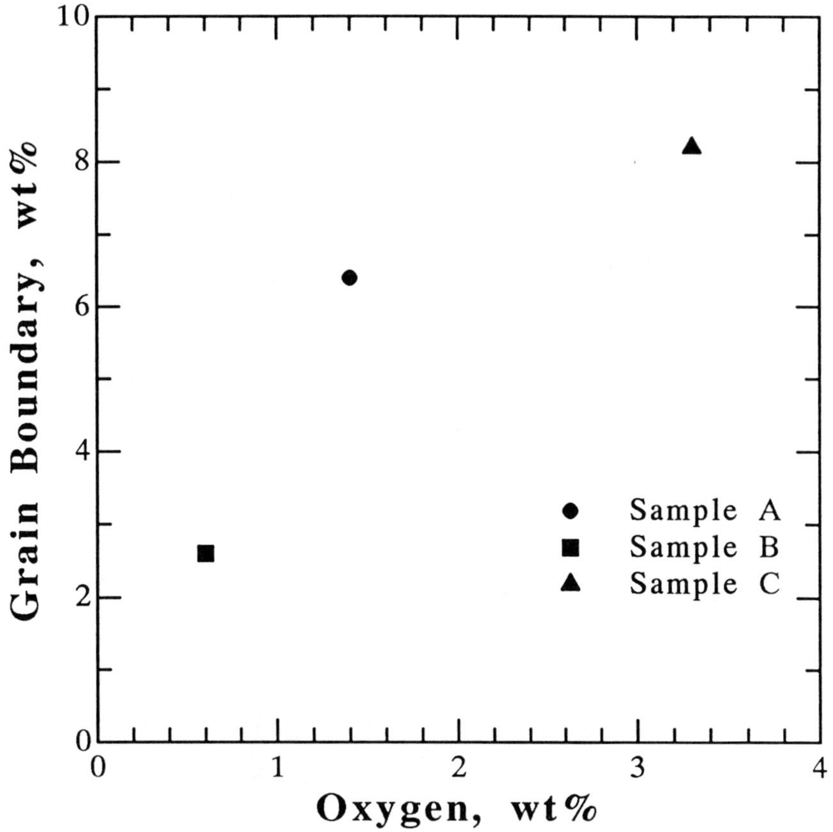

Figure 3. The relationship between the surface oxygen content of starting powder and the amount of grain boundary phases in sintered Si_3N_4. The amount of grain boundary phases in Si_3N_4 increases with increasing the surface oxygen content of the starting powder.

The present method is a nondestructive evaluation of the grain boundary phase and sufficiently sensitive to evaluate small changes in the amounts and kinds of grain boundary phases of Si_3N_4 ceramics.

Figure 4 shows TEM observation of sample A and sample D. The precipitated SiC particles disperse in the grain boundary phase at the Si_3N_4-Si_3N_4 interface and inside of the Si_3N_4 grain.

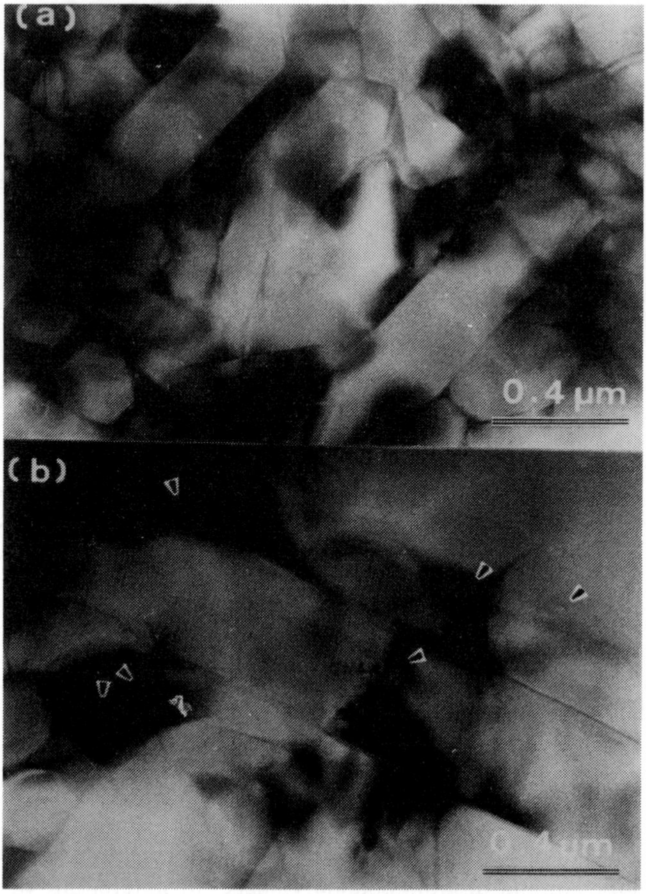

Figure 4. TEM micrographs of Si_3N_4 ceramics made from (a) sample A and (b) Sample D. Arrows in (b) show the precipitated SiC particles that disperse in the grain boundary at the Si_3N_4-Si_3N_4 interface and inside of the Si_3N_4 grain. The SiC particles size is below 30 nm, some of them are as smaller as 5 nm. The SiC structure is confirmed by high resolution microscopy and electron diffraction. [14]

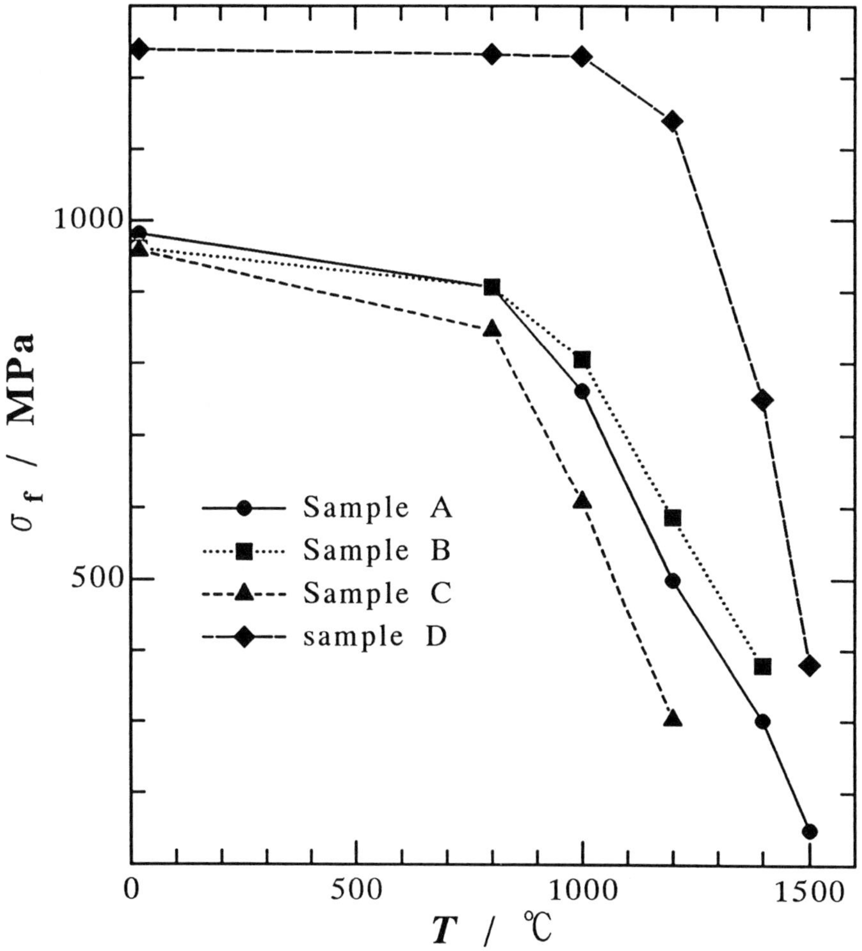

Figure 5. The temperature dependence of the bending strength of samples A, B, C, and D. The symbols for samples are the same as in Figure 1.

Figure 5 shows the temperature dependence of the bending strength of samples A, B, C, and D. Sample D exhibits a gradual loss of the bending strength up to 1200℃. Samples A, B, and C show degradation of the bending strength above 800℃. The bending strength of sample B is the higher than samples A and C at 800℃. The reason is due to the fact that sample B has less grain boundary glassy phase than samples A and C. In the case of the Si_3N_4 made from the carbon coated powder, the bending strength at 1000℃ and 1200℃ were 1230 MPa and 1140 MPa, respectively, which correspond to a decrease of 1 and 9% from the strength at room temperature, respectively. The respective degradation were 23% at 1000℃ and 50% at 1200℃ for the non-carbon coated ones (samples A, B, and C).

The high temperature bending strength is improved by the carbon coating method. When SiC particles disperse at the Si_3N_4-Si_3N_4 interface, a residual tangential compressive stress acts on this interface, and this SiC particles connect both Si_3N_4 grains. SiC particles also disperse in the grain boundary phase, and these particles act as resistance for intergranular fracture, grain boundary sliding and slow crack growth at high temperatures. The grain boundary sliding and slow crack growth are blocked at high temperatures by these mechanisms. Therefore, the carbon coated Si_3N_4 improves the high temperature bending strength.

CONCLUSIONS

Cryogenic specific heat measurements have been conducted for four Si_3N_4 ceramics. The following is concluded from the present work.

(1) The quantity of grain boundary phases of sintered Si_3N_4 can be evaluated by cryogenic specific heat measurements. The present method is a nondestructive evaluation of the grain boundary phase and sufficiently sensitive to evaluate small changes in the amounts and kinds of grain boundary phases of Si_3N_4 ceramics.

(2) The amount of grain boundary phases depends on the content of surface oxygen of the starting powder. The surface oxygen content of the starting powder increases with raising the amount of grain boundary phases in sintered Si_3N_4.

(3) The high temperature bending strength improve by the carbon coated method. The bending strength at 1200℃ was 1140 MPa which correspond to a decrease of 9% from the strength at room temperature.

REFERENCES

1. R. E. Loehman, "Oxynitride Glasses," *J. Non-Cryst. Solids*, **42**, 433-46(1980).
2. N. Hirosaki, A. Okada and M. Mitomo, "Effect of Oxide Addition on the Sintering and High-Temperature Strength of Si_3N_4 Containing Y_2O_3", *J. Mater. Sci.*, **25**[3]1872-76(1990).
3. J. Zeng, O. Yamada, I. Tanaka and Y. Miyamoto, "Hot Isostatic pressing and High-Temperature Strength of Silicon Nitride-Silica Ceramics," *J. Am. Ceram. Soc.*, **73**[4]1095-97(1990).
4. W. E. Lee and G. E. Hilmas, "Microstructural Changes in β-Silicon Nitride Grains Upon Crystallizing the Grain-Boundary Glass," *J. Am.Ceram. Soc.*, **72**[10]191-37(1989).
5. M. K. Cinibulk and G. Thomas, "Grain-Boundary-Phase Crystallization and Strength of Silicon Nitride Sintered with a Y-SiAlON Glass," *J. Am. Ceram. Soc.*, **73**[6]1606-12(1990).
6. G. N. Babini, A. Bellosi and P. Vincenzini, "Factors Influencing Structural Evaluation in the Oxide of Hot-Pressed Si_3N_4-Y_2O_3-SiO_2 Materials," *J. Mater. Sci.*, **19**[11]3487-97(1984).
7. T. Hamasaki and K. Ishizaki, "The Quantitative Evaluation of Grain Boundary Phase by Cryogenic Specific Heat Measurements," pp.53-60 in *Grain Boundary Controlled Properties of Fine Ceramics,* Edited by K. Ishizaki, K. Niihara, M. Isotani and R. G. Ford, Elsevier Applied Science, London and new York, 1992.
8. T. Hamasaki and K. Ishizaki, "Qualitative and Quantitative Evaluation of Silicon Nitride Grain-Boundary Phase by Cryogenic Specific Heat Measurements," *J. Am. Ceram. Soc.*, **77**[4]1101-103(1994).
9. T. Hamasaki, T. Yanai and K. Ishizaki, "Influence of Oxygen content on Si_3N_4 Grain Boundary," in *JFCC International Workshop on Fine Ceramics '94,* The American Ceramic Society, Ohio, USA, to be publish in 1994.
10. M. Kawamoto, C. Ishizaki and K. Ishizaki, "Fluidity-Increasing Behavior of Silicon Nitride Powder by Aqueous Washing", *J. Mater. Sci. Lett.*, **10**[5]279-81(1991).
11. K. Ishizaki, M. Kawamoto and C. Ishizaki, "Improving Flow Behavior of Silicon Nitride Powder by Surface Modification", pp. 655-63 in *Surface Modification Technology IV*, edited by T. S. Sudarshan, D. G. Bhat and M. Jeandin, Minerals, Metals, and Materials Society, Warrendale, PA, 1991.

12. M. Kawamoto, K. Ishizaki and C. Ishizaki, "Characterization of Silicon Nitride Powder by a Temperature-Programmed Desorption Study", pp. 120-24 in Euro-Ceramics, Properties of Ceramics, Vol.2, edited by G. de With, R. A. Terpstra and R. Metselar, Elsevier Applied Science, London and New York, 1989.
13. T. Yanai and K. Ishizaki, "Mechanical Properties of Si_3N_4 ceramics Prepared from carbon Coated Powders", *J. Ceram. Soc. Jpn.* **101**[7]764-768(1993).
14. T. Yanai and K. Ishizaki, "Toughening of Si_3N_4 Ceramics by Nano SiC Particles in Grain Boundary Phase", in *JFCC International Workshop on Fine Ceramics '94,* The American Ceramic Society, Ohio, USA, to be publish in 1994.
15. K. H. Jack, "SiAlON and Related Nitrogen Ceramics", *J. Mater. Sci.*, **11**[6]1135-58(1976).
16. R. E. Loehman, "Preparation and Properties of Yttrium-Silicon-Aluminum Oxynitride Glasses", *J. Am. Ceram. Soc.*, **62**[9-10]491-94(1979).
17. K. Oda and T. Yoshio, "Properties of Y_2O_3-Al_2O_3-SiO_2 Glasses as Model System of Grain Boundary Phase of Si_3N_4 ceramics (Part 1)", *J. Ceram. Soc. Jpn.*, **97**[12]1493-97(1989).
18. R. C. Zeller and R. O. Pohl, "Thermal Conductivity and Specific Heat of Noncrystalline Solids", *Phys. Rev. S*, **4**[6]2029-41(1971).
19. H. M. Rosenberg, *Low-Temperature Solid-State Physics*, p.7, Oxford University Press, London, U. K., 1963.
20. K. Watari, Y. Seki and K. Ishizaki, "Temperature Dependence of Thermal Coefficients for HIPped Silicon Nitride", *J. Ceram. Soc. Jpn.*, **97**[2]174-81(1989), and *J. Ceram. Soc. Jpn. Int.*, **97**[2]170-78(1989).
21. T. Hamasaki, Engineering Doctor Thesis, Nagaoka University of Technology, Nagaoka, Niigata, Japan, 1994.
22. T. Nakamura, *Seramikkusu to Netsu*, p.23, Gihoudou, Tokyo, Japan, 1985.
23. H. Inaba and T. Yamamoto, *Netsu Sokutei*, **10**, 132-45(1983).

NOVEL PROCESS FOR TITANIUM NITRIDE WHISKER SYNTHESIS AND THEIR USE IN ALUMINA COMPOSITES

V. Revankar, R. Hexemer, C. Mroz, D. Bothwell, A. Goel , D. Bray and K. Blakely,
Advanced Refractory Technologies, Inc.,
Buffalo, NY 14207

ABSTRACT

Single crystal ceramic whiskers are viewed as an important reinforcement for ceramic, metal and polymer matrix composites. Titanium nitride (TiN) exhibits high chemical stability in contact with ferrous and nickel alloys, and TiN whiskers (TiNw) are expected to find use in MMC's based on these alloys, as well as in CMC cutting tools for use. Historically the high cost of the whisker materials has prevented their widespread use.

A novel, low cost, scalable process has been developed for the manufacture of titanium nitride whiskers. TiNw - Al_2O_3 composites fabricated using these whiskers are being evaluated as cutting tools inserts for ferrous metals. Characterization of ART's TiNw and initial composite results are presented. The use of TiNw in metal and polymer matrix composites is discussed.

INTRODUCTION

Single crystal ceramic whiskers are an important component in the development of the next generation of materials needed for high temperature and high strength applications. Their principle use is expected to be as the reinforcing constituent in ceramic (CMCs), metal (MMCs) and polymer (PMCs) matrix composites. In CMCs, the matrix material possesses high modulus, low relative weight and excellent high temperature and corrosion resistance. But due to its brittle nature, it is having very poor crack propagation inhibition property. The reinforcement (whisker) is the load bearing constituent because of its higher elastic modulus and therefore restricts crack propagation in the material. Finer diameter fibers/whiskers may provide added advantage in CMCs due to the larger number of whiskers for a given whisker volume. However, finer diameter whiskers are harder to disperse without the formation of whisker agglomerates. This can be partially overcome by having the matrix powder particle size is approximately equivalent to the whisker diameter.[1]

In MMCs, the matrix is ductile and strong in tension with moderate hardness and thermal resistance. The reinforcement material (whisker) should posses high modulus and hardness and in some specialized applications low electrical conductivity and high thermal conductivity. A larger diameter fiber/whisker may provide added advantage in this application. In polymer matrix composites (PMC), the matrix is ductile with low strength, modulus, thermal stability, wear resistance and poor electrical properties. The reinforcement material should have good modulus and strength along with good thermal conductivity.

Titanium nitride whiskers are predicted to have a combination of chemical stability, mechanical strength and electrical properties that surpasses the properties of other ceramic whiskers/fibers. Figure 1 compares TiN whisker properties with other commercially available whiskers and short fibers. These properties are normalized with respect to carbon fiber. TiNw have a tensile strength equal to SiCw and a higher Young's modulus and thermal expansion than SiCw, carbon and alumina fibers. Properties of titanium nitride whiskers are given in Table 1.

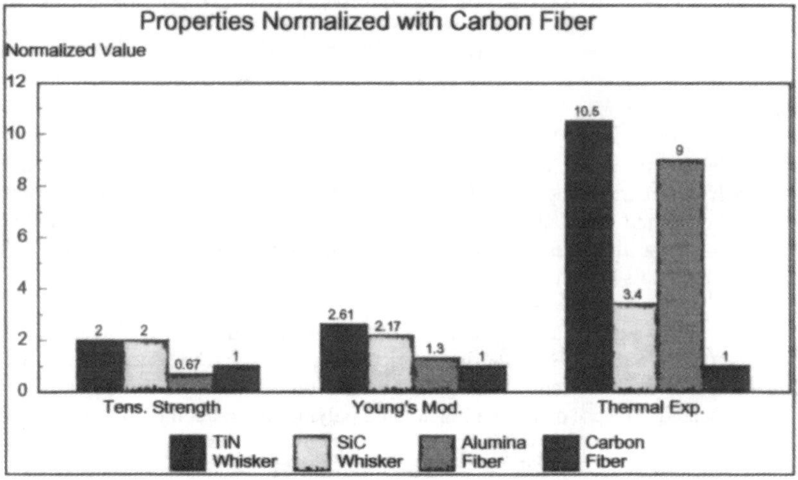

Figure 1. Comparative Properties of Whiskers and Short Fibers.

Table 1. Properties of Titanium Nitride

Chemical Formula:	TiN
Crystal Structure:	Cubic, Rock salt
	Space group 225, Fm3m
Thermal Expansion coeff.	$\alpha = 10.5 \times 10^{-6} / {}^{o}K$
Electrically Conductivity:	μ = 25 microhm-cm
Young's modulus:	E = 590 GPa
Hardness:	2300 - 3000 kg / mm^2
	(Vicker hardness)

Pure titanium nitride has a high melting point (2950 ^{o}C), high hardness (2300 - 3000 kg / mm^2 -Vickers hardness), and high electric conductivity. Thermo-chemical calculations indicate that TiN should be compatible with SiC, Al_2O_3, ZrO_2, Si_3N_4, and TiB_2. TiN has better chemical stability with iron aluminides and nickel aluminides than SiC, and thus may find use in these and other metal matrix composites.[2] Due to its higher thermal and electrical conductivity, TiN is an excellent candidate for PMC applications. In short, these whiskers can be used in CMCs, can be combined with metals for high temperature MMCs, and could improve the performance of PMCs for demanding application (Table 2).

RELATED WORK

There are many references pertaining to the synthesis of TiN powders and films. Generally, titanium nitride synthesis is carried out either by combustion synthesis route [3 - 9] or carbothermal reduction-nitridation route.[10 - 15] However, whisker literature is scarce. Most of the work has been conducted in Japan and the former Soviet Union. Generally, TiN is manufactured via SHS, vapor phase synthesis, carbothermal synthesis or decomposition of organo-metallic compounds. TiN whisker synthesis generally follows one of two routes in literature: catalyzed vapor phase synthesis and thermal decomposition of titanium compounds.

Chemical vapor phase synthesis generally uses the titanium tetra chloride reduction process, followed by reaction with nitrogen gas or nitrogenous compounds.[2, 16 - 21] The deposition is conducted on a catalytic substrate which is kept at an elevated temperature. The catalytic materials generally used for the chemical vapor deposition process (CVD) are Ni, Pd, Pt, Au, Ag, Si, or mullite particles or surfaces.[22] The temperature range varies between 900-1400 ^{o}C depending on the system. Although the process generally produces pure whiskers but production yield is very low resulting in poor economics. Therefore these processes are impractical for production applications.

Table 2. TiN Whiskers

Order	Property	Application/Advantages
1	Thermal expansion is close to that of metals and intermetallic compounds compared to SiC and alumina materials	Reduced stress at whisker/matrix interface during composite fabrication and use
2	Electrically conductive unlike SiC, alumina or alumino-silicate	Can be used as a reinforcement in non conductive polymers, ceramics and glasses to promote electrical conductivity
3	Good chemical stability with ferrous metals and alloys	Can be used as a reinforcement in cutting tools used to machine ferrous metals and alloys (SiC reacts with Fe)
4	Good chemical stability with intermetallics (Fe, Ni , etc.)	Good reinforcement for Ni, Fe, based materials
5	Good chemical stability with ceramics (Co/Ni bonded W_4C, alumina)	Can be used as a reinforcement for W_4C and alumina based cutting tools and wear parts
6	Compatible with most other materials (SiC, Al_2O_3, ZrO_2) and stable towards Si_3N_4 and TiB_2	Could find use in reinforcing these advanced ceramic materials

The other system that has been used for titanium nitride whisker synthesis is thermal decomposition of titanium-alkali-metal salt under nitrogen atmosphere.[23 - 26] In many cases, the salt is prepared via reaction of molten cyanide salt with titania (TiO_2) or TiN and alkali metal oxide.[27] In another process, sodium titanate whiskers are directly converted to titanium nitride whiskers by reaction with ammonia at 1000 ^{o}C. The whiskers are generally polycrystalline in nature.[22, 25] In most processes, reaction temperature is well above 1000 ^{o}C, and reaction time is approximately 15 hours under inert or nitrogen atmospheres. A catalytic surface or material is also required to generate whiskers in most of the cases. Alkali fluro-titanates are also used to produce titanium nitride whiskers in a similar manner.[28] All these processes produce small amounts of whiskers which contain surface faults and defects. None of the above processes are suitable for large scale synthesis.

One interesting patent[29] describes the generation of titanium nitride whiskers using titanium dioxide. The authors generated titanium nitride whiskers using titania, carbonized organic fibers and transition metals or compounds in a graphite tube at 1300 -2100 ^{o}C. A halogen gas was used in association with nitrogen gas to generate brown to golden yellow whiskers with high aspect ratio. This process was further scaled-up at ART but failed to meet the needed economic standards.

WHISKER PRODUCTION APPROACH

Titanium nitride may be synthesized by such methods as sol gel processing, metal nitridation, vapor phase synthesis, carbothermal reduction - nitridation, and self propagating high temperature synthesis. None of the above processes has been reported to result in a scalable process for titanium nitride whisker synthesis.

Initially ART followed the gas phase synthesis approach to manufacture titanium nitride whiskers (AMI/ART process). In this approach, the process is carried out by rapidly heating a mixture of titanium dioxide, fibrous carbon and a catalyst selected from individual combinations of cobalt, nickel, magnesium and calcium. Temperature was maintained at 1300 ^{o}C+/_100 ^{o}C for 60 minutes in a nitrogen and halogen atmosphere. The resultant mass contained titanium nitride whiskers having a uniform diameter of about 0.3 μm to 1.0 μm and an average length of 10 - 30 μm.

In a typical operation, 80 - 85 wt % of finely powdered titanium oxide (rutile) is mixed with 12 - 13 wt % carbonized organic fibers and 1 - 20 wt % powdered metals or metal compound catalysts. The typical whiskers developed are shown in Figure 3.

Whisker composition varied depending on the composition of the initial feed. The main impurities were oxygen and carbon present as unreacted titania, titanium carbide, titanium oxy-nitride or as titanium carbo nitride. The latest results from our new process showed improved whisker characteristics. In the Ti-O-C-N system, the composition of the whiskers can be determined from the following diagram (Figure 4) if the cell dimension is known.

Poor yields, difficulty in scale-up, varied composition and difficulty in separating the titanium nitride whiskers from particulate have lead to the development of an alternative approach to titanium nitride whiskers synthesis.

Currently ART uses a "combination synthesis process" for the production of TiNw. This process involves a combination of several parallel synthesis processes use the advantageous characteristics of each system. This process has several key advantages over which other titanium nitride whisker synthesis processes, especially in the context of ART's existing facilities. The system advantages include rapid synthesis rate, favorable process economics due to low raw material and capital costs and scalability (ART has extensive scale-up and manufacturing expertise and experience).

Figure 3. Whisker synthesized via AMI/ART process

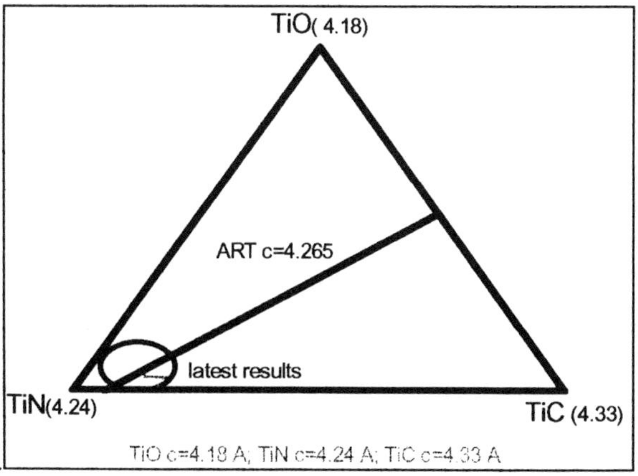

Figure 4. Whisker composition

WHISKER CHARACTERISTICS

Whisker surface morphology was characterized by SEM. Specimens were prepared by ultrasonically dispersing a small quantity of the whisker product in 2-propanol and then depositing a drop of the resulting suspension on an SEM planchet. All microscopy was performed at an accelerating voltage of 5 kV or less. At these low voltages TiN has sufficient electrical conductivity to allow dissipation of electron beam induced charges; therefore, coating was not necessary to render the material electrically conductive. Typical TiNw morphologies are shown in Figure 5.

X-ray diffraction analysis of crude TiNw product showed that the majority of the crystalline material was titanium nitride (TiN) with traces of metallic and oxide material. A typical XRD of the as reacted crude material is shown in Figure 6.

Unlike silicon carbide whiskers, the internal structure of the titanium nitride whiskers is extremely clean without any stacking faults. Under higher magnifications the cubic cross section of the whiskers is clearly visible. Some small whiskers in the crude mass are showing growth balls and blunt ends. Many whiskers have spurs or direction change (by 90°) on the primary whiskers. It appears that there are several different growth mechanisms occurring simultaneously. Most of the whiskers appear to be generated via a VLS or a gas-solid mechanism. These whiskers are golden yellow in appearance with an average diameter of 0.3 - 0.8μm in diameter. The length varies from several microns to 15 microns with an average aspect ratios of 5 - 15. The whisker characteristics are shown in Table 3.

Table 3. Characteristics of ART Whiskers.

Shape	Whisker (aspect ratio > 3)
Surface	Smooth (no faults)
Color	Golden/golden yellow
Diameter	0.3 - 0.8 microns
Length	5 - 15 microns
Aspect Ratio	5 - 15

COMPOSITE APPLICATION

There are numerous published studies concerning whisker reinforced CMCs for applications requiring improved properties, particularly increased fracture toughness and creep resistance at high temperatures. One such application is in the field of cutting tool inserts. Aluminum oxide based cutting tools are very sensitive to thermal cracking as aluminum oxide has a relatively low thermal conductivity and low fracture toughness. This leads to short life in

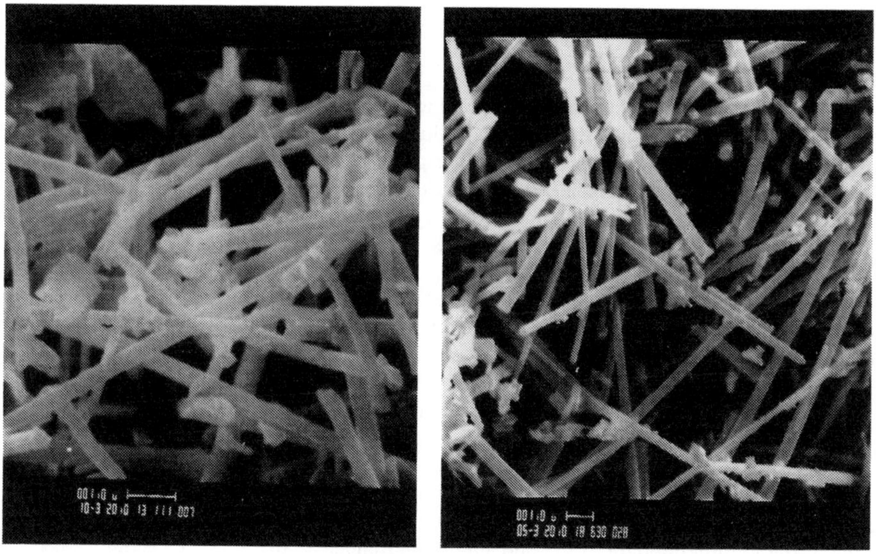

Figure 5. SEM of ART Titanium Nitride Whiskers

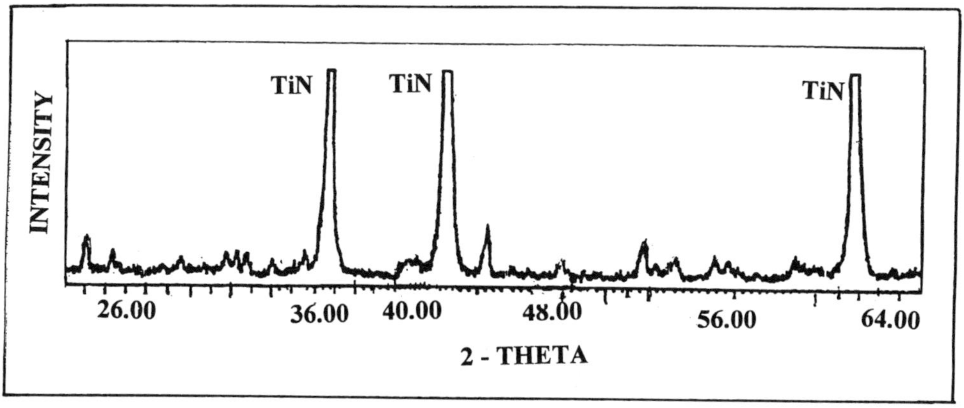

Figure 6. XRD analysis of Crude Titanium Nitride Whisker Product

steel cutting particularly under conditions having short engagement times and varying cutting depths.

Thermal properties have been improved to a certain extent by the addition of titanium nitride powder which improves the thermal conductivity and hardness of the composite material.[30] This results in increased insert life, especially in operations demanding crack resistance. The addition of zirconia (ZrO_2) in the composite results in the transformation of zirconia into metastable zirconium oxide particles during use. This increases both strength and toughness, leading to a more predictable tool life.

Thermodynamic analysis was conducted between the temperature 1300 - 1900 ^oC to identify appropriate composite preparation conditions for titanium nitride whisker reinforced alumina.

Figures 7 and 8 are the thermodynamic stability profiles of alumina and titanium nitride in alumina (70 vol. %) - titanium nitride whiskers (30 vol%) composite fabrication. Figure 9 is the titanium nitride stability profile in the composite in the presence of a small amount of oxygen (0 - 1 wt. % as TiO_2). As is shown in Figures 7 to 9, it is evident that titanium nitride is only stable under certain temperature and pressure operation condition. Optimum hot pressing temperature is under 1500 ^oC - 1525 ^oC, as titanium nitride starts decomposing to form titanium oxides and aluminum titanates above 1550 ^oC. Alumina also starts interacting with TiN at approximately the same temperature and pressure conditions (Figure 8). The pressure used in composite preparation appears to play a key role in thermodynamic stability. As the pressure increases, highest usable while maintaining stability is shifted upwards. Therefore a higher operation pressure is desirable for adequate titanium nitride stability. In the presence of titania the stability is shifted to a lower temperature region (1450 - 1500 ^oC) (Figure 9). These results show that hot pressing should take place below 1500 ^oC and at higher pressures. The titanium nitride material should be free from oxygen as much as possible to avoid reduction in stability.

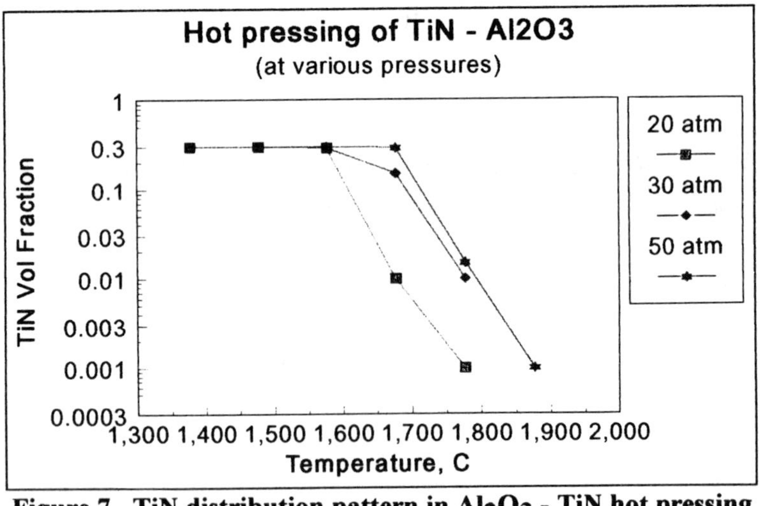

Figure 7. TiN distribution pattern in Al_2O_3 - TiN hot pressing

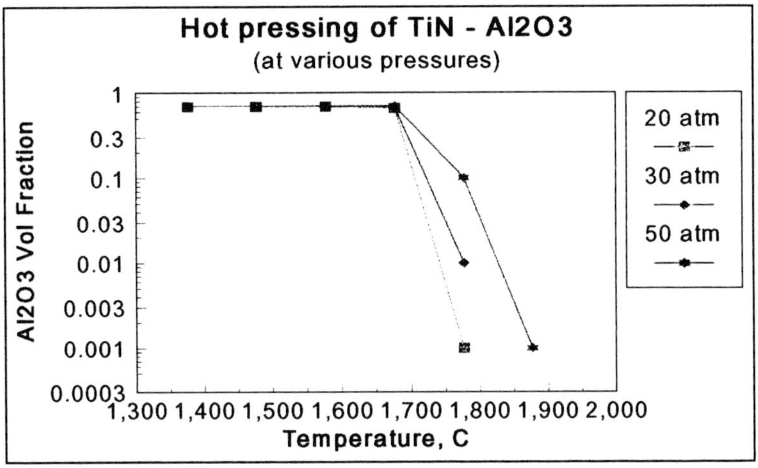

Figure 8. Al_2O_3 distribution pattern in Al_2O_3 - TiN hot pressing

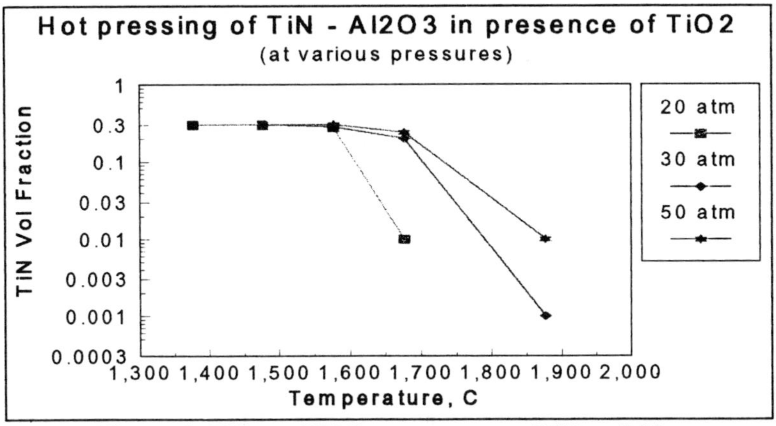

Figure 9. Effect of TiO_2 presence on TiN stability

In experimental studies conducted, suspension processing of Al_2O_3 / TiN whisker composite green bodies was investigated. The primary objective was to maximize final composite density. The matrix phase utilized in the composite sample was high purity α-alumina (Ceralox HPA) with 0.5 % MgO. Al_2O_3 and ZrO_2 (0 - 5 wt %) were added by ball milling with alumina media in distilled water. The suspension pH was adjusted to 9.8 - 10 with ammonium hydroxide and 1 wt.% Darvan dispersant was added. The TiN whisker crude was then added to the slurry and milled for 30 minutes, and the slurry then dried at 100 ^oC.

Hot pressing was conducted at two different temperatures (1650 ^oC and 1500 ^oC) and at 17 atmosphere pressure under nitrogen atmosphere. Hot pressing results are shown in Table 4.

Table 4. Hot Pressing Results of Al2O3 - TiNw Composites

Vol% TiNw Crude	HP Temp. C	HP Time, min	Density (% theoretical)
30 (Grade A)	1650	30	97.6
30 (Batch 1)	1650	30	93.7
30 (Grade A)	1500	30	99.4
30 (Batch 1)	1500	30	94.7
30 (Batch 2)	1500	30	96.9
30 (Batch 3)	1500	30	96.4

These composites exhibited moderate to high densities. SEM analysis of the hot pressed samples showed that all the titanium nitride whiskers did not survive hot pressing at 1650 ^{o}C. This is consistent with thermodynamic calculations shown in Figure 7. 8 and 9. However, analysis of samples hot pressed at 1500 ^{o}C showed that most of the titanium nitride whiskers survived.

Available literature (Brandt and Solna, 1989) describes the properties of alumina - titanium nitride whisker-reinforced composites. Fracture toughness of the composites was measured by Vickers indentation method. An impression is made with a pyramid shaped diamond indenter and k_{1c} is calculated from the length of the cracks which are induced from the corners of the indenter. The k_{1c} values range from 3.9 to 6.1 for titanium nitride particulate reinforced alumina composite and 30 % titanium nitride whisker-reinforced alumina composite. This shows that titanium nitride whiskers significantly increase the composite fracture toughness. The fracture toughness is used to describe the ability of a material to resist mechanical stress without catastrophic failure. Sandvik had evaluated cutting tool life with respect to various forms of titanium nitride reinforced composites. The tools were tested by continuous turning of steel SS2541 at a cutting speed of 0.36 mm rev^{-1}. Under these conditions the life increased from 3 min for titanium nitride particulate reinforced alumina to 16 min for titanium nitride whisker / alumina particulate composites. These results show the potential benefit of using titanium nitride whiskers in ferrous cutting tool applications.

Titanium nitride whiskers also have good chemical stability with ferrous metals and alloys, intermetallics (Fe, Ni, Ti aluminides) and ceramics (Co/Ni bonded W_4C, alumina). They may therefore be used as reinforcements in FeAl, NiAl, and TiAl based materials. Titanium nitride whiskers can also be used as a reinforcement in W_4C based cutting tools. As discussed earlier, titanium nitride whiskers are electrically conductive unlike silicon carbide, alumina and alumino-silicate materials. Therefore they may be highly useful in reinforcing non conductive polymers, ceramics and glasses to promote electrical conductivity.

SUMMARY AND CONCLUSION

ART has developed single crystal titanium nitride whiskers as a reinforcement material for ceramic, metal and polymer matrix composites. The developed process utilizes the relative advantages of several synthesis approaches. The process can be scaled-up and is projected to have very favorable economics.

The titanium nitride whiskers posses an average diameter of 0.3 - 0.8 μm, an aspect ratio of 5 - 15. Titanium nitride whisker alumina matrix composite were fabricated at various conditions and high relative density obtained. Titanium nitride whisker reinforced materials have shown life-time improvements for cutting tool applications. These whiskers may also find widespread use in other applications such as reinforcement of metals and polymers. Further studies are continuing in these directions.

ACKNOWLEDGMENTS

Authors would like to thank Ms. Jayne Spanos for her contribution in SEM analysis.

REFERENCES

1. Krylov, A. V.; Barinov, S. M.; Ivanov, D. A.; Mindlina, N. A.; Patilak, L.; Dusza, J.; Lofaz, F.; and Rudnayova, E.; J. Mat. Sci. Lett., 12, 904, 1993.

2. Nolan, T.; Allard, L.; Coffey, D. W.; Nixdorf, R. D. and Rawline, M. H; J. Am. Ceram. Soc., 74, 2769, 1991.

3. Merzhanov, A. G.; Skhiro, V. M.; Borovinskaja, P.; USSR Patent, 255221, 1971.

4. Kharatyan, S. L; Grigorev, Y. S.; and Merzhanov, A. Z.; Combust. Explos. Shock Waves, 11, 21, 1975.

5. Dunmead, S. D.; Munir, Z. A. and Holt, J. B; Solid State Ionics, 32, 474, 1984.

6. Munir, Z. A., and Holt, J. B.; J. Mat. Sci., 22, 710, 1987.

7. Wriedt, H. A.; and Murray, J. L.; Bull. Alloy Phase Diagrams, 8, 378, 1987.

8. Munir, Z. A.; Amer. Ceram. Soc. Bull., 67, 342, 1988.

9. Deevi, S., and Munir, Z. A.; J. Mat. Res., 5, 2177, 1990.

10. Grey, I. E.; Jones, D. G., and Reid, A. F; Trans. Inst. Min. Metall. C.; 82, 151, 1973.

11. Gray, I. E.; Reid, A. F., and Jones, D. G.; Trans. Inst. Min. Metall. C., 83, 105, 1974.

12. Jones, D. G.; Trans. Inst. Min. Metall. C.; 82, 186, 1973.

13. Coley, K. S; Ph. D Thesis, Imperial College of Science and Technology, 1986.

14. Bourrie, D. B, and Filsinger, D. H; J. Am. Ceram. Soc., 73, 1726, 1990.

15. Terry, B. S.; and Chinyamabobou; Mat. Sci. and Tech.; 7, 842, 1991.

16. Tsuda, H.; Watari, T., and Kato, A; Nippon Kagaku Kaisi, 11, 1847, 1982.

17. Wokulski, Z.; Cryst. Res. Technol., 24, 1067, 1989.

18. Wokulski, Z.; and Wokulska, K.; TMS Conference Proceedings, Chicago, IL, 193, 1989.

19. Guo, W; Ning, X. G.; Zhu, J., and Ye, H. Q; J. Cryst. Growth; 106, 400, 1990.

20. Hubbard, C. R., Nolan, T. A; Allard, L. F; Coffey, D. W., and Padgett, R. A; J. Am. Ceram. Soc., 74, 2769, 1991.

21. Schachner, H; Horlaville, G.; and Richon, D; Minerals Metals and Materials Society Proceedings, 827, 1991.

22. Bamberger, C. S., Powder Diffr., 3, 240, 1988.

23. Hagimura, A; Tamari, N; and Kato, A; Nippon Kagaku Kaishi, 1, 49, 1979.

24. Kato, A. and Tamari, N; J. Cryst. Growth; 49, 199, 1980.

25. Bamberger, C. S.; Allard, L. F; Coffey, D. W; and Bannett, M. R., J. Cryst. Growth., 112, 47, 1991.

26. Bamberger, C. S; U.S. Patent, 5118488, June 2, 1992.

27. Bamberger, C. S., U.S Patent 4883559, April 3, 1989.

28. Tabuchi, M; Kaneko, Y; and Hironichi, I.; J. Ceram. Soc. Japan; 98, 43, 1990.

29. Nixdorf, R. D. and Rawline, M. H; U.S. Patent, 04888084, December 19, 1989.

30. Brandt, G. L, and Solan, Z. D; U.S. Patent, 4867761, September 19, 1989.

MICROSTRUCTURAL DESIGN OF SILICON NITRIDE BY SEEDING WITH ROD-LIKE β-Si_3N_4 PARTICLES

Kiyoshi Hirao, Manuel E. Brito, Masayoshi Ohashi, and Shuzo Kanzaki
National Industrial Research Institute of Nagoya
Hirate-cho 1-1, Kita-ku, Nagoya 462, JAPAN

ABSTRACT

The effect of size, amount, and orientation of seed particles on microstructure development and mechanical properties of seeded silicon nitride was investigated and the relationship between them was established. Seeding with particles in appropriate size and amount allows the development of microstructures where large elongated grains good dispersed contribute to the improvement of fracture toughness retaining a high strength level. Moreover, orientation of seeds particles after tape-casting was effective for improving both fracture toughness and bending strength.

1. INTRODUCTION

It is well known that incorporation of a second phase with an anisotropic shape such as whiskers and platelets is effective for improving fracture toughness of ceramics [1]. However, fabrication of these composites requires hot pressing or hot isostatic pressing technique, an expensive and inefficient method for producing complex components. On the contrary, in the case of silicon nitride ceramics, existence of rod-like β-Si_3N_4 grains, grown in-situ during sintering, leads to high fracture toughness values (K_{IC}). Silicon nitride with K_{IC} of 8 - 12 $MPa \cdot m^{1/2}$ was obtained by a gas pressure sintering technique [2-5]. Such tough silicon nitride exhibits a bi-modal microstructure composed of large elongated grains dispersed in a small grained matrix. In these materials, toughening is generally explained in terms of crack deflection and bridging by the elongated grains. However, the existence of large grains causes a decrease in strength due to the increase in flaw size. It is known that larger grains develop from β-Si_3N_4 particles existing in the raw powder [4], which makes it difficult to control their morphology and distribution.

To overcome this problem, we have developed a novel process to control the microstructure of self-reinforced silicon nitride [6]. The key point of this process is that size and distribution of elongated large grains are controlled by seeds of morphologically regulated β-Si_3N_4 single crystals. The purpose of this study is to investigate the effect of size, amount, and orientation of seed particles on the microstructure and mechanical properties of seeded silicon nitride and characterize the relationship between them.

2. EXPERIMENTAL PROCEDURE

Table I Characteristics of synthesized seed particles.

symbol	diameter (μm)	length (μm)	aspect ratio
A	0.47	1.95	4.2
B	0.96	3.84	4.0
BL	0.96	9.54	10.0
C	1.29	5.14	4.0

Four kinds of β-Si_3N_4 seed particles were prepared by heating powder mixtures of α-Si_3N_4 , Y_2O_3 and SiO_2 at 1850°C for 2h under a nitrogen pressure of 0.5 MPa, followed by acid rinse treatments. Diameter and aspect ratio of seed particles were controlled by the particle size of the starting α-Si_3N_4 powder and a Y_2O_3/SiO_2 ratio, respectively. Details of the preparation procedure and characterization of seed particles are reported elsewhere [7]. Characteristics of the synthesized seed particles are summarized in Table I. These series of seed particles consist exclusively of rod-like pure β-Si_3N_4 single crystal particles and show a uniform size distribution.

The powder mixtures of α-Si_3N_4 , sintering additives and 0, 2 and 5 vol% seed particles were isostatically cold-pressed under 500 MPa. The composition of the additives was 5 wt% Y_2O_3 and 2 wt% Al_2O_3. Besides the above cold-pressing process, for the sake of aligning the seeds, raw powder slurry with an identical composition was tape-cast by the use of a Doctor blade equipment, adjusting the sheet thickness in a range from 130 to 150 [8]. The green sheets were punched in rectangular shape and stacked aligning the casting direction. After calcining, the green compacts were sintered at 1850°C for 6h under a nitrogen pressure of 0.9 MPa.

Sintered specimens were polished and chemical etched for microstructure observation using scanning electron microscopy (SEM). Four point flexural strength measurements were carried out with an inner span of 10 mm, an outer span of 30 mm at room temperature. Fracture toughness was determined by the single-edge-precracked-beam (SEPB) method at room temperature. Measurements of mechanical properties for the tape-cast silicon nitride were done in the direction perpendicular to the casting plane.

3.RESULT AND DISCUSSION

(1) Effect of size and amount of seeds on microstructure and mechanical properties

Figure 1 shows SEM photographs of the microstructure of seeded silicon nitride, compared with the non-seeded specimen (Fig. 1-a). Silicon nitride with a bi-modal

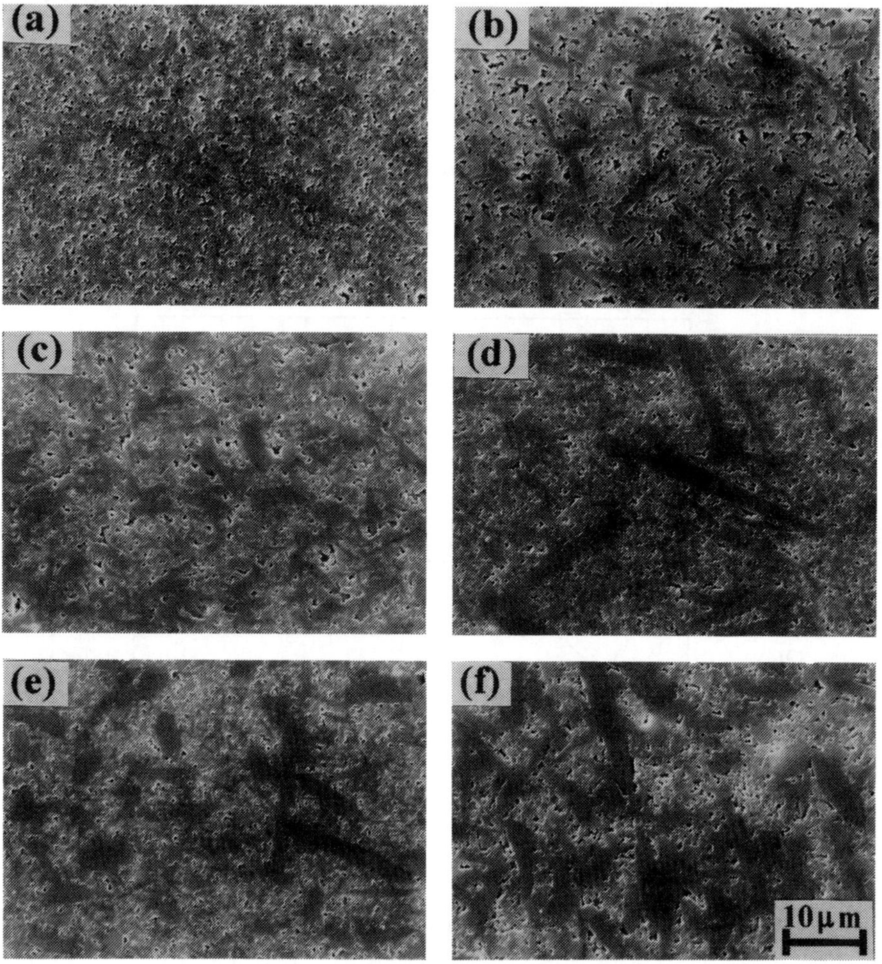

Fig.1 SEM photographs of sintered specimens (a) without seed, with (b) 2vol% A, (c) 2vol% B, (d) 2vol% BL, (e) 2vol% C and (f) 5 vol% C seed.

microstructure composed of small and large elongated grains were obtained in all of the specimens with seeds. Diameter of larger grains increased with an increase in diameter of the seeds (Fig.1-b~e) and amount of them increased with an increase in the amount of the seeds (Fig.1-e, f). These results are consistent with the fact that larger grains in seeded specimens were developed from seeds. This was confirmed by means of the plasma etching technique in a previous study [6]. In cases where amount of seed particles were small, large grains were highly dispersed. A larger amount of seed particles (> 5 vol%) with large diameter caused coalescence of newly developed grains (Fig.1-f).

Figure 2 shows the effect of the seed amount on strength and fracture toughness (K_{IC}) of specimens sintered at 1850°C for 6h. The K_{IC} of about 1-2 MPa·m$^{1/2}$ higher than that of non-seeded specimen was achieved by seeding. Moreover, seeded silicon nitride, except the specimens with 5 vol% BL and 5 vol% C seed, retained a high strength level of about 900-1000 MPa in spite of the existence of larger grains.

In order to characterize quantitatively the relation between microstructure and bending strength for seeded specimens, diameters (d) and lengths (l) of larger

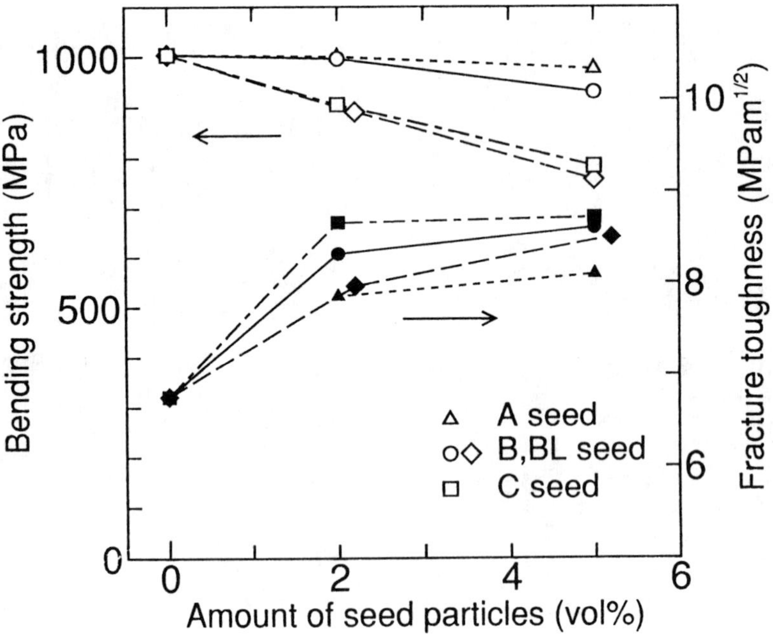

Fig. 2 Relation between amounts of seed particles and mechanical properties of sintered silicon nitride.

elongated grains having larger diameter than the maximum in the non-seeded specimen were measured in an area of about 100 x 100 μm^2 using the SEM photograph of polished and etched surfaces. The area fraction of large elongated grains was determined by the summation of $d \cdot l$ for each grain. Figure 3 shows the relation between area fraction of large grains and bending strength for specimens sintered at 1850°C for 6-12h. There is a marked decrease in bending strength in the vicinity of 30% area fraction, over which large grains begin to contact each other. Low strength of the specimen seeded with large amount seeds should be ascribed to coalescence of larger grains causing a sudden increase in flaw size. The above observations indicate that seeding of β-Si_3N_4 single crystals appropriate in size and amount allows the development of microstructures where large elongated grains with a good dispersion contributes to retaining a high strength level.

The observation of crack propagating behavior suggested that improvement in K_{IC} for seeded specimen is attributable to crack deflection and bridging by larger elongated grains developed from seed particles [6]. Becher has derived the equation for calculating the fracture toughness of whisker reinforced ceramic

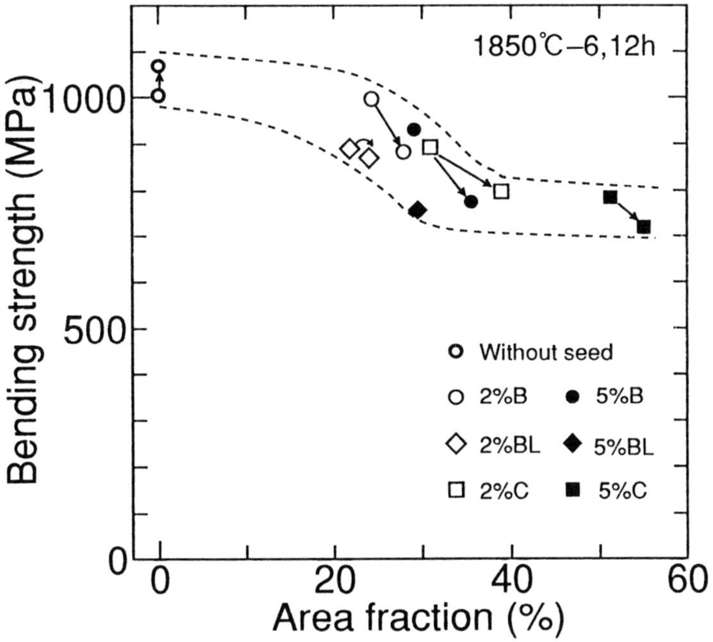

Fig. 3 Relation between area fraction of large elongated grains and bending strength of seeded silicon nitride.

composites under the assumption of a bridging zone at the crack wake[1].

$$K_{IC}^2 - K_{ICm}^2 = R \cdot D \cdot V_f$$

In this equation, K_{IC} and K_{ICm} are fracture toughness of composite and matrix, D and V_f are diameter and volume fraction of whisker, respectively. With regard to the application of this equation to our results on seed silicon nitride, K_{ICm} , D and V_f are regarded as the fracture toughness of a non-seeded specimen sintered at the same condition for seeded specimen, diameter and area fraction of large elongated grains, respectively. The results are shown in Fig. 4. The plots indicated a straight line passing through the origin and, thus, the increasing of fracture toughness for seeded specimen could be quantitatively related to microstructural parameters, that is, diameter and volume fraction of large elongated grains, assuming crack bridging as the dominant toughening mechanism.

In conclusion, the achievement of both high fracture toughness and high strength in Si_3N_4 is made it possible by the development of a bi-modal microstruc-

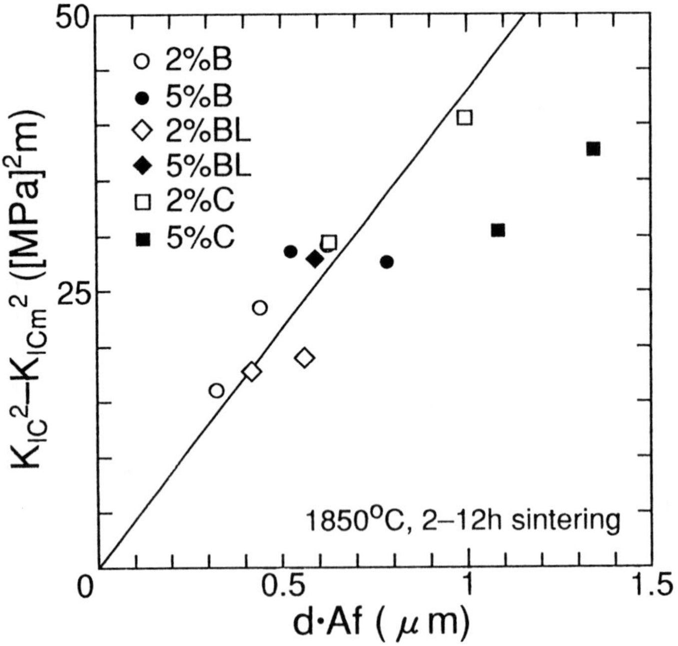

Fig. 4 Correlation between improved fracture toughness and microstructural parameter (d·Af).

ture, where the size of large elongated grains is controlled by seeding.

(2) Effect of seed alignment on microstructure and mechanical properties

Figure 5 shows SEM photographs of polished and plasma-etched surfaces parallel to and perpendicular to the casting plane in the sintered silicon nitride with 2 vol% C seed formed by tape-casting. Large elongated grains appearing in the polished surface perpendicular to the casting plane (Fig. 4-b) were found to be aligned primarily parallel to the casting plane and those appearing in the polished surface parallel to the casting plane (Fig. 4-a) exhibited a rather random orientation, however, lying within the casting plane. Since elongated grains are epitaxially grown from seed particles, the seeds must have aligned during the casting process. The area fraction of large elongated grains in the silicon nitride with 2% C seed formed by cold-pressing was 29%. It is assumed that tape-cast specimens have the same fraction of those grains considering the identical sintering condition. It is noticed that seeding with only 2 vol% particles provided such a large fraction of elongated grains in a two-dimensional arrangement.

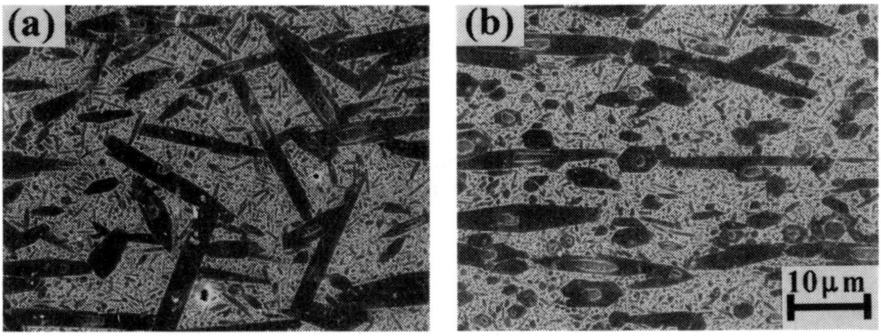

Fig. 5 SEM photographs of polished and plasma-etched surfaces of the seeded and tape-cast silicon nitride (a) parallel to and (b) perpendicular to the casting plane.

Mechanical properties of the tape-cast silicon nitride are shown in Fig. 6, comparing with those of the cold-pressed silicon nitride. The K_{IC} values of the seeded and cold-pressed specimens was about 2 MPa·m$^{1/2}$ higher than specimens without seeds. Tape-cast specimen showed a further improvement of K_{IC} in the plane perpendicular to the casting direction, which values are about 4 MPa·m$^{1/2}$ higher than those of non-seeded specimens. It is considered that the further improvement in K_{IC} is owing to the two-dimensional alignment of large elongated grains perpendicular to the crack plane, which increases the opportunity for crack deflection and bridging.

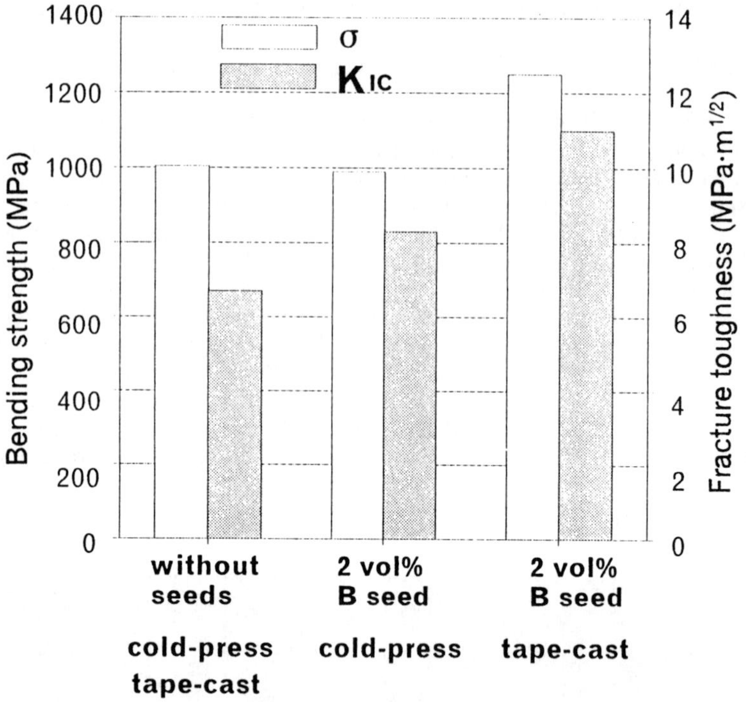

Fig. 6 Comparison of mechanical properties of tape-cast silicon nitride with those of cold-pressed silicon nitride.

Bending strength of seeded and cold-pressed specimen was about 100 MPa lower than the strength of non-seeded specimen, which can be ascribed to the increase in flaw size. On the other hand, seeded and tape-cast specimen had a rather higher bending strength of 1100 MPa. It has been reported that when there exists large amounts of larger elongated grains in silicon nitride, cluster of these grains acts as larger defects to lower the strength [9]. The higher bending strength of tape-cast specimen may be owing to the two-dimensional uniform distribution of large elongated grains, which prevents generation of the cluster building up a three-dimensional network. Also, it should be noticed that a synergistic effect of the high fracture toughness in the plane perpendicular to the crack propagating direction contributes to the high strength.

4. SUMMARY

(1) Silicon nitride with a bi-modal microstructure composed of small and large elongated grains were obtained by adding seed particles. The diameter of larger grains and their amount increased with an increase in diameter and amount of seed particles, respectively. The K_{IC} about 1-2 MPa·$m^{1/2}$ higher than that of non-seeded

specimen was achieved by seeding and the K_{IC} tended to increase with an increase in both diameter and amounts of seed particles. Most of the seeded specimens retained a high strength level, however specimens with larger amount of seed particles presented a low strength.

(2) Silicon nitride with a preferred orientation of large elongated grains could be obtained by tape casting of raw powders slurry with the seed particles. The large elongated grains developed from seeds lay in planes parallel to the casting direction in a two-dimensional distribution. Increased fracture toughness (11.1 $MPa \cdot m^{1/2}$) and bending strength (1100 MPa) were achieved in the direction perpendicular to the grains alignment compared to specimens with a random distribution of elongated grains.

REFERENCES

1) P. F. Becher, "Microstructural Design of Toughened Ceramics," J. Am. Ceram. Soc., **74** [2] 255-269 (1991).

2) E. Tani, S. Umebayashi, K. Kishi, K. Kobayashi, and M. Nishijima, "Gas-Pressure sintering of Si_3N_4 with Concurrent Addition Of Al_2O_3 and 5 wt% Rare Earth Oxide: High Fracture Toughness Si_3N_4 with Fiber-Like Structure," Am. Ceram. Soc. Bull., **65** [9] 1311-15 (1986).

3) C. W. Li and J. Yamanis, "Super-Tough Silicon Nitride with R-Curve Behavior," Ceram. Eng. Sci. Proc., **10** [7-8] 632-645 (1989).

4) M. Mitomo, "Toughening of Silicon Nitride Ceramics by Microstructure Control"; pp.101-107 in Proceedings of the 1st International Symposium on the Science of Engineering Ceramics. Edited by S. Kimura and K. Niihara. Ceramic Society of Japan, Tokyo, Japan, 1991.

5) F. Wu, H. Zhuang, L. Ma, and X. Fu, "Self-Reinforced Silicon Nitride by Gas Pressure Sintering," Ceram. Eng. Sci. Proc., **14** [1-2] 321-332 (1993).

6) K. Hirao, T. Nagaoka, M. E. Brito, and S. Kanzaki, "Microstructure Control of Silicon Nitride by Seeding with Rodlike β-Silicon Nitride Particles," J. Am. Ceram. Soc., **77** [7] 1857-62 (1994).

7) K. Hirao, A. Tsuge, M. E. Brito and S. Kanzaki, "Preparation of Rod-like β-Si_3N_4 Single Crystal Particles," J. Ceram. Soc. Japan, **101** [9] 1071-1073 (1993).

8) K. Hirao, M. Ohashi, M. E. Brito, and S. Kanzaki, "A Processing Strategy for Producing Highly Anisotropic Silicon Nitride," submitted to J. Am. Ceram. Soc.

9) P. F. Becher, H. T. Lin, S. L. Hwang, M. J. Hoffmann, and I. W. Chen, "The Influence of Microstructure on the Mechanical Behavior of Silicon Nitride Ceramics"; pp.147-158 in Silicon Nitride Ceramics. Edited by I. W. Chen, P. F. Becher, M. Mitomo, G. Petzow and T. S. Yen, Material Research Society, Pennsylvania, 1993.

PHASE COMPOSITION CONTROL IN SiC COMPOSITES PREPARED BY REACTIVE-INFILTRATION WITH METAL-SILICON ALLOYS

Leszek Hozer, Jonq -Ren Lee, and Yet -Ming Chiang
Massachusetts Institute of Technology, Cambridge, MA 02139

ABSTRACT

Reactively-infiltrated silicon carbide composites are prepared in the SiC-Al-Si and SiC-Cu-Si systems by using Al-Si and Cu-Si alloys to infiltrate porous carbon preforms. The design process of such materials and control of phase composition is presented and illustrated with microstructural observations. Net-shape processing capabilities and simultaneous joining and reaction are demonstrated.

INTRODUCTION

Fabrication of reaction-bonded silicon carbide by infiltration techniques is a relatively low temperature processing route that has been extensively studied in the past [1-6]. The majority of studies have been concerned with so-called siliconized silicon carbide produced by reacting liquid silicon with a carbon-containing preform. Messner and Chiang [7,8] and Chiang et al. [6] have also demonstrated the use of alloyed melts in order to control the constitution of the secondary phase, using Si-Mo melts to form $MoSi_2$ as a refractory secondary phase. In this work we explore the use of Si-Al and Si-Cu alloys for reactive infiltration, in order to form composites with Al- and Cu-rich secondary phases. An additional role of the alloying is to control wetting and permit infiltration by the normally non-wetting metals without the application of external pressure. We will focus here on the phase constitution developed in SiC composites produced by reactive infiltration. Their processing and physical properties are discussed in more detail elsewhere [6,9]. In addition, the dimensional retention and net-shape processing of prototype components, and the simultaneous joining and reaction-bonding of complex-shape parts, is demonstrated.

PROCESSING

Composites were fabricated by spontaneous, reactive infiltration of microporous glassy carbon preforms prepared after the method of Hucke [4,5]. It is possible to vary the morphology of the preform microstructure (from an interconnected carbon network to spherical particulates), the overall preform density, and the pore size of the preform [6,9].

For the purposes of this work, we chose a spherical particulate microstructure with an average pore size (measured using mercury intrusion porosimetry) ranging from 1 to 3 μm, overall preform densities from 0.80 to 0.65 g/cm^3, and specific glassy carbon density of approximately 1.5 g/cm^3 (measured using mercury porosimetry). The cylindrical preforms had maximum dimensions of 15 mm diameter and 100 mm length, and could be machined using standard tooling.

Machined preforms were mounted in a graphite holder and reaction-infiltrated by dipping into a melt contained in a boron nitride crucible in a vacuum furnace at 10^{-2}-10^{-3} Torr. The infiltration temperature and time were chosen to be high enough to have a single phase liquid. After reactive-infiltration and cooling, the excessive melt was removed from the surface of the samples by either room-temperature chemical etching (mixture of concentrated acids: 70 weight % of HF and 30 weight % of HCl) or high-temperature vaporization in vacuum.

ALLOY COMPOSITION SYSTEMS

Our goal is to introduce a metal-rich secondary phase into the reaction-bonded SiC microstructure. If the reaction proceeds to completion, the compositions and amounts of the individual phases are determined by the overall composition of the composite, and the applicable phase equilibria at the reaction temperature. Thus we targeted overall compositions in the Si-Al-C and Si-Cu-C systems which, based on available phase equilibria, should give two-phase mixtures of SiC and a metal-rich liquid at the processing temperatures (i.e., they lie in two-phase fields). In the Si-Al-C system, we rely on phase equilibria data from several sources [10,11]. For the Si-Cu-C system, the proposed phase

Table I. Melt compositions and infiltration conditions used for metal-silicon-alloy infiltrated samples and phase compositions of reacted composites

	Preform density (g/cm^3)	Melt composition (mol%)	Infiltration temperature (°C)	Infiltration and reaction time (min.)	Final phase composition (vol.%)
1	0.80	15.3% Al + 84.7% Si	1500	60	77% SiC + 13% Al + 10% Si
2	0.80	20.2% Al + 79.8% Si	1500	60	74% SiC + 17% Al + 9% Si
3	0.80	22.1% Al + 77.9% Si	1500	60	73% SiC + 19% Al + 8% Si
4	0.65	28.6% Al + 71.4% Si	1500	5	58% SiC + 24% Al + 18% Si
5	0.65	45.6% Cu + 54.4% Si	1600	30	54% SiC + 46% Cu/Si
6	0.65	51.4% Cu + 48.6% Si	1600	30	59% SiC + 41% Cu/Si

diagram of Warren and Anderson [12] was used. In the absence of any carbon loss or dimensional change during reaction, the overall carbon concentration of the final composite is simply that of the starting preform. The composition of the infiltrating alloy can then be selected according to the phase field desired. As examples, Fig.1a shows the overall composition corresponding to experiment (2) in Table I for the Si-Al-C system, and Fig.1b shows the location of experiment 5 in the Si-Cu-C phase diagram proposed by Warren and Anderson [12]. Based on these data, no ternary compounds were expected in either system, and none were observed in the final composites. The carbon densities and corresponding alloy compositions of a series of experiments are shown in Table I. Several samples were made at each of the compositions shown. The final phase assemblage after cooling is also determined by solidification of the secondary liquid, as discussed later.

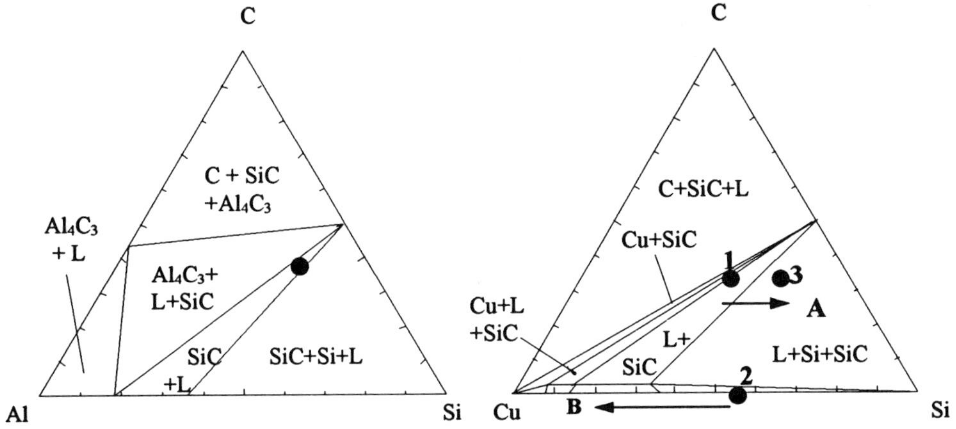

Fig.1. (a) Si-Al-C phase diagram at 900°C (after Ref.10). Composition intended in experiment 2 (Table I) is indicated. (b) Si-Cu-C phase diagram at 900°C proposed in Ref. 12. Composition intended in experiment 5 (Table I) is indicated by point 1. The melt used for infiltration had the composition indicated by point 2. Arrow B shows the shift in composition of the residual liquid phase during reaction. Arrow A denotes an overall composition shift which can occur due to liquid exchange (to point 3, as measured by ICP), as discussed in text.

SiC-Al-Si Composites

The microstructure of a reaction-infiltrated SiC-Al-Si composite is shown in Fig.2a., and its corresponding x-ray diffraction pattern in Fig.2b. The scale (grain size) of the SiC appears to be very close to those seen for silicon infiltrated samples [6,9]. After optimizing infiltration conditions, we did not observe any residual porosity or unreacted carbon. In Fig.2a we can distinguish among three different phases: the darkest-contrast phase is SiC, the intermediate gray phase is an Al-Si eutectic alloy, and the brightest phase is Al. The silicon phase in the final composite (Fig.2b) results from eutectic solidification. No detectable aluminum carbide was seen in any of the composites synthesized. The phase fractions determined by x-ray diffraction, using standards as discussed in [9], are listed in Table I. Depending on the preform density, the SiC volume fraction ranges from ~60% to ~80%. The aluminum volume fraction (including that within the eutectic product) varies from 13% to 24%, and the balance is silicon.

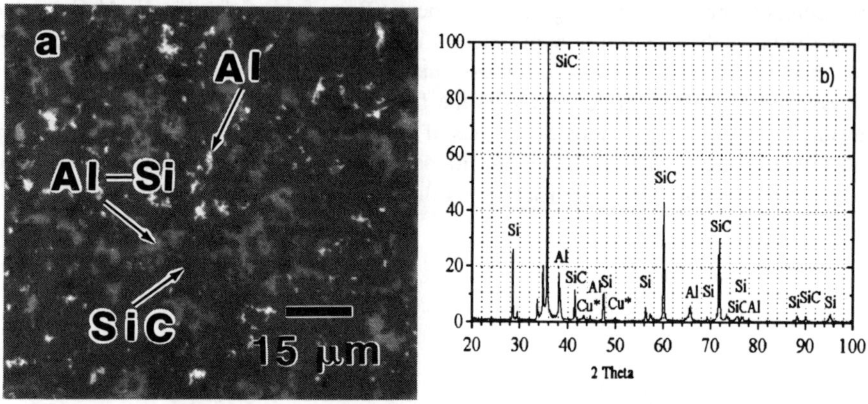

Fig.2. (a) Microstructure of SiC-Al-Si composite reactively-infiltrated using glassy carbon preform of 0.80 g/cm^3 density with 20 mole % Al + 80 mole % Si alloy, viewed with optical microscopy; (b) X-ray diffractogram of the same sample. β-SiC, Al, and Si phases are present. Cu* signal is from sample holder.

SiC-Cu-Si Composites

The microstructure of samples infiltrated with Si-Cu alloys, as shown in Fig.3a, also consist of three principle phases: SiC, Si, and a bright contrast copper-rich phase. The copper-rich phase was analyzed using electron microprobe microanalysis, and found to

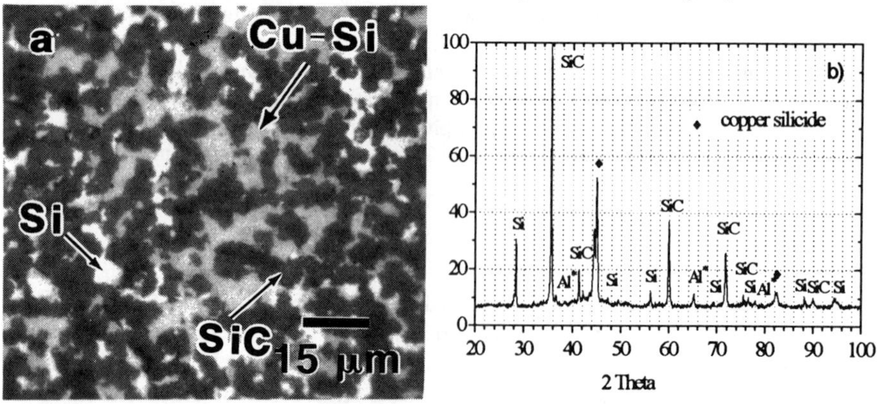

Fig.3. (a) Microstructure of SiC-Cu-Si composite reactively-infiltrated using glassy carbon preform of 0.65 g/cm^3 density with 45 mole % Cu + 55 mole % Si alloy, viewed with optical microscopy; (b) X-ray diffractogram of the same sample. β-SiC, Cu-Si silicide, and Si phases are present. Al* signal is from sample holder.

contain 70±7 atomic % of Cu and 30±7 atomic % of Si. This composition is close to that of the η copper silicide phase [13]. XRD spectra (Fig.3b) do not show the existence of metallic copper in these composites, and show unidentified peaks in which we attribute to η′ copper silicide. The total Si/Cu molar ratio in the composite, as measured using inductively-coupled plasma emission spectroscopy, is higher (2.73) than that in the melt used for infiltration (1.19). The cause of this compositional deviation is discussed below.

Reaction Sequence

Based on the above results, we believe that the composite formation process is as follows. During reactive infiltration, the silicon component of the alloy selectively reacts with carbon to form silicon carbide, leaving a metal-enriched melt within the continuous pore structure of the composite. The secondary phase constitution of the composite is then determined by the solidification behavior of this melt.

As a first approximation, we assume that the liquid supplied to the preform contains silicon in the exact amount needed for full conversion of the carbon skeleton into silicon carbide. The composition for which we can perform this exact calculation is denoted as point 1 in the schematic ternary diagram shown in Fig.1b. The composition of the melt used for infiltration (point 2 in Fig.1b) then has the same Si/metal ratio. As the reaction proceeds, the composition of the remaining liquid contained in the preform becomes enriched in metal (arrow B in Fig.1b). After reaction is completed, the remaining secondary liquid phase is the alloy in equilibrium with SiC at the temperature of reaction. The above results suggest that local equilibrium is quickly achieved in the present systems.

Upon cooling, the residual liquid phase, which contains a minimum of carbon, crystallizes in accordance with the binary metal-silicon phase equilibria. In the case of the Si-Al system some primary Si or Al may first solidify, depending whether the liquid is of hyper- or hypo-eutectic composition; both have been observed. The balance of the liquid then undergoes eutectic solidification. Analytical electron microscopy of the composite has revealed the finely-divided silicon-aluminum eutectic microstructure.

In the Si-Cu system, our reaction temperature is above the melting points of Si and the copper silicides, so it is likely that the SiC co-exists with a single-phase Cu-Si liquid. This liquid also solidifies as essentially a binary system. However, the phases present and the ICP compositional analysis show that the residual melt is more enriched in Si than the intended starting composition, being on the silicon-rich side of the eutectic [13]. The total Si/Cu ratio, in the composite, increases, as indicated by arrow A in Fig.1b.

We believe that the cause of this compositional deviations is liquid exchange between the reacted preform and the excessive liquid remaining in the crucible. This is a consequence of the present experimental procedure, whereby a large excess of alloy is used for infiltration. In this instance the final phase compositions are dependent on the total time of the process, as well as the starting compositions, and must be taken into account if the secondary phase is to be controlled.

DIMENSIONAL RETENTION

Dimensional changes during preform infiltration and reaction were measured over sample gauge lengths of about 60 mm. The dimensions of machined carbon preforms were compared before and after reactive infiltration and removal of the excessive melt (by chemical etching). In Table II we show typical results for the linear shrinkage, $\Delta L/L$. There is a noticeable trend of increasing shrinkage with decreasing carbon density, which is discussed in detail elsewhere [9], but in no case does the shrinkage exceed 0.5%, and for the highest density preforms it is always less than 0.1%. Similar results were observed in composites infiltrated with Si alone, Al-Si, and Cu-Si.

Table II. Typical linear dimensional changes of SiC-Si composites after reactive infiltration

Preform density, (g/cm^3)	Initial length, (mm)	Final length, (mm)	Change of linear dimension, %
0.65	60.45	60.21	-0.41
0.73	61.05	60.89	-0.26
0.80	62.57	62.53	-0.06

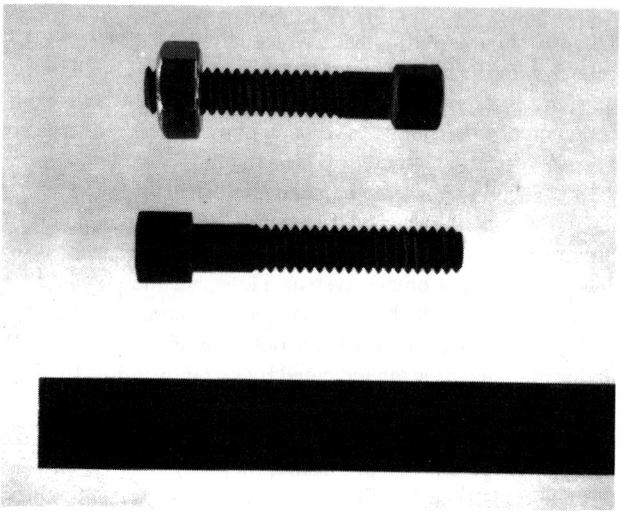

Fig.4. Reactively infiltrated SiC composite component (in the shape of a 1/4" x 20 tpi threaded bolt), showing shape and dimensional retention between the preform and reactive-infiltration and removal of surface residual melt.

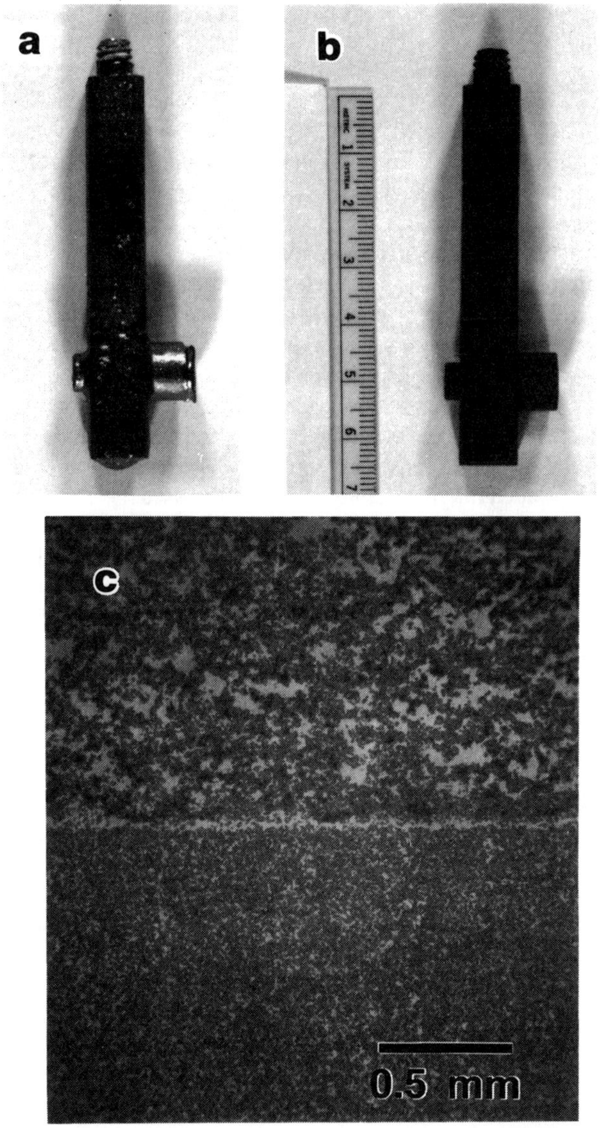

Fig.5. Simultaneous joining and reactive-infiltration. The component was made of two parts (stepped pin and bar), and was infiltrated with silicon after assembling: (a) after reactive-infiltration; (b) after etching to remove excessive melt; (c) shows the microstructure of the interface between the two parts. The thin, interfacial film of residual silicon is believed to be due to machining mismatch. The differing microstructures of the two parts are the result of differing preform densities.

The degree of dimensional retention is far better than those in sintering-based processes, where typical shrinkages are about 30% for fully densified materials. As a demonstration of net-shape capability, we show in Fig.4 an example of a reaction-infiltrated bolt (1/4" x 20 tpi), which remains closely matched to the steel hex nut after reaction and removal of the excess melt.

The small dimensional change during reactive-infiltration also suggests that one can fabricate a complex-shape component from simpler parts machined separately, in a simultaneous reaction-bonding and joining operation. To demonstrate, we machined a cylindrical pin with an outer diameter closely matched to the inner diameter of a drilled hole in a separate preform. To make the interface more easily observable, the two parts were prepared from different carbon densities (0.80 and 0.65 g/cm^3). The assembled component was then reaction-infiltrated with pure silicon. It is shown in Fig.5 after infiltration and after removal of the excess melt. The microstructure of the interface is shown in Fig.5c. A well-bonded interface with a thin interfacial film is visible. This demonstration shows that, in principle, complex parts can be simultaneously joined and reaction-infiltrated, and also suggests that parts with dissimilar microstructures and properties, or possibly graded microstructures, can be processed.

The surface finish retention of reaction-infiltrated composites has been characterized as well [9]. It is found that after removal of residual melt, surface roughness is close to that of the initial machined preform, in which the microstructural scale determines the roughness. For composites prepared from high density (0.80 g/cm^3) preforms, roughness values of R_a=1.2 to 1.5 μm have been achieved.

SUMMARY

Silicon carbide composites in which both the major and the minor phase are interpenetrating, and in which the minor phase consists of metal-rich phases, have been fabricated by pressureless reactive-infiltration of porous glassy carbon preforms using Al-Si and Cu-Si alloys. The secondary phase assemblage is determined by the composition and the cooling history of the residual liquid phase remaining after the primary SiC-forming reaction is complete. In the case of Al-Si melt infiltration, the secondary phases consist of Si-Al eutectic product, along with additional Al or Si depending on whether the secondary liquid is hypo- or hyper-eutectic. In the case of Cu-Si infiltration, the secondary phases are found to be enriched in Si, most likely due to continued liquid exchange with the melt, resulting in crystallization of Si and Cu-Si intermetallic phases.

It is demonstrated that this reactive-infiltration process is capable of producing net-shape components with a linear dimension change of less than 0.1%. The simultaneous joining and reactive-infiltration of complex parts made by assembling simpler multi-part preforms has also been demonstrated.

ACKNOWLEDGMENTS

This work was supported by the Office of Naval Research, Grant Nos. N00014-90-J-1999 and N00014-94-1-0790, Dr. Steven G. Fishman, Program Manager. Shared Experimental Facilities at MIT, supported under NSF/MRL Grant No. DMR 90-22933, were used.

REFERENCES

1 P. Popper, "The Preparation of Dense Self Bonded Silicon Carbide"; p. 209 in *Special Ceramics*, Heywood, London, (1960).
2 C.W. Forrest, P. Kennedy, and J.V. Shennan, "The Fabrication and Properties of Self-Bonded Silicon Carbide Bodies"; p. 99 in *Special Ceramics*, Vol. 5, edited by P. Popper, British Ceramic Research Association, Stoke-on-Trent, (1972).
3 W.B. Hillig, R.L. Mehan, C.R. Morelock, V.J. DeCarlo, and W. Laskow, "Silicon/Silicon Carbide Composites," *Am.Ceram.Soc.Bull.*, 54[12], 1054-56 (1975).
4 E.E. Hucke, "Composite Bodies Comprising a Continuous Framework and an Impregnated Metallic Material and Methods of Their Production," *US Patent No. 3 235 346*, Feb. 15, (1966).
5 E.E. Hucke, "Process Development for Silicon Carbide Based Structural Ceramics," *AMMRC Tech. Rept. TR 83-5*, January (1983) (US Army Materials and Mechanics Research Center).
6 Y.-M. Chiang, R.P. Messner, C.D. Terwilliger, and D.R. Behrendt, "Reaction-Formed Silicon Carbide," *Mat.Sci.Eng.*, A144, 63-74 (1991).
7 R.P. Messner and Y.-M. Chiang, "Processing of Reaction-Bonded Silicon Carbide Without Residual Silicon Phase," *Ceram.Eng.Sci.Proc.*, 9[7-8], 1052-60 (1988).
8 R.P. Messner and Y.-M. Chiang, "Liquid-Phase Reaction-Bonding of Silicon Carbide Using Alloyed Silicon-Molybdenum Melts," *J.Am.Ceram.Soc.*,73[5], 1193-1200 (1990).
9 L.Hozer, J.-R. Lee, and Y.-M. Chiang, "Reaction-Infiltrated, Net-Shape SiC Composites," to appear in *Mat.Sci.Eng.B*.
10 C.A. Handwerker, M.D. Vaudin, U.R. Kattner, and D.-J. Lee, "Interface Reactions and Phase Stability in the Al-SiC System"; pp. 129-137 in *Metal Ceramic Interfaces*, Acta Scripta Metallurgica Proceedings Series, Vol.4, edited by M.Rühle, A.G. Evans, M.F. Ashby, and J.P. Hirth, Oxford, Pergamon Press (1990).
11 I.P. Smyth, "Thermodynamics and Reactivity of Chemically-Derived Silicon Carbide Fibers and Single Crystal Silicon Carbide in Aluminum-Silicon Alloys," Ph.D. Thesis, Massachusetts Institute of Technology, (1991).
12 R. Warren and C-H. Andersson, "Silicon Carbide Fibers and Their Potential for Use in Composite Materials. Part II," *Composites*, 15[2], 101-111 (1984).
13 T.B. Massalski (Ed. in Chief), *Binary Alloy Phase Diagrams*, Sec. Ed., ASM International, Materials Park, OH (1990).

RELATIONSHIP BETWEEN ARCHITECTURE, FLEXURAL STRENGTH, AND WORK OF FRACTURE FOR FIBROUS MONOLITHIC CERAMICS

Dragan Popović, Gene A. Danko, and Kevin Stuffle, Advanced Ceramics Research, Inc., 841 E.47th St., Tucson, AZ 85713
Bruce H. King, and John W. Halloran, the University of Michigan, Ann Arbor, MI 48109-2136

ABSTRACT

Various microarchitectures of fibrous monolithic ceramics were fabricated in the Si_3N_4/BN system. The microstructure consists of high-aspect-ratio polycrystalline cells of Si_3N_4 separated by cell boundaries of BN. Cell boundaries were arranged in several different microarchitectures: uniaxial layup, biaxial layup, woven fabric, and chopped fiber. All types of fibrous monolith microarchitectures fail non-catastrophically in four-point flexure. Architecture - mechanical behavior map, presenting the relationship between specimen architecture and its position on the strength vs. work-of-fracture diagram is shown, and an attempt is made to give basic explanation of the observed properties.

1. INTRODUCTION

Continuous ceramic fiber-reinforced ceramic matrix composites are a new generation of advanced composites with high strength, toughness, and non-brittle fracture characteristics. Although they were expected to become main load-bearing components for many lightweight, high-temperature structural applications such as gas turbine engine components, so far they have found only limited use. The major obstacles are their limited formability and high starting material and fabrication costs. Continuous fully dense ceramic fibers are best suited for shaping into planar bodies, or those of a large radius of curvature, not for thin turbine blades. The "fibrous monolith" concept, introduced by Coblenz, [1] should avoid the above mentioned drawbacks of CFCCs, and bring forth ordinary powder processed ceramics with graceful failure characteristics and strength and toughness values typical for costly composites. The validity of this approach was demonstrated in the literature. [2,3] The major commercial potential of fibrous monoliths lies in their ease of formability, combined with damage-tolerant behavior and low cost.

Fibrous monoliths are multi-phase monolithic ceramics, [4,5] consisting of a primary phase of high-aspect-ratio polycrystalline cells surrounded by cell boundaries of secondary and/or tertiary phase(s) of different polycrystalline material(s). Figure 1. shows a schematic of a two-phase fibrous monolith. Cell boundaries represent regions of lower shear strength interdispersed with high strength cell regions; low shear strength cell boundaries will entrap a propagating crack, thereby inducing crack deflection during fracture. This process will dissipate large amounts of energy and lead to non-catastrophic behavior.

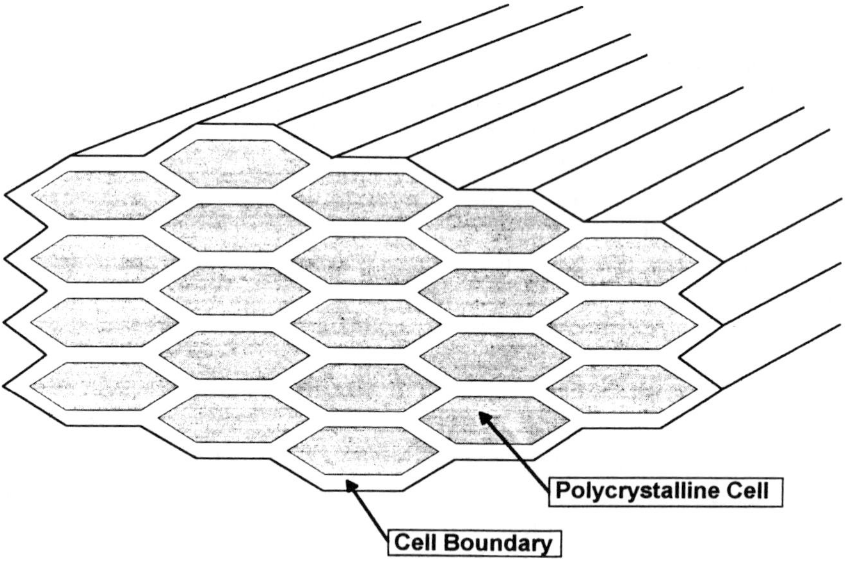

Figure 1. Schematic of a uniaxially aligned two-phase fibrous monolith.

It must be stressed that there are no dense ceramic fibers to start with. The fibrous polycrystalline microstructure of a fired body is an exact replica of a ceramic green body, which is in turn formed from ceramic powder-loaded polymer fibers arranged into a desired microstructural array (uniaxial, biaxial, woven fabric, chopped fiber) and then laminated into a rectangular billet or complex-shape part. "Green" ceramic fibers are the only building blocks of the fibrous monolith, for they consist of a core of primary (cell) polycrystalline powder in a polymer binder, surrounded by shells of secondary + tertiary + ... (cell boundary) polycrystalline powders in a polymer binder. Powders must be sized so that they can be processed in standard thermoplastic extrusion equipment, and they must be thermodynamically compatible.

This paper will show experimental results for several novel two-phase fibrous monolith architectures and materials. Flexural strength, work of fracture, and microstructure - mechanical properties relationships will be presented for Si_3N_4 (cells) / BN (cell boundaries) 0° uniaxial layup as a baseline material, and for a number of new materials: Si_3N_4/BN 0°- 90° biaxial layup, Si_3N_4/BN chopped fiber, and Si_3N_4/BN woven fabric.

2. EXPERIMENTAL PROCEDURE

Slips of various ceramic powders (cell material: Si_3N_4, grade E-10, Ube Industries, Tokyo, Japan, with 9 wt.% Y_2O_3, grade 5600, Molycorp Inc., White Plains, NY, and 3 wt.% Al_2O_3, grade RC-HP-DBM, Reynolds Metals Co, Bauxite, AR, as sintering aids; cell boundary material: BN, grade A-01, or SiC, grade B-10, H.C. Starck, Berlin, Germany) were ball milled in acetone, and then dried using a rotary evaporator. Milled powders were then mixed with thermoplastic polymer binder (ethylene - vinyl acetate copolymer, EVA 470, DuPont, Wilmington, DE) at about 50 volume % loading of ceramic using a high-shear mixer (Plasticorder, C.W. Brabender, New Hackensack, NJ), and the viscosity of the mixture (extrudate) was adjusted by adding lubricant until the desired torque reading was obtained on the machine. Extrudates containing various ceramic powders (cell or cell boundary materials) were then assembled into controlled geometry feed rods, consisting of about 85 vol.% of the cell (Si_3N_4) extrudate in the core and about 15 vol.% of the cell boundary (BN or SiC) extrudate in the shell. Single fiber or seventy-filament rods were then co-extruded preserving the cross-sectional geometry of the feed rod. Diameter of both single green fiber and individual filaments in the green multi-filament rod were about 250-260 μm. Some fifteen-filament rods were extruded; the diameter of individual filaments was about 615 μm. Details of this fabrication process are reported elsewhere. [5,6]

Preparation of ceramic green bodies proceeded as follows: co-extruded single filament was either chopped and then laminated into rectangular billets, or woven into 3-D mats which were then laminated into rectangular billets. Co-extruded multi-filament rods were cut to desired lengths, aligned inside the die into either 0° uniaxial or 0°-

90° biaxial arrays, and then laminated into rectangular billets. The billets were then loaded into a hot-pressing die, binder was burned out in a controlled-atmosphere furnace, and billets were hot-pressed under 25MPa uniaxial pressure until ram compaction stopped. At this point, near full density was achieved.

Hot-pressed billets were 114.3 mm long by 76.2 mm wide by about 4 mm thick. The long axis of the billet was denoted direction x, the short axis direction y, and the thickness was denoted direction z. All billets were machined according to MIL-STD 1942A configuration B four-point flexure bars. Flexural strength was measured using a fully articulated four-point bend fixture with a 20 mm inner span and a 40 mm outer span at a crosshead speed of 0.5 mm/min. Stress values were calculated from elastic beam equations.

3. RESULTS AND DISCUSSION

Figure 2. summarizes experimental results for the samples presented in this paper - the architecture / mechanical behavior map, showing the relationship between specimen architecture and its position on the flexural strength vs. work-of-fracture diagram. It shows individual samples as points, and then the regions of predominant behavior based on sample architecture. All silicon nitride/boron nitride samples were tested in two orientations, with long axis of the test bar parallel and perpendicular to the long axis of the billet. All of the samples exhibited non-brittle failure. Only the 0° uniaxial billet, with cells parallel to the long axis of the billet, exhibited significantly different results in the two orientations. These samples, uniaxial small and large cell (cells parallel to test bar long axis, or on-axis) and uniaxial off-axis (cells parallel to test bar short axis), were also the strongest and the weakest of all the samples tested. This is what one would expect, having in mind that uniaxial on-axis samples had all of the strong cells along the test bar length, whereas uniaxial off-axis samples had no continuous strong cells along the test bar length. Other silicon nitride / boron nitride samples (0/90 biaxial, woven fabric, chopped fiber) were also tested in both orientations, but their strength values did not differ as nearly as much as uniaxial on-axis samples, and were therefore contributed to the scatter in data, and not to the fundamentally different behavior. The fact that strength values were so similar is not surprising, because 0°/ 90° biaxial billet and woven fabric billet had the same amount of strong cells in x and y directions, and the orientation of strong cells in the chopped fiber billet was random.

The work of fracture values were calculated from the load - displacement graphs as the area under the curve after the first load drop, which takes into account only the non-elastic portion of it. Close inspection of the Figure 2. reveals that 0/90 biaxial

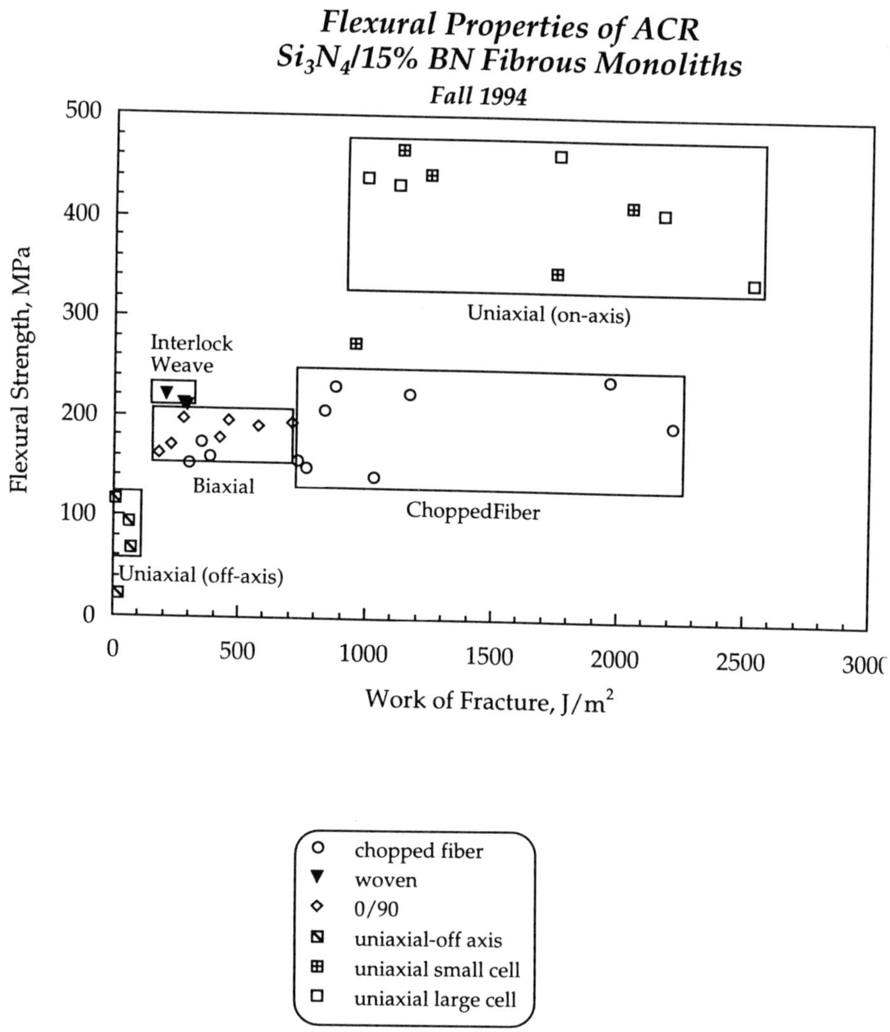

Figure 2. Relationship between strength and work of fracture for various silicon nitride / boron nitride fibrous monolith architectures.

architecture region lies between regions of uniaxial on-axis and uniaxial off-axis specimens. At this early stage of the study, it was believed that the scatter of data was too large to perform a statistical analysis and calculate standard deviations of strength and work of fracture values. It is believed that the better control of the production process would yield specimens that are grouped closer together on the strength vs. WOF diagram, and then this dependence of mechanical behavior on specimen architecture might become even more pronounced. Even at this early stage the relationship between architecture and mechanical properties is still quite interesting because it suggests that strength and work of fracture are directly proportional to the amount of strong phase in the given direction, for a given architecture.

Further inspection of the Figure 2. reveals that the chopped fiber samples about the same strength but generally higher work of fracture than 0/90 biaxial samples, whereas the woven fabric samples ("Interlock Weave" - 7 filament tows, 20% of tows in +/- 45° bias along x axis) had higher strength but lower work of fracture than most biaxial samples. The fact that the orientation of strong cells in the chopped fiber samples was random in the middle of the specimen but more aligned close to the edge, due to the friction against the die wall during lamination, would probably explain relatively large data scatter in this group. Higher work of fracture of the chopped fiber samples than of the biaxial samples could be due to the fact that there is more mechanical interlocking among randomly oriented strong cells than among uniform layers of 0° and 90° orientation. Higher fracture strength of the woven fabric samples than of the biaxial samples might, on the other hand, be due to the fact that it has more regular and repetitive mechanical interlocking, which could also be responsible for the lower work of fracture. Inspection of the Figure 3., SEM micrograph of the fractured woven fabric sample, shows that some 45° reinforcement actually acted as a preferred path for the crack to propagate. This suggests that some other types of weave might be more suitable for structural applications. However, a closer look in the actual mechanism of fracture and possibly theoretical modelling would be necessary to prove actual relationships between architecture, fracture strength, and work of fracture, which is beyond the scope of this paper.

4. CONCLUSIONS

It is suggested that strength and work of fracture are directly proportional to the amount of strong phase in the given direction, for uniaxially and biaxially aligned silicon nitride / boron nitride fibrous monoliths. For other fibrous monolith architectures in the same material system, it appears that random mechanical interlocking, as in the chopped fiber specimens, leads to higher work of fracture without change in strength values, and that very regular mechanical interlocking, as in the woven fabric specimens, leads to higher strength and lower work of fracture. However, more experimental work needs to be done to prove the two latter claims, due to imperfect processing and resulting relatively high data scatter at this early stage of the project. Theoretical modelling of fracture behavior would also help to understand the relationship between microarchitecture and mechanical properties and it would help to determine proper architecture for a particular use.

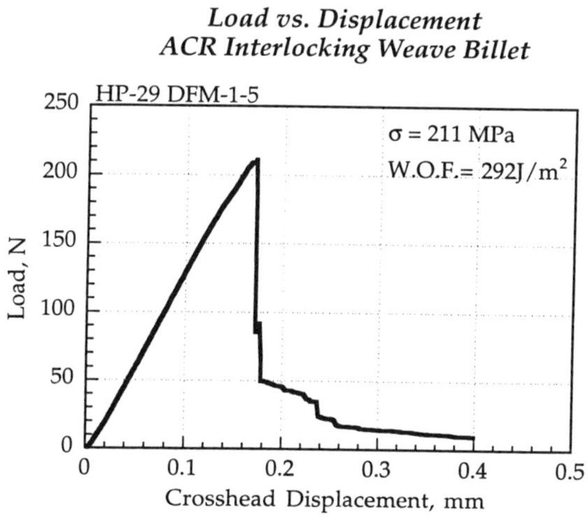

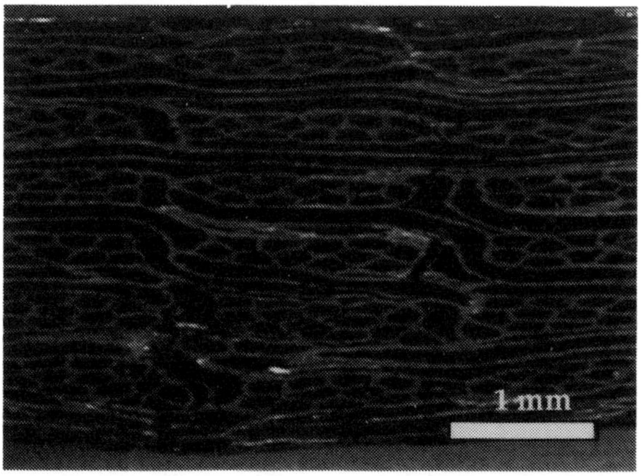

Figure 3. Silicon nitride / boron nitride woven fabric fibrous monolith, load vs. displacement diagram and side view of the broken test bar.

5. REFERENCES

[1] W. S. Coblenz, "Fibrous Monolithic Ceramics and Method for Production," U.S. Pat. No. 4 772 524, September 20, 1988.

[2] S. Baskaran and J. W. Halloran, "Fibrous Monolithic Ceramics: III, Mechanical Properties and Oxidation Behavior of the Silicon Carbide/Boron Nitride System", *J. Am. Ceram. Soc.*, **77** [5] 1249-55 (1994).

[3] S. Baskaran, D. Popovic, G. Zywicki, C. Arens, and J. W. Halloran, "Silicon Nitride and Silicon Carbide Fibrous Monolithic Ceramics," proc. of the Silicon Based Structural Ceramics Symposium, Am. Ceram. Soc. Pacific Rim Meeting, Nov. 7-10, 1993, Honolulu, Hawaii.

[4] S. Baskaran, S. D. Nunn, D. Popovic, and J. W. Halloran, "Fibrous Monolithic Ceramics: I, Fabrication, Microstructure, and Indentation Behavior," *J. Am. Ceram. Soc.*, **76** [9] 2209-16 (1993).

[5] D. Popovic, "Fibrous Monolithic Ceramics," Master's Thesis, Materials Science and Engineering Department, the University of Michigan, September 1993.

[6] G. Hilmas, A. Brady, U. Abdali, G. Zywicki, and J. Halloran, "Fibrous Monoliths: Non-brittle Fracture from Powder Processed Ceramics," submitted to *Mat. Sci. & Eng. B.*, 1994.

ACKNOWLEDGEMENTS

The authors are grateful for research support by the Advanced Research Projects Agency and Office of Naval Research under Contract No. N00014-93-1-0302.

PROCESSING AND CHARACTERIZATION OF ALUMINUM NITRIDE / METAL COMPOSITES

Koji Atarashiya
Hokkaido University, Sapporo 060, Japan

ABSTRACT

Composites of the system AlN-M were prepared using ultrafine Al and M particles by the method of reaction sintering. A mixture of the raw materials was pressed by a cold-isostatic pressing (CIPing) under 100 MPa, followed by the reaction sintering at 1273 K in flowing nitrogen under no pressure. Techniques of ultrafine-particle mixing and sintering method with no crystal grain growth were developed. The structure and composition of the composite were examined by XRD, EPMA, and FTIR. The properties of the composite such as density, thermal expansion coefficients, micro-hardness etc. were measured.
The composites prepared by this process had about 15-35 % porosity, but structural soundness was accomplished.

INTRODUCTION

Aluminum nitride has good electrical insulating properties and high thermal conductivity. Thus, the usages as an insulated substrate and a heat-sink material are expected. The composite obtained by this work are applicable to the heat-sink material of an energy-conversion device. Here, a metal dispersoid M in the composites contributes to increase cooling efficiency.
Generally, sintering of AlN must need extremely high temperature such as above

2173 K with sintering aids [1,2], but reaction sintering of AlN using an ultrafine Al particle was easy at 873 K for 84.6 ks [3]. Thus, reaction sintering of composites was also easily accomplished.

.EXPERIMENT

An ultrafine Al particle and an ultrafine M particle were prepared by means of the arc-plasma processing. The ultrafine Al particle 0.42 μ m in average diameter was partially AlN whose composition was ca. 30 wt% AlN. Ultrafine Ni and W particles were of coarse pure substances 0.7 μ m and 2.7 μ m in average diameter respectively.

Because the raw materials were very fine it was hard to mix them homogeneously. Thus, the mixing of them was carried out in butanol with ultrasonic wave. A complete mixture of raw materials was dried in vacuum. Then, the mixture was put in a cylindrical plastic tube and was pressed by CIPing under 100 MPa. Green compacts were sintered at 873 K for 43.2 ks in flowing nitrogen followed by more heating at 1273 K for 43.2 ks under no pressure. Then, structures and compositions on cross sections of composite blocks were examined by XRD and EPMA. Density of sintered composite was measured by Archimedean method. Furthermore, the powders of the raw materials and the sintered composites were examined by FTIR. Micro Vickers' hardness and thermal expansion coefficient of each composite were also measured.

RESULTS AND DISCUSSION

The composite blocks of AlN-Ni and AlN-W systems were prepared easily. Micro-structures of the composite blocks are shown in Fig. 1 and Fig. 2. Nickel and W-dispersoids whose grain sizes did not increase by sintering were observed. About matrices of AlN, we can see well-developed bonding between the particles, but no densification occurred. The composite blocks had about 35% porosity, but the pores were very small and uniformly distributed. Relative density, linear

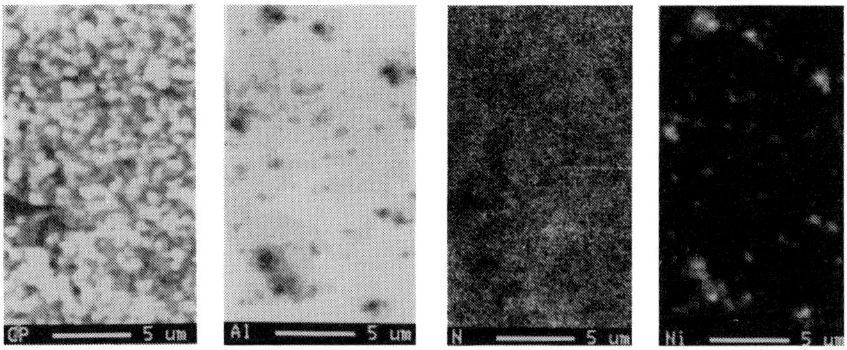

Fig.1 Micro-structure and image analysis of AlN-12.8 vol% Ni composite by EPMA.

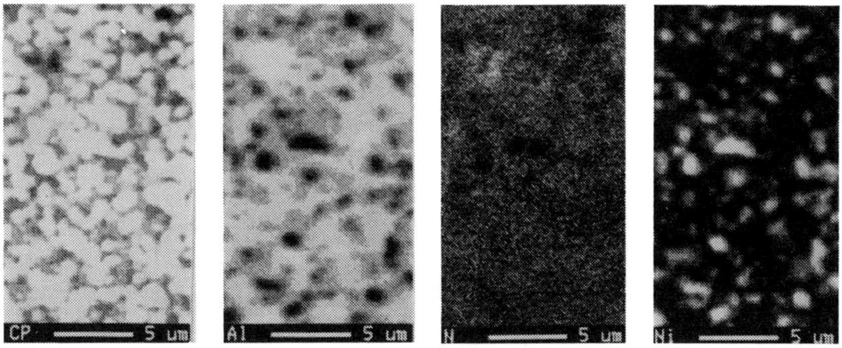

Fig.2 Micro-structure and image analysis of AlN-25.5 vol% Ni composite by EPMA.

thermal expansion coefficient and micro Vickers hardness of composite blocks are shown in Fig. 3 to Fig. 6. These physical properties smoothly changed with compositions. Law of mixture was clearly effected concerning the thermal expansion coefficient. The differences between apparent density and bulk density shows the content of an open pore. The composite of 30.7 vol.% Ni had an only open pore, but in the other parts open pore decreased and closed pores gradually increased. Thus, each pure composition i.e. AlN and Ni had an only closed pore. The influence of composition on micro Vickers' hardness is observed. The low

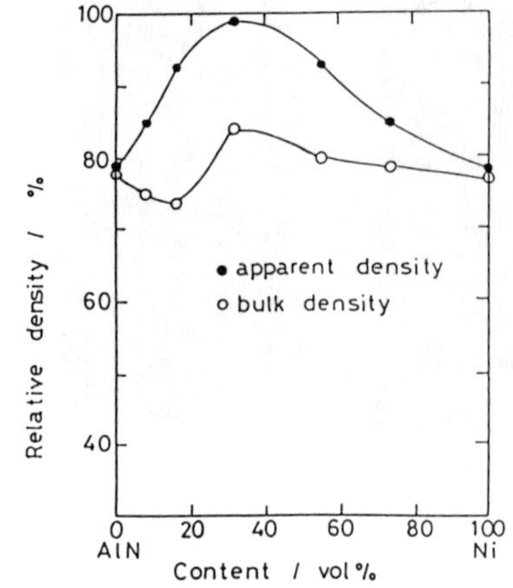

Fig.3 Relative density of AlN-Ni composite.

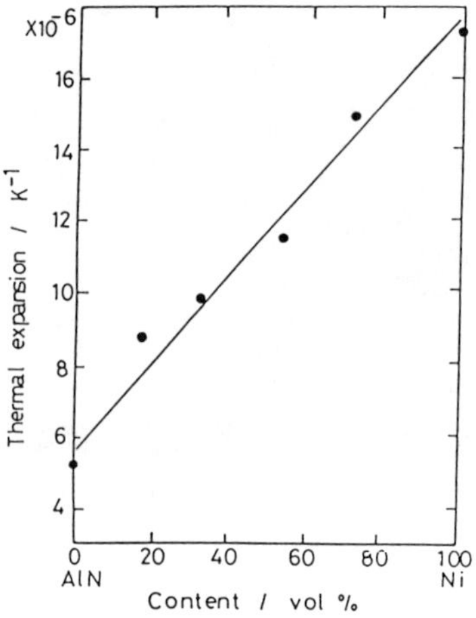

Fig.4 Thermal expansion coefficient of AlN-Ni composite.

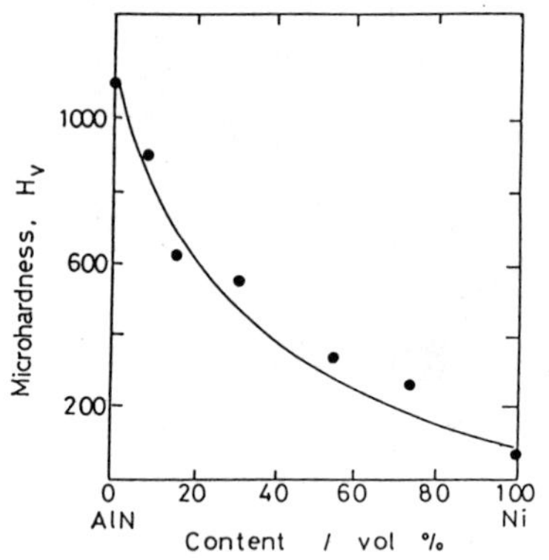

Fig.5 Micro Vickers' hardness of AlN-Ni composite.

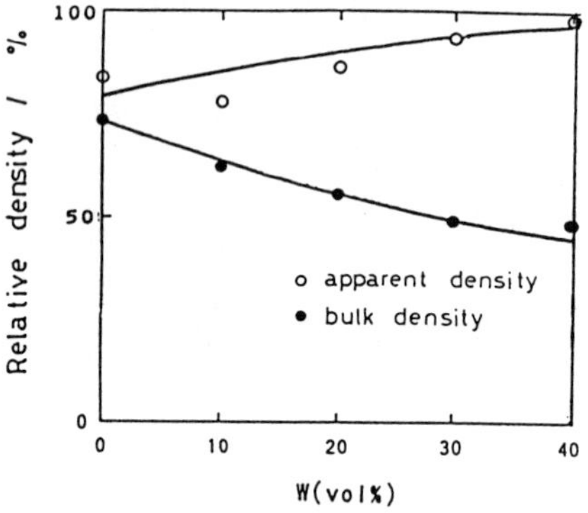

Fig.6 Relative density of AlN-W composite.

hardness value is caused by the presence of several pores. The FTIR spectrum of ultrafine Al particle (30 wt% AlN) is shown in Fig. 7. After sintering in nitrogen a new AlN phase with Al_2O_3 is formed and its spectra are given in Fig. 8. The former shows a broad band with a maximum at 878 cm^{-1}, and the later shows also

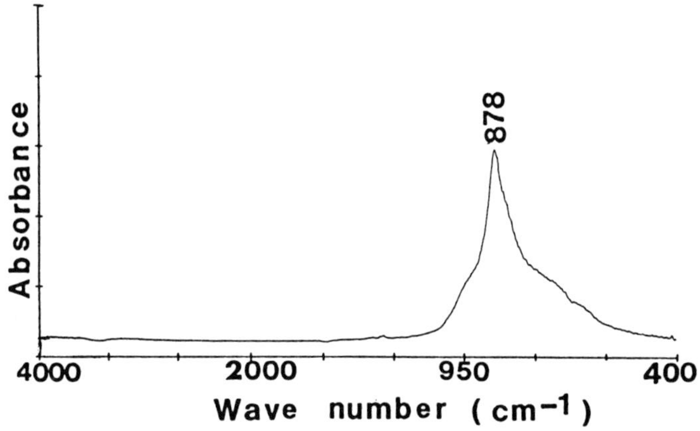

Fig.7 Fourier-transformed IR spectrum of ultrafine Al particle (30 wt% AlN).

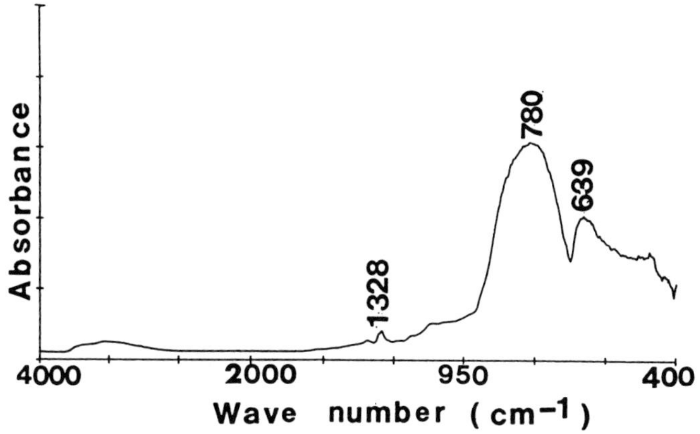

Fig.8 Fourier-transform IR spectrum of AlN sinter.

broad band with two peaks at 780 cm^{-1} and 639 cm^{-1}. They show characteristic vibration bands of Al-N at 878 cm^{-1} and 780 cm^{-1}, and of Al-O at 639 cm^{-1} [4,5]. A band at 1328 cm^{-1} in the sintered AlN sample shows the vibration of Al-NH where the -NH group vibrates as a single mass [4]. The bands of ultrafine AlN particle and nano-structured sinter were shifted into large wave numbers.

CONCLUSION

The main conclusions are as follows:

(1) Composite blocks of the system AlN-M were easily prepared by powder metallurgy under no pressure using an ultra fine Al particle.

(2) Metal-dispersoid grains in the composite blocks did not grow and did not react with any species.

(3) The Ni composite blocks had about 15-35% porosity and their micro-hardness was low to be caused by pores. But, the structural soundness of composite was good because the pores were very small and uniformly distributed, thus the composite blocks did not crush up by cutting and polishing.

(4) Relative density and thermal expansion coefficient of the composite blocks were also measured.

(5) Fourier-transform infrared characterization of an ultrafine Al particle (30 wt% AlN) and sintered block of AlN was carried out. Vibration bands of Al-N, Al-O and Al-NH were observed.

REFERENCES

[1] N.Mizutani, K.Shinozaki and T.Yagi: Sintering and Grain Boundary Phenomena of Y_2O_3 Doped AlN Ceramics: Ceram. Trans., 33, pp. 307-320, (1993)

[2] K.Komeya, H.Inoue and A.Tsuge: Effect of Various Additives on Sintering of Aluminum Nitride (in Japanese): J. Ceram. Soc. Japan, 89, pp. 330-336 (1981)

[3] K. Atarashiya, Y. Ishida, T.Nagai and M.Uda: Joining of AlN Plate and Composite of the System Al-AlN Using Ultrafine Al Particle: Ceram. Trans., 34, pp. 263-269, (1993)

[4] M. Hoch, T. Vernardakis and K. M. Nair: Ultrafine Powders of Aluminum Nitride, Silicon Nitride and Sialon: an Infrared Study: Sci. Ceram., 10, pp. 227-233, (1994)

[5] L.Ponthieu, P.Grange and B.Delmon: Proposal of a Composition Model for Commercial AlN Powders: J. Europ.Ceram.Soc., 8, pp. 233-241, (1991)

DEVELOPMENT OF A MULTI-STEP FORCED CHEMICAL VAPOR INFILTRATION PROCESS

W.M. MATLIN,* D.P. STINTON,** T.M. BESMANN,** Y.G. ROMAN,*** P.K. LIAW*
*University of Tennessee, Knoxville, TN 37966-2200
**Oak Ridge National Laboratory, Oak Ridge, TN 37996-2200
***Center for Technical Ceramics - TNO, AN Eindhoven, The Netherlands.

ABSTRACT

Using GTCVI, a finite element modeling code, a multi-step forced chemical vapor infiltration process was developed to infiltrate woven fiber bundle ceramic preforms. The fine pores within fiber bundles are uniformly infiltrated during the first few hours of the process. Reactant gas concentrations are then increased to more rapidly fill the larger porosity between bundles and layers. Processing times were more than halved, while maintaining high, uniform densities.

INTRODUCTION

The fabrication of Nicalon (Nippon Carbon, Tokyo, Japan) fiber-reinforced silicon carbide ceramic-ceramic composites by isothermal, isobaric chemical vapor infiltration (ICVI) is currently commercially established. However, due to surface overcoating and the resulting slow infiltration of relatively thick preforms, the production of > 3 mm cross-section components is difficult. An alternative approach developed at the Oak Ridge National Laboratory utilizes thermal gradient and forced convection chemical vapor infiltration (FCVI) to dramatically reduce processing times, with requirements of less than 24 hours for preforms 1.3 cm thick. [1]

FCVI has been generally performed in one step where gas flow rate, reactant concentration, outlet pressure, and hot-side preform temperature are held constant for the duration of the run. In previous attempts to reduce infiltration times the effects of pressure, reactant concentration, and hot-side temperature were investigated. It was concluded that increasing temperature, total gas flows, and reactant concentration did result in faster deposition times. Unfortunately, reductions in processing time corresponded with reductions in overall density [2]. This was due, in part, to the fact that more rapid deposition rates overcoated the fiber bundles and resulted in higher residual porosity within the bundles. In addition, steep axial density gradients resulted in large pressure drops while a portion of the preform remained relatively porous.

A multi-step process has been proposed as a way to reduce processing time for the infilitration of Nicalon preforms with SiC without sacrificing high densities. Typical preforms are fabricated by stacking layers of woven Nicalon. As a result of this fiber architecture, micrometer-sized pores exist within the bundles and hundred-plus micrometer channels lie between bundles and layers. Initially, deposition rates will be slow enough such that overcoating of fiber bundles will not occur . Once all the bundles are infiltrated, processing conditions will be changed to more rapidly infiltrate the channels between bundles. Previous work on the initial processing conditions allowed uniform infiltration of the pores between fibers [3]. This paper will focus on the second stage where channels need to be efficiently infiltrated.

MODELING

GTCVI [4], a finite-volume modeling code developed specifically for the FCVI process, was used to simulate the infiltration of pores between bundles. A plain-weave cloth layup yielded a 40 percent fiber loading. For this fiber architecture, an initial preform density of 66 percent was determined to correspond to the point where fiber bundles are completely infiltrated. The hot-side temperature was fixed at 1200 °C, the highest temperature that can be used without degrading the Nicalon fibers. The thermal gradient, hydrogen gas flow, reactant molar ratio and outlet pressure were varied using a response surface approach where 16 cube points and 8 star points were used. The center point was at 1200 °C, 545 cm³/min total flow, 0.0827 methlytrichlorosilane (CH_3SiCl_3) molar ratio (CH_3SiCl_3 moles/ total gas moles), and 101 kPa outlet pressure. Intervals of 100 °C, 100 cm³/min, 0.025, and 25 kPa, respectively, were used to generate the cube and star conditions (Table 1). At each of these points GTCVI was used to simulate the process. Final density and infiltration time were then curve fit and processing trends observed.

EXPERIMENTAL PROCEDURE

Two preforms were infiltrated using conditions determined from the earlier work on fiber bundle filling, i.e. a hot-side temperature of 1200 °C, a cooled-side temperature of 950 °C, hydrogen flow of 500 cm³/min, CH_3SiCl_3 flow of 45 cm³/min, and an outlet pressure of 101 kPa. The runs were terminated after 4 h and the preforms were removed from the furnace. The onset of overcoating of the bottom of the preforms was observed and as in previous experimental runs this resulted in low final densities. To prevent this from occurring the preforms were inverted, the peforated lid was removed, and the cooled-side preform temperature was reduced to 750 °C. The first preform was then reinfiltrated using a CH_3SiCl_3 flow of 45 cm³/min and the second preform was reinfiltrated using a CH_3SiCl_3 flow of 83 cm³/min. Each of the preforms were sectioned in half. Metallographic samples were machined from one half and flexure bars from the other (2.5 mm by 2.5 mm by 25 - 40 mm). Geometric densities were determined for each of the flexure bars. Theoretical densities were then calculated using a density of 2.55 g/cm³ for the Nicalon fiber and 3.21 g/cm³ for the SiC matrix.

RESULTS

Modeling

The GTCVI results (Table 1) were used to generate second order response surfaces for processing time (Figure 2) and density (Figure 3). Increasing temperature was determined to have the greatest effect on reducing processing time, while increasing concentration, pressure and flow were much less important, but still contributed to reduced processing times. Decreasing temperature, pressure, and flow tended to increase density, while increasing CH_3SiCl_3 concentrations increased density. No variable dominated the control of density. The result observed is that reducing processing time is in direct conflict with increasing density for all variables except CH_3SiCl_3 concentration.

Experimental

The preform that utilized a 45 cm³/min CH_3SiCl_3 flow had a second stage infiltration time of 10 hours, resulting in a total process time of 14 hours and a final average density of 80.5 %. When the CH_3SiCl_3 flow was increased to 83 cm³/min, second stage infiltration time was reduced and the part infiltrated in a total time of 11.3 hours, with the final density increased to 84.6 % (Figure 3).

Table 1. GTCVI model results for response surface analysis of the second stage process. A partially infiltrated preform density of 66 % was used.

File Name	Response Surface Intervals				Temp. (°C)	Pres. (kPa)	Hydrogen Flow (cm³/min)	MTS Molar Ratio	Time (hours)	Density (%)
tst401	-1	-1	-1	-1	1100	75	419.52	0.0577	29	88.1
tst402	1	-1	-1	-1	1300	75	419.52	0.0577	9	87.0
tst403	-1	1	-1	-1	1100	125	419.52	0.0577	24.5	85.7
tst404	1	1	-1	-1	1300	125	419.52	0.0577	7.5	84.2
tst405	-1	-1	1	-1	1100	75	607.63	0.0577	22.5	86.7
tst406	1	-1	1	-1	1300	75	607.63	0.0577	6.5	85.1
tst407	-1	1	1	-1	1100	125	607.63	0.0577	17	82.9
tst408	1	1	1	-1	1300	125	607.63	0.0577	6	82.9
tst409	-1	-1	-1	1	1100	75	397.26	0.1077	24	88.5
tst410	1	-1	-1	1	1300	75	397.26	0.1077	7	87.9
tst411	-1	1	-1	1	1100	125	397.26	0.1077	11	87.9
tst412	1	1	-1	1	1300	125	397.26	0.1077	5.5	84.4
tst413	-1	-1	1	1	1100	75	575.39	0.1077	18	86.6
tst414	1	-1	1	1	1300	75	575.39	0.1077	5.5	87.0
tst415	-1	1	1	1	1100	125	575.39	0.1077	14.5	83.8
tst416	1	1	1	1	1300	125	575.39	0.1077	4	82.3
tst417	0	0	0	0	1200	100	499.76	0.0827	10.5	85.9
tst418	0	0	0	0	1200	100	499.76	0.0827	10.5	85.9
tst419	-2	0	0	0	1000	100	499.76	0.0827	56.5	87.0
tst420	2	0	0	0	1400	100	499.76	0.0827	4.5	84.9
tst421	0	-2	0	0	1200	50	499.76	0.0827	12.5	88.2
tst422	0	2	0	0	1200	150	499.76	0.0827	8	82.1
tst423	0	0	-2	0	1200	100	316.64	0.0827	15	88.0
tst424	0	0	2	0	1200	100	683.26	0.0827	8	84.0
tst425	0	0	0	-2	1200	100	527.00	0.0327	19	86.4
tst426	0	0	0	2	1200	100	472.52	0.1327	8.5	86.6
tst427	0	0	0	0	1200	100	499.76	0.0827	10.5	85.9

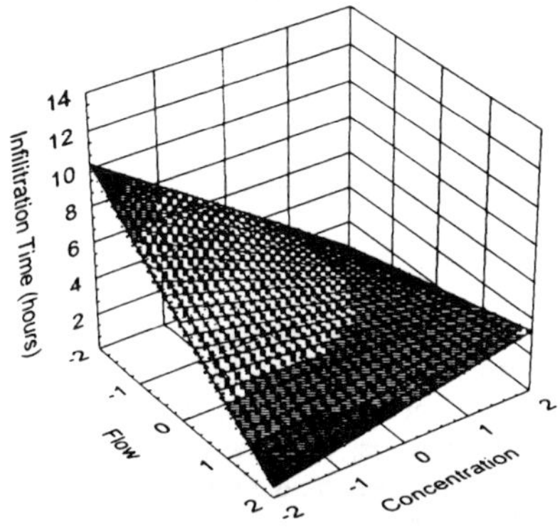

Figure 1. Time response surface based on GTCVI model results for second stage infiltration.

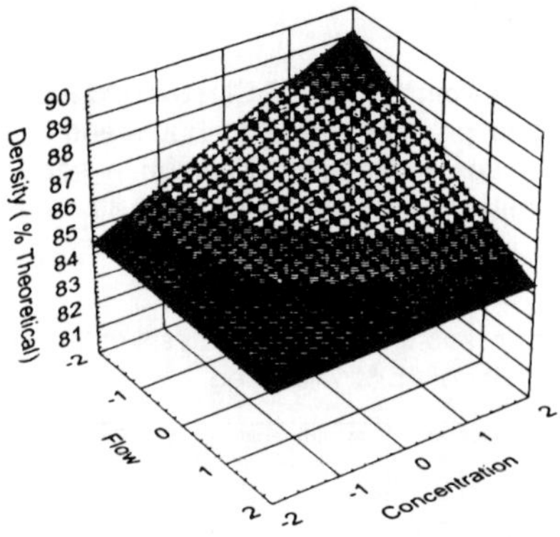

Figure 2. Time response surface based on GTCVI model results for second stage infiltration.

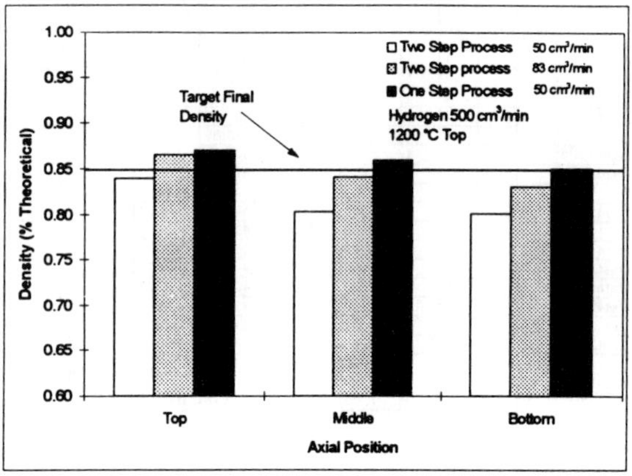

Figure 3. Experimental density for fully infiltrated preforms.

DISCUSSION

Reductions in processing time while maintaining high overall densities was made possible by tailoring the thermal gradient so that reactant depletion is offset by increased temperature. This results in a uniform deposition rate throughout the preform, thereby allowing all bundles to be filled simultaneously. Having the bundles well-infiltrated in the first portion of the run allows deposition rates to be increased in the second stage, avoiding overcoating of partially infiltrated bundles.

A uniform, partially infiltrated density of 66 % was used in the model to simulate uniform first stage infiltration. In the second stage the model predicted lower processing times and higher densities with increasing reactant concentration. This trend was confirmed in the experimental runs. For the 45 cm³/min CH_3SiCl_3 flow the model predicted an infiltration time of 14.5 hours and a final density of 85 % whereas the experimental run took 14.0 hours and produced a final density of 83 %. For the 83 cm³/min CH_3SiCl_3 flow the model predicted an infiltration time of 10.75 hours and a density of 87 %, whereas the experimental run lasted 11.3 hours and resulted in a final density of 85 %.

Finally, based on the model results further reductions in processing time should be possible by modifying the processing equipment. The current injection system uses a vapor controller that regulates gas flow based on thermal conductivity. This setup prevents heating the gas stream flowing into the furnace. As a result low H_2:MTS are not possible. In addition, in the current process cooled-side preform temperature can not be continuously regulated. Process system modification is currently underway that is anticipated to resolve both of these problems.

CONCLUSION

Processing time for a 1.3 cm thick preform was reduced from approximately 24 hours to 11.3 hours while maintaining an average final density greater than 85 %. By tailoring the thermal gradient to result in uniform deposition, processing conditions were achieved whereby final density was increased at the same time processing time was reduced. This is a major improvement over the one-step process where higher final densities were always associated with longer processing times. Since the GTCVI process model correlated closely with the experimental results, it should be relatively easy to scale up the process by using the model to duplicate these conditions within much larger preforms.

ACKNOWLEDGMENTS

Research sponsored by the U. S. Department of Energy, Office of Fossil Energy, Advanced Research and Technology Development Materials Program, under contract DE-AC05-84OR21400 with Martin Marietta Energy Systems, Inc.

REFERENCES

1. D. P. Stinton, T. M. Besmann, and R. L. Lowden, J. Am. Ceram. Soc. **67** (2), 350-355, (1988).

2. R. Lowden, A. Caputo, D. Stinton, T. Besmann, M. Morris, "Effect of Infiltration Conditions on the Properties of SiC/Nicalon Composites", Oak Ridge National Laboratory Report, No. ORNL/TM-10403, (1987).

3. W. M. Matlin, D. P. Stinton, T. M. Besmann, P. K. Liaw,"Optimization of Bundle Infiltration in the Forced Chemical Vapor Infiltration Process, ", in Ceramic Matrix Composites - Advanced High Temperature Structural Materials, Materials Research Society Symposium Proceesings, Vol. 365. Edited by R.A. Lowden, J.R. Hellmann, M.K. Ferber, S.G. DiPietro, and K.K. Chawla, Materials Research Society, Pittsburg, PA, in press.

4. T. Starr, A. Smith, T. Besmann, J. McLaughlin, and B. Sheldon, in High Temperature Matrix Composites, edited by R. Naslain, J. Lamon, and D. Doumeingts (Woodhead Pub. Ltd., Cambridge, pp. 231-240 (1993).

CHARACTERIZATION OF DEFECT EVOLUTION DURING SINTERING

David H. Phillips
Advanced Research and Applications Corporation
514 East Dayton-Yellow Springs Road
Fairborn, OH 45324-6432
(513) 879-4200

John J. Lannutti
The Ohio State University
477 Watts Hall
2041 College Avenue
Columbus, OH 43210
(614) 292-3926

ABSTRACT

Research on firing of powder-based compacts has primarily focused on important processing variables and their effects on overall density. While providing some insight into sintering behavior, overall compact densities give little or no information concerning density variations within a compact. For example, one cannot distinguish between a compact of uniform 50 percent theoretical density and one containing a mix of low density defects and high density regions averaging to 50 percent theoretical density. This information is critical in attempting to minimize strength loss and dimensional changes due to flaws generated during forming. X-ray computed tomography (CT) has been employed to quantitatively locate, measure and monitor the influence of density gradients and defects during firing. Experimental methodology and results of the study will be presented. Application of this technique to composites and other materials will also be discussed.

INTRODUCTION

Consistent manufacturing by dry pressing requires a green body which is free of density gradients in order to ensure uniform densification and shrinkage upon firing. The effects of firing a green body which is not free of these pressing defects can vary from loss of tolerance, to the development of regions of high porosity, to partial or total loss of component integrity via defect mechanisms such as end capping[1, 2].

Current technologies employed to minimize the effects of density variations range from using extremely high pressures, to overfiring, to the use of substantial amounts of organic additives. However, these strategies are not without associated economic penalties. High pressures lead to increased die wear and added tooling cost. Overfiring can add significantly to energy costs and production time. The use of larger amounts of organic additives calls for additional raw materials, longer burnout cycles, and further investment in pollution control technology.

CT offers the potential to establish a better understanding of the link between density variations which develop during dry pressing and flaws observed in fully fired compacts. Attaining this knowledge will help bridge the gap between the numerous theoretical sintering models in existence and the results of cumbersome trial-and-error experimentation. This knowledge may also help optimize/minimize additive systems and firing schedules, thereby, improving compact quality and reducing rejects. All of this will help ceramic component manufacturers fabricate parts via dry pressing to closer tolerances without costly post-firing machining.

COMPUTED TOMOGRAPHY

X-ray computed tomography (CT) is a nondestructive, noninvasive, noncontact inspection technique. CT measures the transmitted intensity of a finely collimated beam of radiation passing through an object. The CT scanner used in this research has a second-generation configuration which employs a stationary fan beam radiation source, a stationary linear detector array, and a rotate-translate handling system. During data collection, the object to be scanned is translated across the fan beam, rotated by a small angle, and translated back across the beam. This procedure is repeated until views have been obtained over approximately 180° of angular motion.

To further illustrate the process by which CT images are generated, it is useful to compare CT and conventional film radiography. Figure 1 shows a schematic representation of the two techniques. In radiography, the transmitted X-rays are projected onto a sensitized film creating an X-ray opacity map or shadowgraph of the object. Herein lies the most significant disadvantage of radiography. Because a 3-D object is being projected upon a 2-D surface, all object features perpendicular to the radiation source are superimposed. This limits the ability of radiography to spatially locate and size internal features. CT on the other hand, extracts a cross-sectional map of the object's internal structure preserving spatial information.

After the data has been collected, it is processed by a computer to reconstruct the multiple views into a map of CT numbers. These numbers are closely related to mass density, and as a result, CT is often said to generate a map of internal density. CT images are cross-sectional views of the object along the plane of the fan beam. Because CT images are densiometrically accurate (much like a photographic negative), complete morphological and part characterization information can be obtained without the need for physical sectioning.

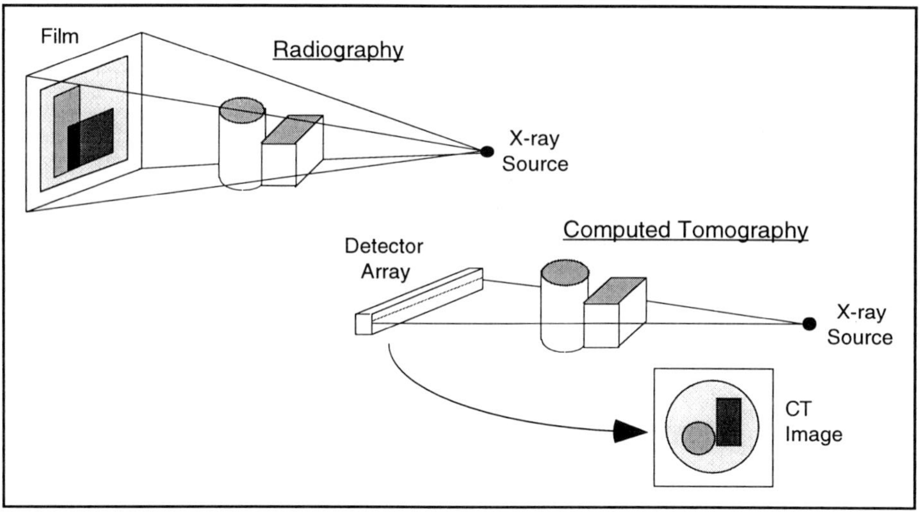

Figure 1. Comparison of CT and conventional film radiography.

EXPERIMENTAL PROCEDURES

Cylindrical compacts were single-action uniaxial dry pressed from as-received ZnO powder (J.T. Baker, Inc., Phillipsburg, NJ). The pellets formed were 12.7 mm (0.5 inch) in diameter and had an average green state aspect ratio of 0.75. Pressing was performed in an unlubricated steel die at approximately 100 MPa (15 ksi). A small laboratory tube furnace was used to fire the pellets. In order to evaluate sintering at intermediate stages, the pellets were put into a pre-heated furnace for a short period of time and then air quenched prior to CT analysis. The furnace was pre-heated to 1100 °C and a compact was placed at the edge of the hot-zone for one minute to equilibrate with the environment. The pellet was then slid into the hot-zone of the furnace for a predetermined amount of time. At the completion of firing the pellet was slid back to the edge of the hot-zone so that cooling could begin without subjecting the compact to excessive thermal shock. Finally, the pellet was removed entirely from the furnace and cooled to room temperature before scanning.

CT ANALYSIS

Figure 2 shows one of the ZnO compacts at four stages of processing: in the green state, and then after firing for 5, 10 and 20 minutes. The pellet is positioned in each image such that the stationary ram is at the bottom and the moving ram is at the top, acting downward. In each image, black signifies regions of lowest density and white regions of highest density. The green state image has been windowed individually to highlight a low density defect centrally located in the pellet. The images of the fired compact have been collectively windowed for direct comparison.

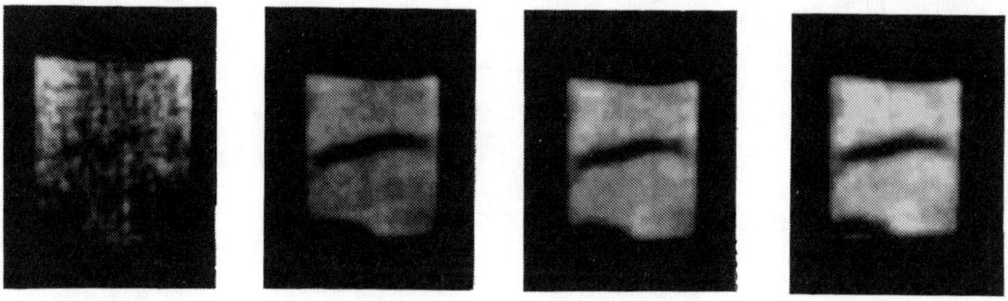

Figure 2. CT images of cylindrical ZnO compact with central defect.

The green state image shows density gradients which are characteristic of a classical shape (i.e., a circumferential high density region near the moving ram and a circumferential low density region near the stationary ram). These density gradients are primarily the result of die wall friction during dry pressing. Additionally, the green state image shows evidence of a low density defect centrally located within the pellet. Although the normal circumferential density gradients are themselves significant defects, it is the centrally located low density zone which will be the focal point of this paper.

To evaluate the magnitude of this defect, the density of the defect is compared to a 2 mm thick rectangular reference region on either side of it. Table I shows the results of this analysis. In the green state the defect is only 4.0% less dense than the reference region. However, after only five minutes of firing, what appeared to be a minor defect in the green state is 16.1% less dense than the reference region. Additional firing increases this difference.

Table I. Results of CT analysis.

	Green	5 min.	10 min.	20 min.
Defect Mean	2.89	4.07	4.21	4.28
Neighboring Region Mean	3.01	4.85	5.04	5.18
Percent Difference	4.0 %	16.1 %	16.5 %	17.4 %
Defect Width (in mm)	—	1.25	1.25	1.25
Defect Area (in mm^2)	4.69	7.29	6.10	5.87

DEFECT SIZING

Two techniques were used to measure and monitor the size of the defect. Density line-outs were taken along the centerline of the pellet in each image to measure the change in width of the defect. A density line-out plots the CT density versus position for each image pixel it crosses. Figure 3 shows this line-out analysis.

To obtain a single valued measurement of the defect width the full-width half-max (FWHM) value of the density line-out is determined. Table I lists the measured defect widths. One will note that for each of the fired pellets this value is nearly identical. This is most likely an artifact of the moderate resolution CT system employed. Since the system resolution (approximately 1 mm) is very near the defect width, in-plane partial volume effects blur the edges. Use of a higher resolution CT system may show that the width of the defect varies more than can be seen with the current system.

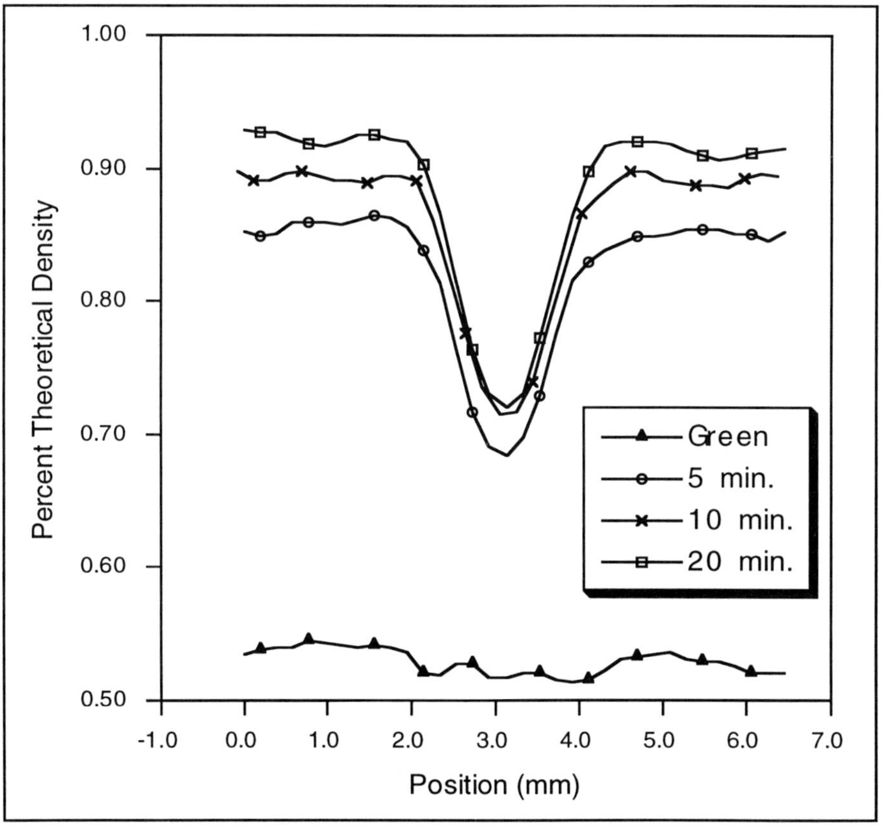

Figure 3. Density line-outs through defect.

An interesting observation which can be made from the density line-outs is that the edges of the defect appear to become more defined as sintering progresses. Notice the gradual drop from the high density plateau to the low density valley in the line-out of the pellet fired for five minutes. In comparison, this drop off in the line-outs for the pellet after it has been fired for 10 and 20 minutes is slightly more severe. This is an indication that the edges of the defect have become sharper.

A second technique for sizing the defect is to determine its area in the CT image. This is accomplished by thresholding the image to determine the defect boundaries, and then evaluating the number of image pixels within that boundary. Table I shows the results of this analysis. Measurement of the area provides evidence that the defect begins to heal as sintering progresses. Initially, the defect grows significantly; however, it then shrinks upon further firing. Re-examination of the CT images suggests that this healing occurs both near the surface of the pellet and near the center. Taking into account all of the results of the CT analysis, we can conclude that what was initially a minor low density pressing defect evolves into a significant fired state flaw. As sintering progresses, the flaw becomes more severe in magnitude and exhibits more defined boundaries, but shrinks in overall size.

Measurements of the defect width and area may be significantly improved through the application of advanced edge detection algorithms which function at a sub-pixel level. Current algorithms are capable of detecting edges within 1/5 of a pixel.

CONCLUSIONS

Computed tomography has been successfully employed to quantitatively measure the magnitude and size (width and area) of an internal defect centrally located in a dry pressed ceramic compact. The evolution of the defect was monitored throughout firing of the compact.

CT has also been applied to a variety of materials and processes from functionally gradient materials[3] to ceramic and metal matrix composites[4]. As an example, consider the task of monitoring densification patterns in chemical vapor deposition (CVD) processed carbon/carbon composite disks. Performing CT analysis after each successive CVD run has proven to be a very successful technique for quantitatively evaluating the effectiveness of the CVD process. The process can be optimized for a specific component by determining the minimum number of CVD runs required to ensure full densification. These same concepts can be applied to optimizing the firing of ceramic components.

ACKNOWLEDGMENTS

This work was completed with funding from the Edward Orton Jr. Ceramic Foundation, Westerville, OH. The authors also wish to thank Dr. Thomas J. Moran of the Wright Laboratory Materials Directorate at Wright-Patterson AFB for his generous support of CT scanner time under Air Force contract #F33615-89-C-5618, "NDE X-ray Computed Tomography Applications Research".

BIBLIOGRAPHY

1. D. H. Phillips and J. J. Lannutti, "X-ray Computed Tomography for the Testing and Evaluation of Ceramic Processes," *Am. Ceram. Soc. Bull.*, **72** [11] 69-75 (1993).

2. D. H. Phillips, "Quantitative Characterization of Density Gradients in Ceramic Processing by X-ray Computed Tomography," Master's Thesis, The Ohio State University (1995).

3. D. P. Miller, J. J. Lannutti and R. N. Yancey, "Functionally Gradient $NiAl/Al_2O_3$ Structures," *Proceedings of the 16th Annual Conference on Composites and Advanced Ceramic Materials,* 13[7-8] Part 1 of 2, pp. 365-373, Cocoa Beach, FL (1992).

4. R. N. Yancey and S. J. Klima, "High Resolution Computed Tomography of Advanced Composite and Ceramic Materials," *Proceedings of the 15th Annual Conference on Composites and Advanced Ceramic Materials,* 12[9-10] Part 2 of 2, pp. 2029-2041, Cocoa Beach, FL (1991).

Novel Synthesis

Solid Freeform Fabrication Using Gas Precursors: Selective Area Laser Deposition

J. V. Tompkins, B. R. Birmingham, H. L. Marcus
The Center for Materials Science and Engineering
The University of Texas at Austin

Abstract

Two gas-based Solid Freeform Fabrication techniques, Selective Area Laser Deposition (SALD) and Selective Area Laser Deposition Vapor Infiltration (SALDVI), can produce unique structural materials. By controlling the interaction of a laser beam with precursor gases, these techniques synthesize advanced materials such as silicon carbide and silicon nitride directly into parts with high density while also benefiting from the advantages of Solid Freeform Fabrication.

INTRODUCTION

Solid Freeform Fabrication using material additive techniques has become an established technology for rapid prototyping to reduce manufacturing cycle times [1-5]. Existing techniques offer three dimensional parts suitable for planning and design or as forms or molds. A smaller number of processes address the need for the direct production of structural parts. The following describes two evolving Solid Freeform Fabrication techniques: Selective Area Laser Deposition, and Selective Area Laser Deposition Vapor Infiltration. These approaches involve the interaction of a precursor gas with a laser beam to selectively create solid material to build a desired part. Since the number of gaseous chemical precursors continues to grow, Solid Freeform Fabrication using gas precursors may prove capable of producing a wide range of materials, including materials which current processes cannot readily produce. This suggests that gas phase approaches may have special advantages in producing structural materials, and may compete economically with traditional methods in the case of materials, such as ceramics, which resist machining.

SELECTIVE AREA LASER DEPOSITION

Selective Area Laser Deposition (SALD) describes the interaction of a laser beam with a precursor gas on a solid surface to create solid material only on those areas which the laser beam scans [6]. Moving the beam via computer control according to a predefined description of an object, such as a CAD file, builds up material into the appropriate geometry for the object.

As a result of the gas phase origin of the material, structural parts made by SALD enjoy advantages in material properties over those made by other SFF approaches and in some cases over those made by conventional manufacturing. Although gas deposited material may have a level of porosity, it does not exhibit the inherent porosity associated with the voids found in a powder bed. Also, since the material is synthesized directly into a solid piece, SALD can shape materials with high melting points or materials which decompose before melting, such as silicon nitride, at moderate temperatures without the use of binders or hot isostatic pressing. Producing these materials in a pure form preserves their resistance to high temperatures and chemical environments, as well as hardness and other properties which suffer from the addition of sintering agents.

Experimental Approach

Fig. 1a shows the equipment configuration currently used for SALD. A vacuum chamber equipped with a transparent window provides environmental control while allowing the laser beam to pass through to the substrate. In order to scan the desired areas of the substrate, computer controlled mirrors or a computer controlled x-y table direct the motion of the laser beam. Containers filled with liquid or vapor precursors provide the reactants for the process. For liquids, evaporation in the container provides a source of vapor at pressures up to the equilibrium vapor pressure. Multiple gas ports in the chamber allow mixing of various gases.

The laser beam passes through one or more lenses to provide a spot size consistent with the resolution required in making parts and in getting the required energy density; a 5 W beam focused to a 0.5 mm diameter gives a flux of approximately 25 MW/m^2. Reported research in other applications has indicated that these power densities may support a thermal plasma at the deposition point [7]. The use of a long focal length creates a waist in the beam as opposed to a point focus, making the spot size relatively constant for small distances from the focal plane. In some cases, a defocused beam is used to give a larger spot size, when more gradual thermal gradients or larger deposition schemes are desired. Fig. 1b shows a schematic of deposition from the gas phase.

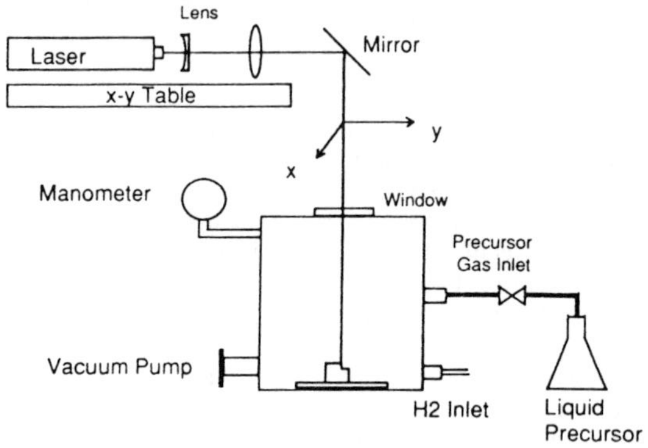

1a

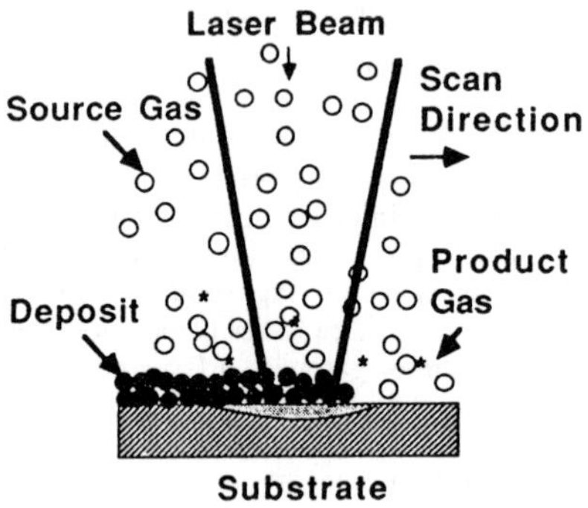

1b

Fig 1a: Gas control and laser scanning configuration. Fig 1b: Physical and chemical processes occurring at the surface (Zong).

The small size of the deposition point also dictates the mass and heat transport phenomena around the deposition site. As the reaction occurs, fresh reactant can diffuse in three dimensions through a solid angle of 2π, as opposed to one dimensional diffusion perpendicular to the surface in the case of general chemical vapor deposition over a large area. Since the area available for diffusion increases as the square of the distance from the deposition point, diffusion can occur at rates high enough to make SALD of production parts feasible.

Chemistry

Most elements and many compound materials serve as candidates for SALD, since many precursors have emerged for use in electronic materials synthesis and of films with desirable properties. In addition to inorganic compounds with low boiling points, such as metal halides or hydrides, many groups of organometallic compounds offer volatile precursors for nearly every element of interest. These include, among others: acetylacetonates, carbonyls, and metallocenes. Chemists have tailored metallorganic precursors to contain chosen elements and have desirable decomposition temperatures, and in some cases act as single source precursors providing two different elements for compound materials. Initial work on SALD has focused on relatively innocuous, non-corrosive, non-pyrophoric precursors which provide sufficient vapor pressure at room temperature. Previously reported SALD results included work with acetylene and methane and resulted in carbon shapes [8].

The SALD research described in this paper uses tetramethylsilane ($Si(CH_3)_4$), primarily in view of its potential products: ß-silicon carbide [9], α-silicon nitride (by addition of ammonia), as well as the possibility of silicides such as cobalt silicide in conjunction with other gases in the future. It was hoped that the large energy of formation which gives these materials their hardness and stability would also provide a driving force for rapid deposition rates. Tetramethylsilane (TMS) is not pyrophoric, is not corrosive, has low toxicity, and provides a vapor pressure of nearly 600 Torr at room temperature. TMS decomposes according to the following overall reaction:

$$Si(CH_3)_4\ (g) \Rightarrow \beta\ SiC + 3\ CH_4\ (g)$$

In addition, side reactions and byproducts are possible. Among these, the formation of hydrogen, solid elemental carbon, and various hydrocarbon gas byproducts may occur. The growth of silicon nitride from TMS and ammonia does not appear in the literature; however, one might expect an overall reaction to take the form:

$$3\ Si(CH_3)_4\ (g) + 4\ NH_3\ (g) \Rightarrow \alpha\ Si_3N_4 + 12\ CH_4\ (g)$$

Calculations of the free energies of formation for both nitride and carbide in the range of temperatures estimated for deposition show that carbide is favored for very low ammonia pressures (< 0.001 Torr), while nitride is favored for higher ammonia pressure.

Deposition Rate

Deposition rates observed to date for silicon carbide have reached 2.7 mm/min, sufficiently large to support the application of SALD to structural parts. The enhanced reactant concentration at the surface allowed by three dimensional feed may account for the high growth rate, though the likely presence of a thermal plasma as indicated by incandescent gas over the deposition site may also play a role. The above rate exceeds that seen in most CVD applications by several orders of magnitude [10]. To measure deposition rate, a stationary beam at a fixed intensity produces a rod with a length that is easily measured, and the rate is directly determined based on the elapsed time. Rates on two dimensional deposits are more difficult to obtain, since as material accumulating adjacent to the beam begins to absorb heat, the temperature of the deposition point decreases visibly. Rod growth, on the other hand, provides a nearly steady state growth temperature once the rod has grown only a short distance from the underlying substrate. The substrate material used affects the initial heat transfer conditions observed, as powder substrates are easily heated to the deposition temperature (~1500C) with 2-4 W, and the temperature of the deposition point decreases as solid material accumulates. A solid graphite substrate, on the other hand, requires a high power level, 50-75 Watts, to initiate growth. This is due to a combination of high reflectivity and high conductivity. After material begins to accumulate on the smooth graphite surface, the laser power may be decreased to 15-20 watts as the conductive path length to the substrate increases.

Growth rates for silicon nitride have not been completely characterized relative to TMS pressure and ammonia pressure. However, they roughly match those for silicon carbide. Increasing ammonia pressure noticeably decreases the deposition rate, possibly reflecting displacement of TMS by hydrogen and nitrogen byproducts from decomposition of excess ammonia.

Currently, the inability to accurately measure the temperature of the deposition point complicates basic modeling of the growth rates. An anticipated Arrhenius dependence on temperature makes even small errors in temperature measurement an obstacle to accurate predictions of deposition rates or kinetic studies.

Ability to Vary Structure and Composition

SALD offers a unique ability to control the geometry and/or composition of the interior of structural parts. Since feed material comes from the gas, complex internal cavities or voids may be accommodated (for internal cooling, for example). Furthermore, one may vary the material properties with position inside the part. For example, controlling TMS pressure dictates the grain size of deposited silicon carbide. Fig. 2a shows decreasing grain size of silicon carbide with increasing TMS pressure as measured by XRDS. This may be explained by an increased nucleation rate due to the higher TMS concentration. The Scherer equation [11] indicates that grain size ranges from amorphous to greater than 500 nm.

Another way to control the properties of the material is to change the chemical species in the chamber while the beam is scanning. The addition of ammonia to TMS demonstrates this by suppressing silicon carbide deposition in favor of silicon nitride. Fig. 2b shows an XRDS showing two phase silicon nitride/silicon carbide formation in the presence of 50 Torr of TMS and 44 Torr ammonia. For a TMS pressure of 50 Torr, single phase ß-silicon carbide forms at ammonia pressures below 36 Torr, while single phase α-silicon nitride dominates at ammonia pressures above 50 Torr, with two phase deposits occurring in the intermediate range. The ability to codeposit two phases likely results from near complete consumption of the ammonia when it is present in an amount lower than its stoichiometric ratio. This allows a graded transition between carbide and nitride by gradually varying the ammonia pressure, while a large step in ammonia pressure from below 36 Torr to over 50 Torr produces an abrupt transition.

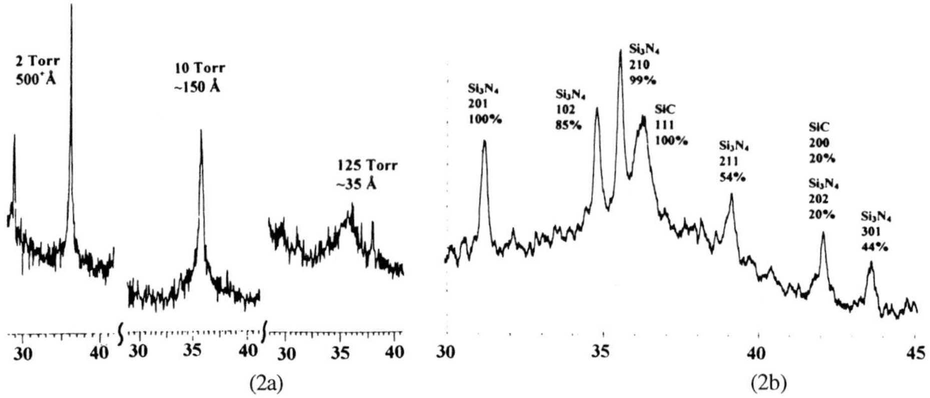

Fig 2a: Variation in grain size vs. TMS pressure as measured by XRDS. Fig 2b: Codeposition of α-silicon nitride and ß-silicon carbide by addition of ammonia.

Weak Interfaces Between Overlapping Scans

The temperature gradient at the edges of the deposition area suggests a possible complication in SALD. Any given point scanned on the surface will see a ramping up of temperature, a maximum temperature while inside the beam, and a decreasing temperature and exposure to heated gases while proximate to the departing beam. At temperatures occurring between those in the center of the scan and the temperature of the rest of the substrate an undesired solid phase may form. Alternatively, heated gas compositions from the deposition point may come in contact with unheated substrate areas near the deposition point and may precipitate an undesired byproduct. Auger analysis of silicon carbide deposited from TMS shows a surface with high carbon concentration at the surface. This may be interpreted as a deposition of carbon occurring after the temperature has dropped below the temperature needed for silicon carbide deposition. If this phenomenon occurs during overlapping scans, weak interfaces in the material might form. As a representative solution to this problem, the particular chemistry of the undesired material (carbon) was considered, and the addition of hydrogen was employed to prevent its formation. After the addition of 20 Torr of hydrogen, SEM micrographs showed the absence of the undesired second phase. The following equations characterize the reactions:

$Si(CH_3)_4 \Rightarrow SiC + nC + 2nH_2 + (3-n)\,CH_4$		(TMS only)
$nC + mH_2 \Rightarrow nCH_4 + (m-2n)\,H_2$		(Hydrogen added)
$Si(CH_3)_4 \Rightarrow SiC + 3\,CH_4$		(Desired reaction)

In general, the approach to solving the problem of weakened interfaces by a layer of second phase material will require looking at the specific chemical characteristics of the phases and gases present.

Material Hardness

Micro-hardness tests on SALD material provide a comparison with standard materials produced by conventional means to determine the suitability of this approach for production of structural parts. Possible complications include porosity, mechanical defects, poor stoichiometry, and thermal stresses.

To determine material strength, Knoop micro-hardness tests were performed to determine inherent material strength, and SEM micrographs of polished cross sections were used to compare porosity and defect size between materials synthesized under different conditions. Knoop micro-hardness tests on ß-silicon carbide from SALD were compared with both published values for ß-silicon carbide and tests performed on a commercially sintered α-silicon carbide standard with identical load and surface preparation (300g/1 um diamond paste). Hardness values in highly dense regions typically exceeded both published values and those obtained from the commercial standard, probably due to the reduced grain size which raises the hardness values. Significant peak broadening in the XRDS patterns supports this hypothesis. Hardness values up to 4300 KHN were observed, vs. 2700 published for ß-silicon carbide [12] and 2300 obtained on the standard α-silicon carbide sample. One standard deviation for 20 hardness measurements on the standard was 170 KHN. Limited data was obtained on the SALD samples due to small sample cross sections and interference from areas having defects or porosity. Fig. 3 shows SEM cross sections of a silicon nitride rod which yielded hardness readings of 1100 KHN under the same conditions.

Preferential Growth

The need to heat the surface of the substrate in the deposition area from above with the laser beam as opposed to heating from below as is the case in conventional chemical vapor deposition causes a significant limitation to the growth of multiple layers of material: heating with the beam results in heat flow downward into the substrate. When a projection or irregularity, however small, is generated, the slight increase in the thermal conduction path length causes the projection to be slightly hotter than the planar surface of the material. This excessive heating increases the deposition rate causing areas with projections to grow faster than planar or concave areas. Techniques to reduce this problem include heating the entire substrate to decrease the heat flow into the substrate and the corresponding temperature gradient, as well as selecting a deposition chemistry which does not increase exponentially above the normal deposition temperature. An example of the latter might be use of a gas mixtures containing hydrogen or halides to promote a reverse reaction at higher temperatures, etching away the projections as they try to grow.

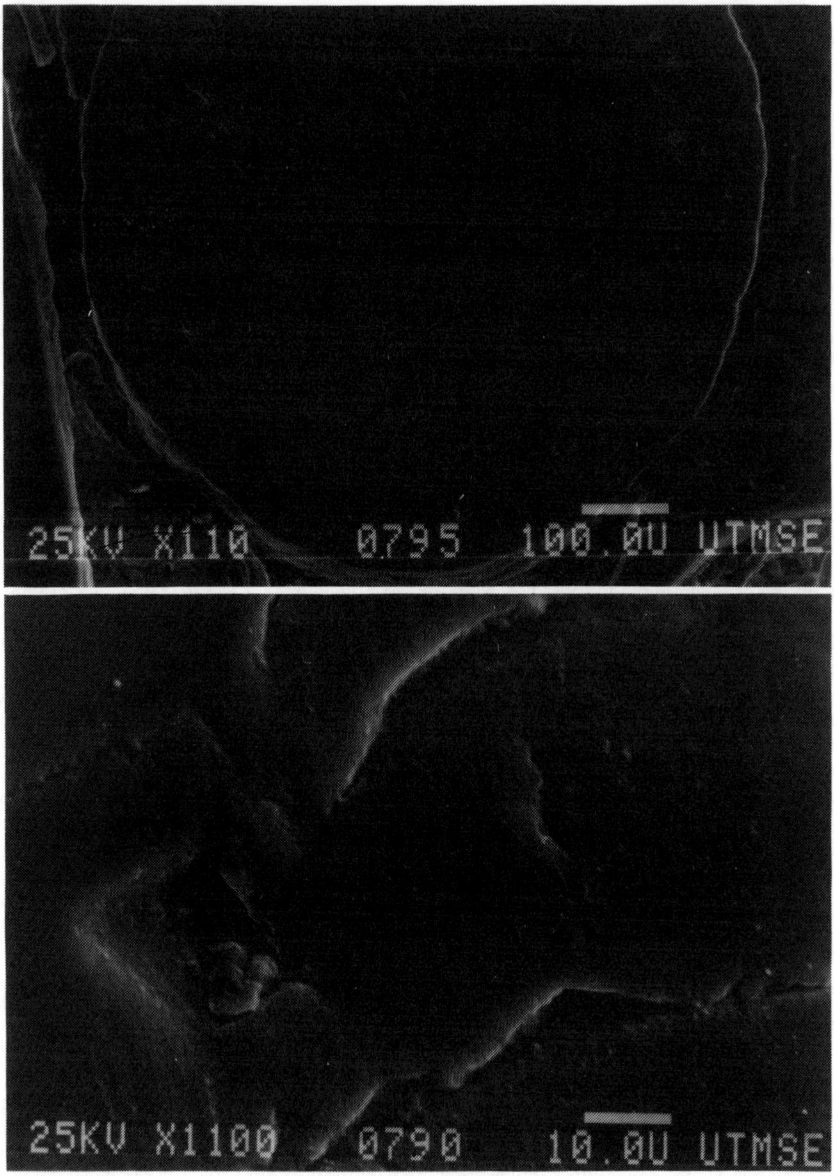

Fig. 3: SEM micrographs of SALD silicon nitride rod cross section produced with 50 Torr TMS and 200 Torr ammonia.

Thermal Shock

Thermal shock is an important consideration in scanning a point of high temperature across a ceramic material. This becomes more important in considering material with sizable defects, such as porosity. Early efforts at producing multi-layer material failed when the scanning intended to produce an overlying layer caused the first layer to disintegrate. Subsequent efforts with slower scan speeds appear to have eliminated the problem. By decreasing scan rate, the thickness of the underlying material increased, probably giving greater strength, and the thermal shock of the slower moving beam was less intense. For the conditions studied, a rate of 400 um/sec failed to produce integral layers, while rates of 50 um/sec showed no cracking in the material. Beam size and intensity, as well as growth rates, will have a strong influence on the maximum scan rate used, as will the material being deposited (metal, ceramic, etc.)

SELECTIVE AREA LASER DEPOSITION VAPOR INFILTRATION

Selective Area Laser Deposition Vapor Infiltration (SALDVI) is a solid freeform fabrication (SFF) process that combines the SALD concept discussed above with simultaneous localized chemical vapor infiltration (CVI) of a powder bed to produce three-dimensional shapes. The process uses a scanned laser to locally heat and infiltrate a thin layer of powder with material produced by the pyrolytic decomposition of a gas precursor, Fig. 4. Consecutive layers of

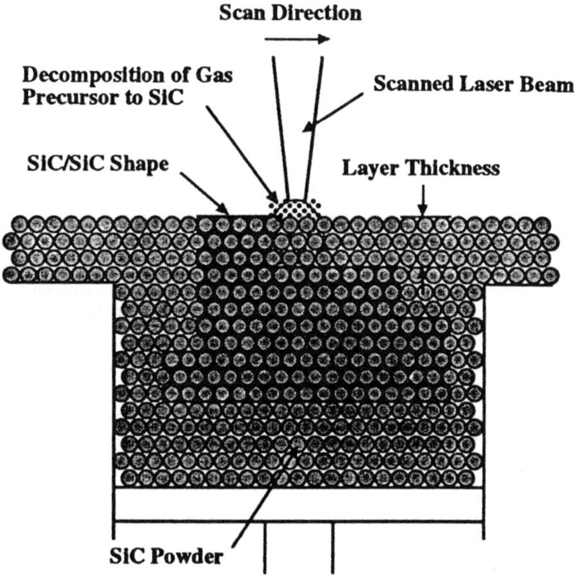

Figure 4: Selective Area Laser Deposition Vapor Infiltration (SALDVI)

powder are infiltrated and bonded together in this fashion to build up a shape. The finished part has a composite structure consisting of the starting powder particles contained within a matrix of vapor deposited material. A variety of metal/metal, metal/ceramic, and ceramic/ceramic structures may be produced by SALDVI using the gas precursors described earlier for SALD in combination with appropriate powder bed materials.

In addition to its diverse material possibilities, the SALDVI process has the potential to overcome size constraints inherent to traditional chemical vapor infiltration (CVI) techniques and produce 'large' chemical vapor infiltrated shapes. Typical CVI processes have difficulty producing 'large' shapes because of surface closure. The porous outer surface of a shape being infiltrated tends to close before its interior is fully dense, thus preventing full densification of the shape. More complex CVI processes such as forced flow reverse gradient and microwave heating are being developed specifically to address this problem[13,14]. However, the SALDVI process avoids part size limitations altogether by infiltrating consecutive thin layers to build shapes. This makes it possible to produce shapes with maximum dimensions limited only by machine size.

The experimental SALDVI arrangement is similar to that described for SALD. The principal differences are the addition of an alternate laser and the installation of a powder handling system located inside the environmental chamber. The powder handling system, which uses technology similar to that developed for the selective laser sintering process[15], provides the multiple layers of powder that are selectively infiltrated to produce a shape. The addition of a second laser, a 150W cw Nd-YAG laser (1.06μm wavelength), provides a wavelength alternative to the 25watt cw CO_2 laser (10.6μm wavelength) used for most of the SALD work. The shorter wavelength Nd-YAG laser has certain advantages when using specific organic gas precursors and ceramic powder substrates that are discussed later.

SiC/SiC SALDVI

The primary materials used for characterization of the SALDVI process were silicon carbide powder and the TMS gas precursor. The objective was to produce SiC/SiC ceramic composite shapes. SiC powder was the chosen substrate material because of the material's desirable engineering properties and its 2560°C decomposition temperature [16]. SiC's high decomposition temperature makes it stable under the laser beam, thereby minimizing any melting or decomposition of the powder that would complicate the vapor infiltration process. TMS was used because of its demonstrated ability to produce SiC in a laser deposition process.

Research to date has focused on two important aspects of the process. Beam interactions with precursor and substrate are considered important because they determine the distribution of power delivered to the process. Power distribution is the major variable determining temperature distribution and the resulting decomposition/infiltration behavior. Infiltration behavior has also been addressed to determine basic infiltration rates and the resulting densities. This establishes scan rate and powder layer thickness parameters that will ultimately determine part build rate and process feasibility.

Laser beam interactions with the precursor gas(es) and the substrate material are important for both SALD and SALDVI. Attenuation of the beam by gas absorption directly reduces energy available to drive the precursor decomposition reaction. TMS and ammonia absorb infrared energy through molecular vibration/photon interactions [17,18]. The absorption coefficients of TMS and ammonia were determined experimentally by directing the SALDVI CO_2 and Nd-YAG laser beams through a chamber filled to various pressures of each precursor gas. Laser power leaving the chamber was measured using a standard power meter and compared to the power entering the chamber. Experimental data were fit to the Beer-Lambert relationship,

$I = I_o \exp(-\alpha x p)$ = transmitted intensity, in watts
where
I_o = initial intensity, in watts,
α = absorption coefficient, in $Torr^{-1} cm^{-1}$,
p = pressure, in Torr and
x = absorption length, in centimeters.

The absorption coefficients α for the 10.6μm wavelength laser are $3.62 \cdot 10^{-4}$ $Torr^{-1}cm^{-1}$ for TMS and $1.35 \cdot 10^{-3}$ to $1.33 \cdot 10^{-4}$ $Torr^{-1}cm^{-1}$ for ammonia. No absorption was detectable by either precursor for the 1.06μm wavelength laser. The results were used to adjust power output at the laser to compensate for absorption in the gas to maintain the desired power at the point of deposition. It is noted that ammonia absorption of the CO_2 laser beam was both severe and irregular. This made the Nd-YAG laser the laser of choice when using ammonia as a gas precursor.

Reflection, absorption and transmission of the beam by the substrate material are also important to both SALD and SALDVI because they determine distribution of laser power in the substrate and the resulting temperature profile. These beam interactions are controlled by both intrinsic material properties and substrate geometry, i.e., surface roughness for SALD and powder shape and size for SALDVI. Several models are reported [19,20,21] that describe radiation transfer through a powder bed. One model developed specifically to determine reflectance and absorptance for laser interaction with a powder substrate predicts higher absorptance values for powdered material when compared to solid material due to light trapping between the particles [22]. The model assumes opaque behavior for the solid particle, leading to a prediction of total beam absorption within one particle diameter of the surface. Many of the ceramic materials being considered for SALDVI are not opaque to certain infrared energies. As an example, the absorption coefficients K for SiC at 10.6μm and 1.06μm wavelengths are 1.369 and 0.00187 μm^{-1} respectively [23]. The resulting characteristic absorption lengths, I/I_o=0.37, are 0.73μm for the CO_2 laser and 534.7μm for the Nd-YAG laser. Clearly for this example, transmission must be taken into account when predicting reflectance and absorptance values of powdered material and the Nd-YAG laser is used. Reflectance values for Si, SiC and Si_3N_4 powders have been experimentally determined using a typical integrating sphere arrangement [24], and preliminary results support the importance of transmission.

Infiltration behavior was characterized by a series of static single layer infiltrations using the CO_2 laser and TMS/H_2 precursor. Time and powder size were the main variables of the study. Three powder sizes, 0.6,15 and 100 micron particle size, were used to explore the extremes of the variable. 'Best' infiltration as determined by SEM microscopy of layer cross-sections was achieved using the 15 micron material. The 0.6 micron material sealed off at the surface almost immediately. This is attributed to high deposition rates at the surface due to the large surface area for reaction combined with resistance to infiltration of reactants due to small pore size. The 100 micron powder showed limited infiltration at the surface and very little penetration into the powder. This is attributed to the relatively small surface area available for reaction. The 15 micron material demonstrated significant infiltration to depths exceeding 300 microns in times as short as 10 seconds. Extrapolation of this rate results in a shape build rate of $2mm^3$/minute when using a single scanned beam having 0.5 mm diameter. Rates could easily be increased by using multiple beams or varying beam size.

Static beam multi-layer tests were run, again using the CO_2 laser and TMS/H_2 precursor to determine optimum layer thickness for infiltration of the 15 micron material. Thin layers are desirable in terms of final part resolution because layer thickness establishes resolution in the build direction. It is also assumed that thinner layers will more rapidly infiltrate to full density. However, there is a minimum layer thickness limited by powder flow characteristics. Flow characteristics are dependent on a variety of properties including powder size, shape and surface chemistry. The 15μm SiC powder chosen for this study flows well and is easily spread to layer thicknesses approaching 100 microns. Figure 5 shows the result of one multi-layer test which used a 135 μm layer thickness. Knoop micro-hardness testing performed on the most dense region of the sample determined a hardness of 2100 KHN (300g loading and 1 um surface polish). This is comparable to bulk SiC and SALD SiC hardness values discussed previously, suggesting that optimization of the process can result in high performance material.

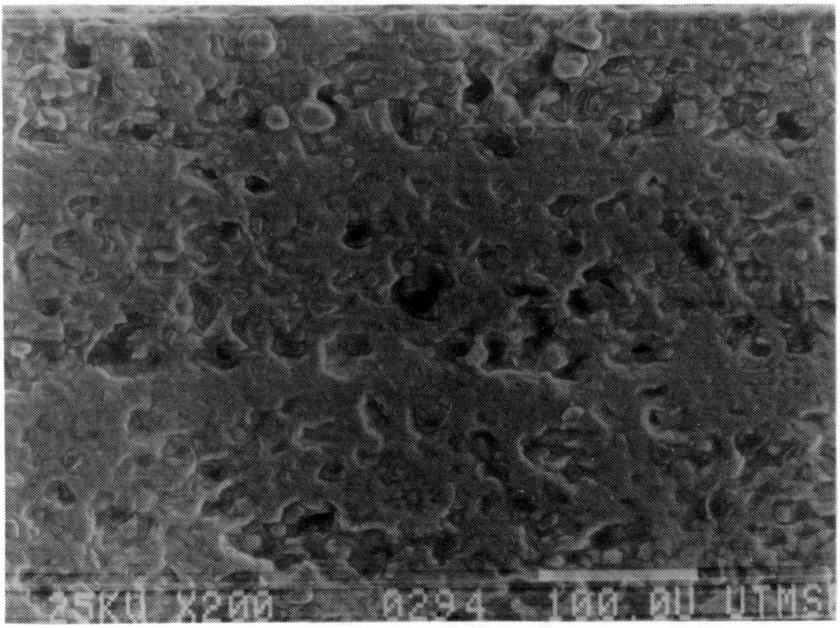

Figure 5: Cross-section of multi-layer infiltration shape made using static CO_2 beam, 20 Torr TMS and 20 Torr H_2.

Scanned beam multi-layer tests were run to complete the proof-of-concept phase of the work. Multi-layer shapes were produced; however, the scanned beam accentuated two processing difficulties previously discussed for SALD that must also be addressed for the SALDVI process. The moving temperature gradient created by the scanned beam can produce secondary undesirable phases and induce excessive thermal stresses. The secondary phases can result in poor overall or inter-layer material properties, and the thermal stresses can lead to part failure during or after fabrication.

The codeposition of undesired secondary phases, discussed previously for SALD, is further complicated for SALDVI by the addition of a three dimensional temperature and reactant/byproduct gas composition gradient into the powder substrate. This amplifies the importance of developing a deposition chemistry which is relatively temperature independent and robust with respect to reactant depletion or by-product accumulation. For example, experiments indicate infiltrating with Si_3N_4 from TMS and ammonia requires higher ammonia pressures than determined for SALD to avoid ammonia depletion and resulting carbide formation. In other systems, secondary gases can be added that retard the formation of the undesirable material as in the deposition of silicon carbide. Hydrogen addition has reduced carbon formation during the current work with the TMS precursor.

Solutions to the thermal stress problem described earlier for SALD are also applicable to SALDVI. Slower scan rates or biasing the substrate will reduce the thermal gradient. Another possible solution can be found by looking at the transmission behavior of materials versus laser wavelength. It is plausible that use of a laser/powder combination that has high transmittance will favor volume heating over surface heating, thereby reducing the temperature gradient. Figure 6 shows the cross-section of a SiC/SiC multi-layer shape produced using the Nd-YAG laser (high transmission through SiC). The material infiltration was more uniform possibly indicating that the temperature gradient was reduced.

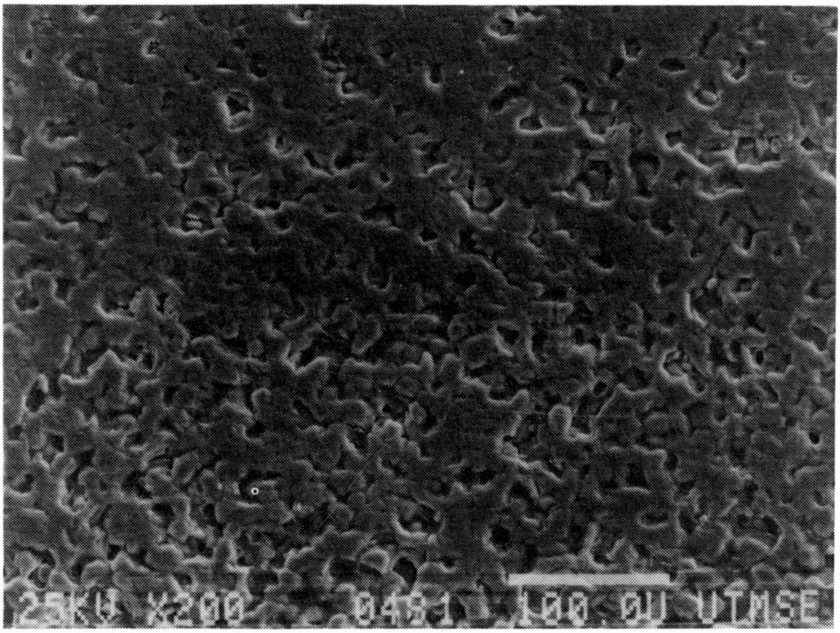

Figure 6 Cross-section of multi-layer SiC/SiC shape made using Nd-YAG laser, TMS/H_2 gas.

CONCLUSION

Research efforts into both SALD and SALDVI have demonstrated the capability to directly synthesize dense silicon carbide objects using tetramethylsilane. Both techniques also offer unique abilities in producing materials such as silicon carbide/silicon nitride composites, and they appear to offer the possibility of rapidly producing parts with a very wide range of materials in the future.

ACKNOWLEDGMENTS

The authors gratefully support the support of ONR grant N00014-92-J-1514 and Texas ATP grant 351.

References

1. *Proceedings of the Solid Freeform Fabrication Symposium*, edited by Harris L. Marcus, J. J. Beaman, Joel W. Barlow, David L. Bourell, and Richard H. Crawford, ISSN 1053-2153, The University of Texas at Austin, Austin, Texas, 1990, 1991, 1992, 1993, and 1994.
2. *Proceedings of the International Workshop on Rapid System Prototyping,* Editors, R. P. Chartoff, A. J. Lightman and J. A. Schank, University of Dayton Rapid Prototyping Development Laboratory, 1990, 1991, 1992, 1993, and 1994.
3. *Proceedings of the European Conference of Rapid Prototyping,* Editors, I. Campbell and P. Dickens, Nottingham University, 1992 (ISBN 0951975900), 1993 and 1994.
4. Paul Jacobs, *Fundamentals of Stereolithography*, Society of Manufacturing Engineers, Dearborn, MI, 1992.
5. Harris L. Marcus and David L. Bourell, "Solid Freeform Fabrication", Advanced Materials and Processes, Vol 9 1993, p 28-35.

6. Zong, G., Yves Jacquot, W. Richards Thissell, and H. L. Marcus, "Solid Freeform Fabrication Using Selective Area Laser Deposition," *Plasma and Laser Processing of Materials*, edited by K. Upadhya, The Minerals, Metals, and Materials Society, 1991.
7. *Laser-Induced Plasmas and Applications*, edited by Leon J. Radziemski and David A. Cremers, Marcel Dekker, Inc., New York, NY, 1989, pp 169-170.
8. Zong, G., J. V. Tompkins, W. R. Thissell, E. Sajot, and H. L. Marcus, "Processing Problems Associated With Gas Phase Solid Freeform Fabrication Using Pyrolytic Selective Area Laser Deposition", *Proceedings of the Solid Freeform Fabrication Symposium,* edited by Harris L. Marcus, Joeseph J. Beaman, David L. Bourell, and Richard H. Crawford, The University of Texas at Austin, Austin, Texas, August 12-14, 1991, p 271.
9. Figueras, A., S. Garelik, J. Santiso, R. Rodroguez-Clemente, B. Armas, C. Combescure, R. Berjoan, J.M. Saurel, R. Caplain, "Growth and Properties of CVD-SiC Layers Using Tetramethylsilane", Materials Science and Engineering, Volume B11, No.s 1-4, January 15, 1992, pp 83-88.
10. *Ibid.*
11. *X-ray Diffraction in Crystals, Imperfect Crystals, and Amorphous Bodies,* A. Guinier, W. H. Freeman and Co., San Francisco, CA, 1963, p 124.
12. David W. Richarson, *Modern Ceramic Engineering: Properties, Processing, and Use in Design* , Marcel Dekker INC, New York, NY 1992 p. 359.
13. S.M. Gupte and J.A. Tsamopoulos, "Forced-Flow Chemical Vapor Infiltration of Porous Ceramic Materials," *Journal of the Electrochemical Society*, 137[11], 1990, pp. 3675-3682.
14. D. Gupta and J. Evans, "A Mathematical Model for Chemical Vapor Infiltration With Microwave Heating and External Cooling,"*Journal of Materials Research*, 6[4], 1991,pp. 810-818.
15. C. Deckard and J.J. Beaman, "Progress in Selective Laser Sintering," (Paper presented at the 15th Conference on Production Research and Technology, Berkeley, California, 9-13 January 1989, pp. 623-629.
16. *Binary Alloy Phase Diagrams*, 2nd edition, T.B. Massalski editor-in-chief, Ohio, ASM International, 1990, pp. 882-883.
17. *Aldrich Library of Infrared Spectra*, edited by C.S. Pouchert, 1970.
18. J.S. Haggerty and W.R.Cannon,"Sinterable Powders from Laser Driven Reactions,"*Laser Induced Chemical Processes,* edited by J.I. Steinfeld, Plenum Press, New York,1981, pp.176-184.
19. Y.S. Yang, Howell, J.R., Klein, D.E., "Radiative Heat Transfer Through a Randomly Packed Bed of Spheres by the Monte Carlo Method,"*Journal of Heat Transfer*, 105,1983,pp325-332.
20. Chan, C.K. and Tien, C.L., K. Argo and J.M. Smith, "Radiative Transfer in Packed Spheres," *Journal of Heat Transfer*, 96, 1974, pp. 52-58.
21. Van der Held, E.F.M., "The Contribution of Radiation to the Conduction of Heat," *Applied Science Research*, A3, pp237-249, (1952); A4, 1953, pp. 77-99.
22. Sun, M., "Physical Modeling of the Selective Laser Sintering Process," Ph.D. dissertation, The University of Texas at Austin, Austin, Texas, 1991, pp. 22-28.
23. *Handbook of Optical Constants of Solids*, Edited by E.D. Palik, Academic Press, Orlando,1985, pp. 587-595.
24. *Thermophysical Properties of Matter*, edited by Y.S. Touloukian and D.P. DeWitt, Volume 8, I/P Plenum, New York, 1972, pp. 34-36.

MODELING, SINTERING, MICROSTRUCTURE, AND MECHANICAL PROPERTIES OF RBAO CERAMICS

Suxing Wu, Shaun P. Gaus*
Helen M. Chan, Hugo S. Caram*, and Martin P. Harmer

Materials Research Center
Department of Chemical Engineering*
Lehigh University
Bethlehem, PA 18015

ABSTRACT

A macroscopic model of the RBAO process, which considers the interaction of oxygen diffusion through a porous body and reaction, has been developed. The model shows reasonably good agreement with the TG-data in the literature. The paper also reports the sintering, microstructure and mechanical properties of ZrO_2-containing RBAO ceramics. Final densities of >99% TD and fracture strengths of >800 MPa were achieved in these materials.

1. INTRODUCTION

Since the invention of the RBAO (Reaction Bonding Aluminum Oxide) process at the Technical University Hamburg-Harburg some five years ago, this technique has been attracting increased interest. The RBAO process utilizes the oxidation reaction of intensively milled and compacted Al/Al_2O_3 powder mixtures to make monolithic alumina ceramics[1-7]. It has been demonstrated that ZrO_2 additions can significantly improve the final microstructure and mechanical properties of RBAO ceramics[4,5]. During heat treatment in an oxidizing atmosphere, the volume expansion associated with the oxidation reaction partially compensates for the sintering shrinkage. Although the shrinkage can be reduced by increasing the green density[4], linear shrinkages of >5% still occur in this system. The volume fraction of Al is limited to ~0.60[7] and the volume expansion associated with the oxidation reaction of Al (28%) is insufficient to completely compensate for the sintering shrinkage. However, the RBAO process can be modified by incorporating other metal and ceramic additives which exhibit larger volume expansions on oxidation[3], in order to achieve net shape forming. The ratio of final and green density, ρ/ρ_o, is the compensation factor that must be satisfied for zero shrinkage:

$$\rho/\rho_o = 1 + \Sigma \upsilon_i V_i \qquad (1)$$

where υ_i and V_i are the volume expansion and volume fraction of the additives including Al, respectively. As a good example, SiC (υ_{SiC}=1.08) has been incorporated in Al/Al_2O_3 mixtures to make low-to-zero shrinkage mullite and mullite composites[3,6].

The current work presents a macroscopic model of the RBAO process along with preliminary results from a parametric study of the model and reports further work on ZrO_2-containing RBAO ceramics in terms of sintering, microstructure and mechanical properties.

2. MODELING OF THE RBAO PROCESS

A macroscopic model of the pure RBAO process, where $4Al+3O_2 \rightarrow 2Al_2O_3$ is the only reaction, has been developed. The model considers the interaction of diffusion through the porous body and reaction, while ignoring the microscopic details of the body. Assuming a second-order reaction, the reaction rate, r, can be written as:

$$r = kC_{O_2}C_{Al} \qquad (2)$$

where C_{O2} and C_{Al} are the oxygen and Al concentrations, respectively. k is the reaction rate constant, which may be expressed according to the Arrhenius equation as:

$$k = A\exp(-E / RT) \qquad (3)$$

where A is a constant known as the frequency factor, E is the activation energy, R is the gas constant, and T is the absolute temperature.

Under the isothermal conditions expected in this relatively slow process, only differential material balances have to be considered. The differential material balance on the reacting gaseous component (continuity equation for O_2) contains a term arising from the transport by effective diffusion through the pore system and a reaction term:

$$D_{e,O_2} \cdot \frac{\partial^2 C_{O_2}}{\partial z^2} = 3r \qquad (4)$$

and the continuity equation for the reacting metal, Al, is

$$\frac{\partial C_{Al}}{\partial t} = -4r \qquad (5)$$

where $D_{e,O2}$ is the effective diffusivity of oxygen, and z is the position in the reacting system[8]. The solution to these equations using standard boundary conditions gives the concentrations of the gas and the metal as a function of time and position in the body. One can then calculate the conversion profile of Al and the overall weight gain of the sample.

Figs. 1-3 show the results of a parametric study of the frequency factor, A, conducted on the model. The values of the frequency factor used in Figs. 1, 2, and 3 are 25, 330, and 150,000, respectively. Part a of each figure shows the progression of the Al conversion profile with time as a function of the dimensionless length of the body. The center of the body is given as a value of zero for the dimensionless length, while the outer edges of the body are given by ±1. The Al conversion profile corresponds directly to the composition gradient that develops across the sample during oxidation. The Al conversion profile, and thus the composition gradient, depends critically on the rate of reaction relative to the rate of gas phase diffusion. Part b of each figure shows the % weight change of the body as a function of time. The model is unable to describe the evaporation of fugitive species in the TG data. Therefore, the process is modeled from the beginning of the oxidation, which corresponds to a temperature of ~380°C[4]. The experimental TG-data presented in the figures are from the literature[4]. The heating cycle used for the TG measurement is given in Fig. 4.

It can be seen from Fig. 1(a) that at a low value of the frequency factor (25) the aluminum is uniformly consumed throughout the body. This indicates that the rate of reaction is slow relative to the rate of gas phase diffusion. That is, the process is reaction controlled. The % weight change of the body as a function of time for a frequency factor of 25 is shown in Fig. 1(b). The reaction is seen to proceed at a much slower rate than the experimental TG data.

Fig. 3(a) shows that at a high value of the frequency factor (150,000) the reaction occurs in an external layer leaving an internal unreacted core that shrinks as the reaction progresses. In general, a steep composition gradient, as seen in Fig. 3(a), indicates a high rate of reaction relative to the rate of gas phase diffusion. This is because a high reaction rate promotes an instantaneous reaction as the gas penetrates into the body, thus producing a steep gradient between reacted and unreacted regions. Fig. 3(b) shows the % weight change as a function of time for a frequency factor of 150,000. The reaction is seen to proceed at a faster rate than the experimental TG data.

It can be seen from Fig. 2(a) that at a moderate value of the frequency factor (330) a smoother Al conversion profile develops in the body. The process in this case is controlled by a combination of reaction and diffusion, somewhere in between the two extremes shown in Figs. 1 and 3. Fig. 2(b) shows reasonably good agreement between the model and the experimental TG data.

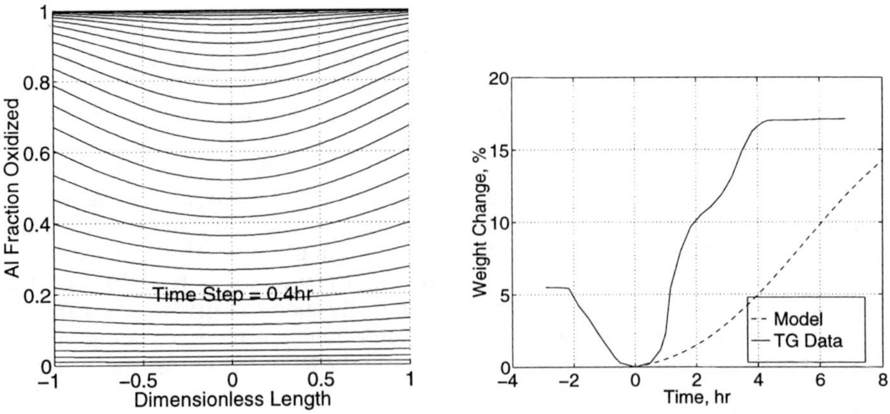

(a) Al conversion profile (b) % weight change with time

Fig. 1. RBAO model at a frequency factor, A, of 25. Reaction controlled case.

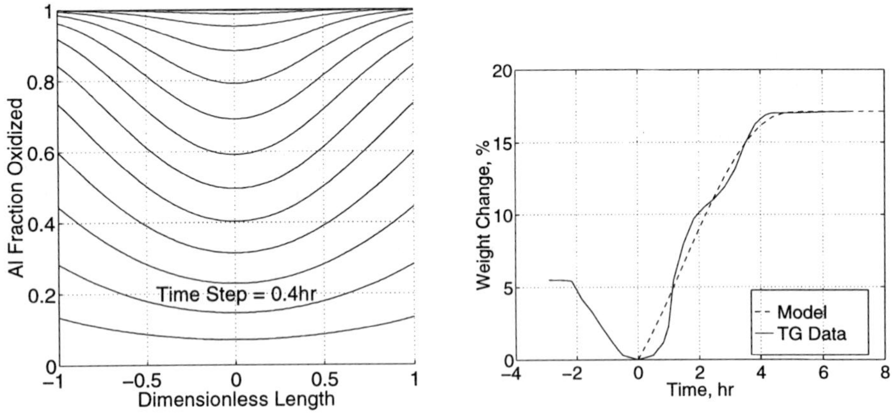

(a) Al conversion profile (b) % weight change with time

Fig. 2. RBAO model at a frequency factor, A, of 330.

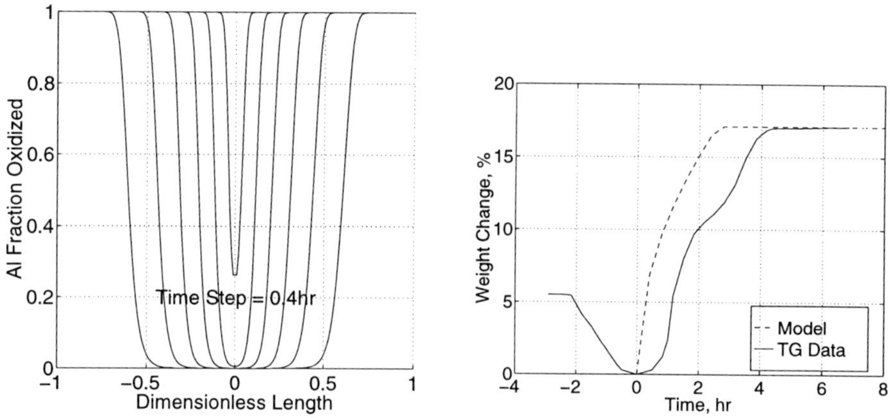

(a) Al conversion profile

(b) % weight change with time

Fig. 3. RBAO model at a frequency factor, A, of 150,000. Diffusion controlled case.

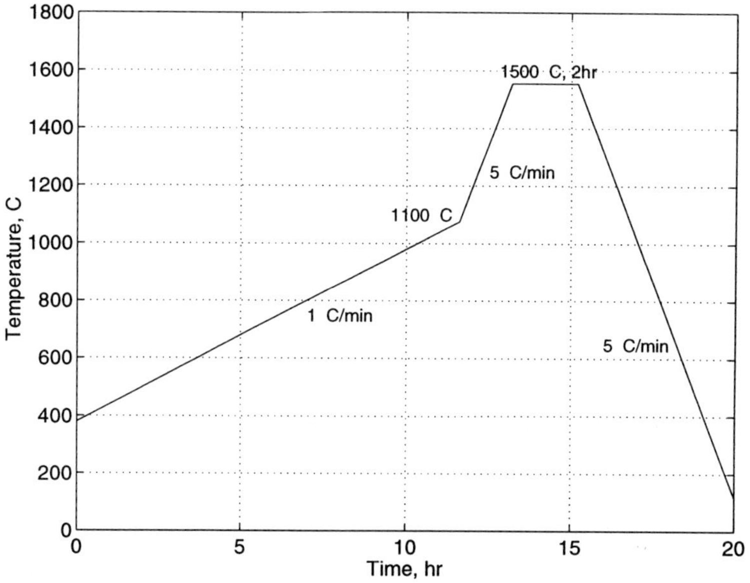

Fig. 4. Heating cycle used for TG measurements (ref. 4) and modeling.

In order to further develop the model, various input parameters, such as effective oxygen diffusivity and reaction constants, need to be experimentally determined. The model will then be used to predict the effect of various process parameters on the reaction behavior of the system.

3. SINTERING, MICROSTRUCTURE AND MECHANICAL PROPERTIES OF ZrO_2-CONTAINING RBAO CERAMICS

The two powder compositions listed in Table I were used to study sintering behavior and mechanical properties. The 45-10 composition was attrition milled in acetone for 5 hrs and 7 hrs, respectively, while the 45-20 composition was milled for 6 hrs. We shall designate these powders as 5h-, 6h-, and 7h-sample accordingly. After rotovap drying, the powder mixtures were uniaxially pressed in a graphite die at 44 MPa to form disks of 25 mm in diameter and 4 mm in thickness and then isopressed at 300 MPa. The disks were then heated in a box furnace in air. The sintering temperature and time were varied between 1450 - 1550 °C and 10 - 300 min, respectively.

Table I. Powder compositions used in this study

Composition	vol% Al*	vol% Al_2O_3**	vol% ZrO_2***
45 - 10	45	45	10
45 - 20	45	35	20

* ~20 µm, >99%, Aldrich Chemical Company, Inc.

** AKP-30, 0.3 - 0.5 µm, Sumitomo Chemical Co., Ltd. Tsukuba, Japan

*** TZ-3Y, Tosoh Co., Tokyo, Japan

Reaction bonded disks were ground and polished to 1-µm finish for mechanical testing. Samples for strength measurement were broken in biaxial flexure using a 2-mm-radius flat with the specimen loaded on a three-ball support of radius 9.37 mm[9]. Microstuctures of the RBAO ceramics were examined by scanning electron microscopy (SEM).

Data showing the final density of 7h-samples as a function of sintering time at 1550°C are shown in Fig. 5. After only 30 min sintering, a final density of 98.6% TD was achieved. The highest density presented in Fig. 5 is 99.6% TD, corresponding to 60 min sintering. Longer sintering times (> 2 hours) result in desintering which may be explained in terms of pore accumulation[10] associated with grain growth in the final stage of sintering.

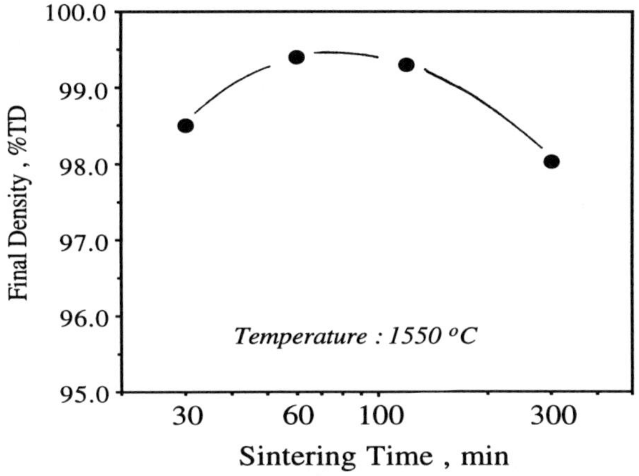

Fig. 5. Final density of 7h-sample as a function of sintering time at 1550°C.

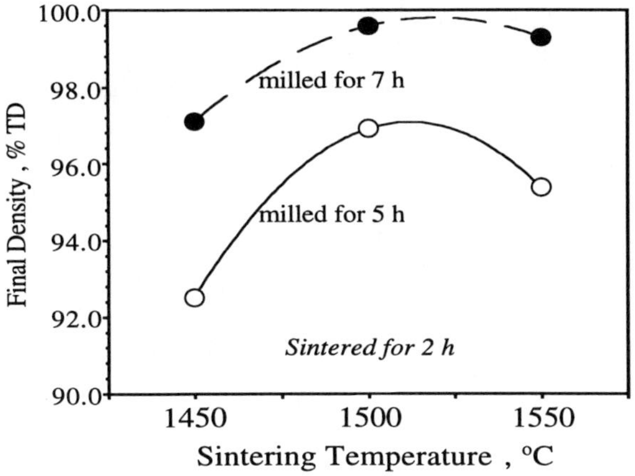

Fig. 6. Final densities of 5h- and 7h-samples as a function of sintering temperature.

Data showing final density as a function of temperature and milling time for the 45-10 composition are given in Fig. 6. It is seen from Fig. 6 that the final density increases from 1450°C to 1500°C, and then decreases at 1550°C in both samples. It is also seen that increased milling time, and hence reduced particle size, enhances densification. Since the particle sizes of the originally added submicron

Al_2O_3 and ZrO_2 powders in the mixtures (cf. Table I) remain almost constant for prolonged milling times, the size effect (7h-samples in Fig. 6) is entirely due to the higher extent of comminution of the ductile Al phase. Smaller Al particle size not only improves microstructural homogeneity of the green body, but also enhances the oxidation reaction. It has been shown that RBAO precursor powders milled for less than 5 hrs result in inhomogeneous microstructures which retain the porous agglomerate structures seen in the green state[11]. Therefore, fine precursor powders are favorable for achieving dense and uniform microstructures. The microstructure of the 7h-sample sintered at 1550°C for 2 h is shown in Fig. 7. In spite of the prolonged sintering time, the average grain sizes of Al_2O_3 and ZrO_2 are ~0.6 μm and ~0.2 μm, respectively. The body was sintered to almost full density (99.3% TD) without post-HIPing.

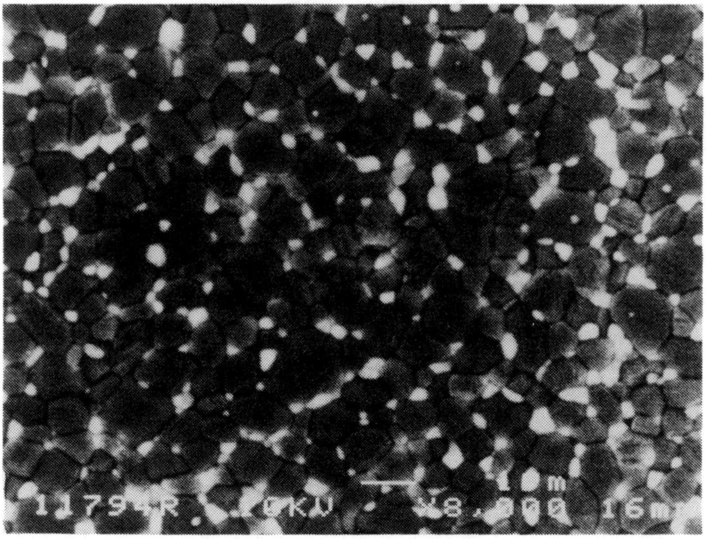

Fig. 7. SEM micrograph showing the microstructure of 7h-sample sintered at 1550°C for 2 hrs.

Data showing fracture strength of the 45-20 composition samples milled for 6 hrs and sintered at 1550°C as a function of sintering time are given in Fig. 8. The strength increases slightly with firing time from 800 MPa after 30 min to 820 MPa after 60 min. The Vicker's hardness of these samples are ~18 GPa which is comparable to that of Al_2O_3.

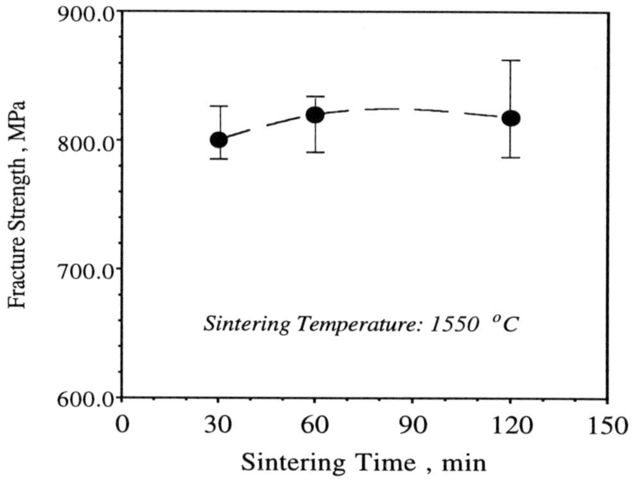

Fig. 8. Fracture strengths of 6h-samples (45-20 composition) sintered at 1550°C as a function of sintering time.

4. CONCLUSIONS

(1) A macroscopic model of the RBAO process has been developed, which has shown reasonable agreement with the experimental data in the literature.

(2) ZrO_2-containing RBAO ceramics of high densities (> 99 % TD) and nano-microstructure are readily producible. The materials exhibit superior fracture strength: 800 - 820 MPa.

(3) Longer milling times and hence smaller Al grain sizes of precursor powders enhance final density.

(4) Prolonged sintering times at temperatures >1500°C result in desintering.

ACKNOWLEDGMENTS

The authors wish to thank US Office of Naval Research (ONR) and US Department of Energy (DOE) for financial support. Thanks are also due to Sumitomo Chemical Co. Ltd for supplying Al_2O_3 powders.

References

1. N. Claussen, N. A. Travitzky, and Suxing Wu, "Tailoring of Reaction-Bonded Al2O3 (RBAO) Ceramics", *Ceram. Eng. Proc.*, **11**, 806-20 (1990).

2. Suxing Wu, A. Gesing, N. A. Travitzky, and N. Claussen, "Fabrication and Properties of Al-Infiltrated RBAO-Based Composites", *J . Euro. Ceram. Soc.*, **7**, 277-281 (1991).

3. Suxing Wu and N. Claussen, "Fabrication and Properties of Low-Shrinkage Reaction-Bonded Mullite", *J. Am. Ceram. Soc.*, **74**, 2460-2463 (1991).

4. Suxing Wu, D. Holz, and N. Claussen, "Mechanisms and Kinetics of Reaction-Bonded Aluminum Oxide Ceramics", *J. Am. Ceram. Soc.*, **76**, 970-980 (1993).

5. N. Claussen , Suxing Wu, and D. Holz, "Reaction Bonding of Aluminum Oxide (RBAO) Composites: Processing, Reaction Mechanisms, and Properties", *J. Euro. Ceram. Soc.*, **14**, 97-109 (1994).

6. Suxing Wu and N. Claussen, "Reaction Bonding and Mechanical Properties of Mullite/SiC Composites", *J. Am. Ceram. Soc.*, **77**, 2898-904 (1994).

7. D. Holz, Suxing Wu, S. Scheppokat, and N. Claussen, "Effect of Processing Parameters on Phase and Microstructure Evolution in RBAO Ceramics", *J. Am. Ceram. Soc.*, **78** (1995). In press.

8. G. F. Froment and K. B. Bischoff, "Chemical Reactor Analysis and Design, John Wiley & Sons, Inc., 1979.

9. D. B. Marshall, "An Improved Biaxial Flexure Test for Ceramics," Am. Ceram. Soc. Bull., **59**, 551-53 (1980).

10. Junhong Zhao and M. P. Harmer, "Effect of Pore Distribution on Microstructure development: II, First- and Second-Generation Pores", *J. Am. Ceram. Soc.*, **71**, 530-39 (1988).

11. D. Holz, M. Geerken, Suxing Wu, R. Janssen, N. Claussen, "HIP of Reaction-Bonded Alumina (RBAO) with Various Ratios of Open-to-Closed Porosity", in *Proceedings of HIP '93,* Antwerp, Belgium, Edited by L. Delaey, H. Has, and W. Kaysser. Elsevier, 425-33 (1993).

CONTINUOUS SINTERING OF SELF-REINFORCED SILICON NITRIDE CONTAINING REDUCED AMOUNTS OF SINTERING AIDS

D. E. Wittmer, J. J. Conover, V. A. Knapp, and B. E. Rodely
Southern Illinois University
Carbondale, IL 62901

C. W. Miller, Jr.
Centorr/Vacuum Industries
Nashua, NH 03063

ABSTRACT

Seeding of Si_3N_4 to promote the in-situ formation of elongated β-Si_3N_4 during sintering was investigated for continuously sintered Si_3N_4 compositions. Si_3N_4 formulations containing 2 or 4 wt% Al_2O_3 and 6 or 8 wt% Y_2O_3, with and without the addition of 5 wt% β-Si_3N_4 seed, were continuously sintered for 30 to 90 min in 1 atm of flowing N_2, over the temperature range of 1700 to 1750 °C. Room temperature four-point flexural strength was slightly lower and fracture toughness was only marginally improved by the addition of β-Si_3N_4 seed. XRD and microstructural results indicated that the addition of β-Si_3N_4 seed does promote the in-situ formation of β-Si_3N_4 at lower sintering times and temperatures

INTRODUCTION

Advanced ceramics are continuing to be of great interest for replacement of metals, especially for severe conditions involving high temperatures, wear, and corrosion. Typical applications include: high-speed cutting tools, pump seals, bearings, high-temperature furnace fixtures, wire and thread guides, drawing dies, and other common wear-intensive parts.[1,2] Although sufficiently high strength can be obtained for most of the advanced ceramics, improvements in fracture toughness are still desired. For Si_3N_4, the elongated β-Si_3N_4 formed during sintering has long been known to produce inherent in-situ toughening. Recently.

efforts have been successful in improving fracture toughness through the enhancement of the growth and morphology of the β-Si_3N_4 during sintering.[3-8] This has been accomplished through tailoring the liquid phase formers used for liquid phase sintering to also affect β-Si_3N_4 phase development and through the addition of β-Si_3N_4 seed as a means of controlling the resultant sintered microstructure.

Prior research by Wittmer, Doshi and Paulson[5] on seeding Si_3N_4 involved conventional over-pressure sintering of A2Y6-Si_3N_4[1] at 1880 C for 4 h under 2 MPa of N_2. The addition of 5 wt.% of β-Si_3N_4 seed to this composition was found to increase the fracture toughness (by the controlled flaw method) from 8-10 MPa-$m^{1/2}$ to 10-13 MPa-$m^{1/2}$, while only slightly reducing the flexural strength. Both the diameter and length of the β-Si_3N_4 were increased by seeding. Seeding was also found to increase the toughness for continuously sintered A4Y13, sintered at 1675 to 1725 °C for 30 to 90 min. under flowing N_2 at ambient pressure, while not appearing to have much effect on the flexural strength.[8]

In this work, the effect of β-Si_3N_4 seeding for microstructural enhancement of continuously sintered compositions, containing reduced amounts of sintering additives, was investigated. Three Si_3N_4 compositions (using Ube E-10 Si_3N_4 from Ube Industries America, NY, NY) were evaluated: A4Y6, A2Y8, and A2Y6 with and without the incorporation of 5 wt.% β-Si_3N_4.

PROCEDURE

The same procedures used by Wittmer, Doshi and Paulson[5] were used to prepared the β-Si_3N_4 seed and make additions to the respective formulations. The basic process involved, aqueous turbomilling using Si_3N_4 milling media, followed by pressure casting, drying, and isopressing at 310 MPa. Continuous sintering took place in a CVI Model 6-BF belt furnace (Centorr/Vacuum Industries, Inc., Nashua, NH) at 1700 to 1775 °C for 90 min. under flowing N_2 at ambient pressure. Closed BN boats without sacrificial powders were used for sintering. Following sintering, the immersion density and weight loss were determined and standard test bars were machined from the sintered billets.

---

[1]A refers to Al_2O_3, Y refers to Y_2O_3 and the number following the letter refers to the weight %.

Four-point flexural strength was determined by ASTM standard techniques[2] (five bars per sintering condition) and fracture toughness was determined by several indentation methods (three bars per sintering condition).[9-11] Portions of each sintered composition were ground to -325 mesh and the residual α-Si_3N_4 content was determined by the XRD techniques reported by Gazzara and Messier[12]. For microstructural evaluation, test bars were first diamond polished to 0.25 μm and then microwave plasma etched according to the techniques developed by Knapp et al.[13] After etching, the bars were coated with Au-Pd and evaluated using standard SEM techniques.

RESULTS AND DISCUSSION

Density and Weight Loss

All of the compositions were easily sintered to closed porosity with weight loss generally less than 2 wt.%. The A4Y6 composition with and without the β-Si_3N_4 seed sintered to theoretical density, while the A2Y8 and A2Y6 compositions sintered to a maximum of 98% of theoretical density. The β-Si_3N_4 seed additions to the A2Y8 and A2Y6 formulations slightly reduced the density (by about 2-3 % of theoretical) compared with the unseeded formulations. It is believed that seeding produced a higher growth rate of the elongated β-Si_3N_4, thereby creating more entrapped porosity

Four-Point Flexural Strength

Table I summarizes the average four-point flexural strengths as a function of sintering temperature for all of the formulations continuously sintered for 90 min. These results show that high flexural strength was obtained for all formulations at significantly lower sintering temperatures and times, compared to conditions required by conventional overpressure sintering methods. For some of the formulations, the highest strength was obtained at the intermediate temperature of 1725 °C, while for others it was obtained at the highest sintering temperature.

In general, the average four-point flexural strengths of the unseeded formulations were somewhat higher than the corresponding seeded formulation. This would be

---

[2]Type-B specimens, ASTM Vol. 15.01, C-1160-91

expected if grain growth were enhanced by seeding; which would also be expected to improved fracture toughness.

Fracture Toughness

Due to a length restriction for this paper, only the fracture toughness results obtained by the strength method [11] will be reported and discussed.

The expected significant improvements in measured fracture toughness were not observed for any of the formulations investigated, and only marginal improvement in the measured fracture toughness may have been obtained by seeding. As seen from the results in Table II, the range of fracture toughness values for the seeded formulations was statistically the same as that of the unseeded ones. Only the seeded A2Y6 formulation had higher or equivalent fracture toughness for all sintering conditions.

It is interesting to note that higher fracture toughness values were determined for sintering conditions yielding the highest flexural strengths for all formulations with the exception of the seeded A4Y6. This may be an indication of how strongly the strength controls the fracture toughness value determined by the strength method.

XRD and Microstructure

Although there was not much evidence that seeding had any effect on the measured fracture toughness of the compositions evaluated, there was XRD and microstructural evidence that seeding did promote β-Si_3N_4 phase development and grain growth. Table III shows that the presence of the β-Si_3N_4 seed promoted the $\alpha \rightarrow \beta$ transformation at shorter sintering times and had the greatest effect on the A2Y6 formulation. This enhanced transformation would be expected to produce larger grain size which is normally associated with improved toughening. Larger, more elongated grains were observed for all of the seeded formulations. Examples of the observed microstructural development are given in Figures 1 through 6.

Figure 1 shows that the A4Y6 without seeding developed uniform elongated β-Si_3N_4 phase, compared with the much larger β-Si_3N_4 grains developed in the seeded A4Y6 (Figure 2). Through close examination of Figure 2, the original β-Si_3N_4 seed can be observed within several of the large β-Si_3N_4 grains. It is clear that seeding does promotes the formation and growth of the β-Si_3N_4 but this may actually lead to pore entrapment in the process. This would explain the lower densities and flexural strengths obtained for the seeded formulations.

For the A2Y8 formulation, seeding appears to have an even greater effect on the β-Si_3N_4 development. The unseeded composition (Figure 3) shows that grain growth is much more developed compared with the unseeded A4Y6 (Figure 1); however seeding (Figure 4) produced even greater grain growth. The microstructure of the seeded A2Y8 reveals several large, low aspect ratio β-Si_3N_4 grains, several intersected grains, and residual porosity. Although the grain growth appears to be enhanced, low aspect ratio microstructures are generally associated with reduced toughness.

The unseeded A2Y6 composition (Figure 5) appears to have significantly reduced β-Si_3N_4 grain growth and a major portion of the microstructure is very fine grained. Seeding (Figure 6) does appear to increase the β-Si_3N_4 development and the microstructure appears to contain less fine grained structure. Therefore, it would be expected that the seeding could produce much greater toughening effects for the lower sintering aid composition.

CONCLUSIONS

β-Si_3N_4 seed was added to Si_3N_4 formulations, containing Al_2O_3 and Y_2O_3 sintering aids, as a means of enhancing in-situ toughening during continuous sintering. Room temperature four-point flexural strength was slightly lower and fracture toughness was at best only marginally improved by the addition of β-Si_3N_4 seed. XRD and microstructural results indicated that the addition of β-Si_3N_4 seed does promote the in-situ formation of β-Si_3N_4 at lower sintering times and temperatures, although the measured fracture toughness results did not confirm the definable improvements expected.

ACKNOWLEDGMENTS

This work was funded by the U.S. Department of Energy, as part of the Ceramic Technology Project of the Propulsion System Materials Program under Materials Development Program, under contract DE-AC05-840R21400 with Martin Marietta Energy Systems, Inc. Appreciation is also expressed to Robert K. Olsen at Centorr/Vacuum Industries for his technical assistance.

REFERENCES

1. R. N. Katz, "Application of Silicon Nitride Based Ceramics in the US," Silicon Nitride Ceramics, Scientific and Technical Advances, MRS Symposium Proceedings, Edited by Chen, Becher, Mitomo and Yen, MRS V287, 197-208 (1993).

2. H. Miao, L Qi, D. Ma, and Z. Jiang, "Studies on Si_3N_4 Ceramic Cutting Tool Materials and Their Applications," ibid 507-12.

3. A. Pyzik, D. Carrol and C. Hwang, "The Effect of Glass Chemistry on the Microstructure and Properties of Self-Reinforced Silicon Nitride," ibid 411-16.

4. G. Wötting, H. Feuer and E. Gugel, "The Influence of Powders and Processing Methods on Microstructure and Properties of Dense Silicon Nitride," ibid 133-46.

5. D. E. Wittmer, D. Doshi and T. E. Paulson, "Development of β-Si_3N_4 for Self-Reinforced Composites," 4th International Symposium on Ceramic Materials & Components for Engines, Eds. R. Carlsson, T. Johansson and L. Kahlman, Elsevier, Amsterdam, 594-602 (1992).

6. F. Wu, H. Zhuang, L. Ma and X. Fu, "Self-Reinforced Silicon Nitride by Gas Pressure Sintering," Ceram. Eng. Sci. Proc., **14** [1-2] 321-22 (1993).

7. M. Ohashi, S. Kanzaki, H. Tabata, "Effect of Additives on Some Properties of Silicon Nitride Ceramics," J. Mat. Sci., **26** 2608-14 (1991).

8. D. E. Wittmer, T. E. Paulson, D. Doshi and C. W. Miller, Jr., "Improved Processing and Cost Effective Sintering of Si_3N_4," SAE Publication P-256, 16 7-170 (1992).

9. R. F. Cook and B. R. Lawn, "A Modified Indentation Toughness Technique," J. Am. Ceram. Soc., **66** [11] C200-201 (1983).

10. G. R. Anstis, P. Chantikul, B. R. Lawn and D. B. Marshall, "A Critical Evaluation of Indentation Techniques for Measuring Fracture Toughness: I, Direct Crack Measurements," J. Am. Ceram. Soc. **64** [9] 533-38 (1981).

11. P. Chantikul, G. R. Anstis, B. R. Lawn and D. B. Marshall, "A Critical Evaluation of Indentation Techniques for Measuring Fracture Toughness: II, Strength Method," J. Am. Ceram. Soc. **64** [9] 539-43 (1981).

12. C. P. Gazzara and D. M. Messier, "Determination of Phase Content of Si_3N_4 by X- Ray Diffraction Analysis," Am. Ceram. Bull., **56** [9] 777-80 (1977).

13. V. A. Knapp, D. E. Wittmer, J. J. Conover and C. W. Miller, Jr., "Microwave Plasma Etching of Si_3N_4," Ceramic Engineering and Science Proceedings, **15** [5] 1118-1127 (1994).

Table I. Summary of average four-point flexural strength results as a function of sintering temperature for Si_3N_4 continuously sintered for 90 min.

Temperature	Average Four-Point Flexural Strength (MPa)					
(°C)	A4Y6	A4Y6-S	A2Y8	A2Y8-S	A2Y6	A2Y6-S
1750	910	870	1050	930	920	880
1725	950	760	950	980	850	920
1700	890	750	950	970	780	820

S-indicates 5 wt.% β-Si_3N_4 seed addition

Table II. Summary of average fracture toughness (strength method) results as a function of sintering temperature for Si_3N_4 continuously sintered for 90 min.

Temperature	Average Fracture Toughness (MPa-$m^{1/2}$)					
(°C)	A4Y6	A4Y6-S	A2Y8	A2Y8-S	A2Y6	A2Y6-S
1750	5.7	5.7	6.9	6.5	6.5	6.5
1725	6.7	6.2	6.6	6.7	6.1	6.7
1700	6.6	6.6	6.5	6.7	5.8	6.2

S-indicates 5 wt.% β-Si_3N_4 seed addition

Table III. Results of XRD analyses.

Temperature /Time (°C/min)	Residual α-Phase Content (%)					
	A4Y6	A4Y6-S	A2Y8	A2Y8-S	A2Y6	A2Y6-S
1750/90	nd	nd	nd	nd	nd	nd
1725/90	nd	nd	nd	nd	nd	nd
1700/90	nd	nd	nd	nd	nd	nd
1700/60	nd	nd	nd	nd	4	nd
1700/30	nd	nd	14	12	17	5

S-indicates 5 wt.% β-Si_3N_4 seed addition
nd-not detected

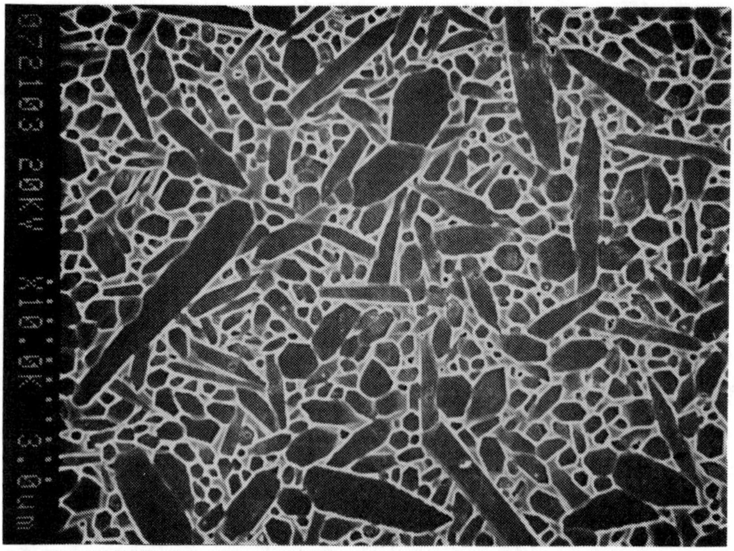

Fig 1. SEM of A4Y6-E10 Si_3N_4 after continuous sintering at 1750°C and 90 min. The microstructure shows uniformly dispersed, elongated β-Si_3N_4 grains in a fine grained matrix.

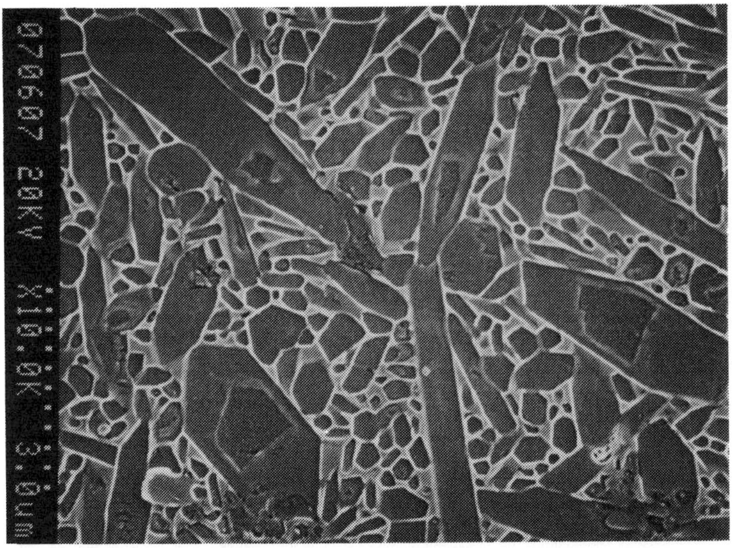

Fig 2. SEM of A4Y6-E10 containing 5 wt% β-Si_3N_4 seed after continuous sintering at 1750°C and 90 min. The microstructure contains larger β-Si_3N_4 grains than observed for the unseeded formulation.

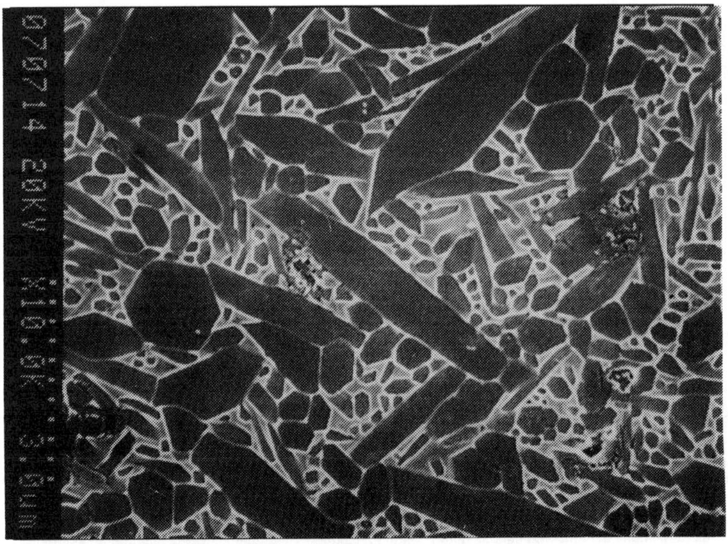

Fig 3. SEM of A2Y8-E10 Si_3N_4 after continuous sintering at 1750°C and 90 min. The microstructure contains some large β-Si_3N_4 grains in a fine grained matrix.

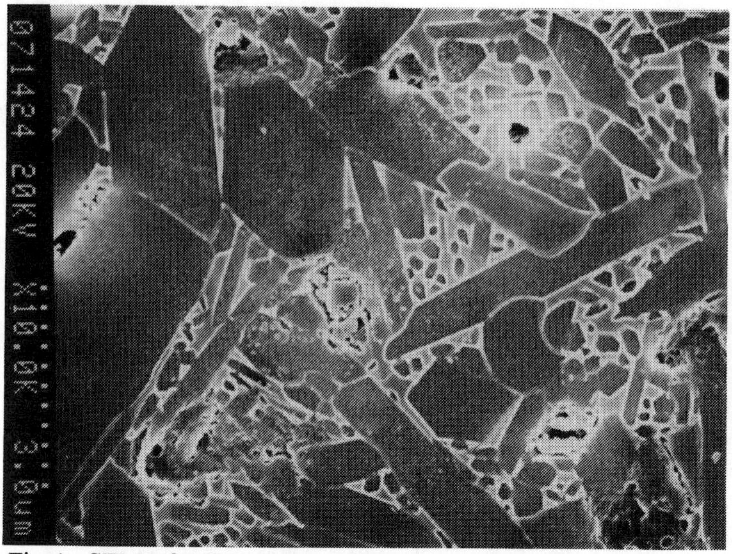

Fig 4. SEM of A2Y8-E10 containing 5 wt% β-Si_3N_4 seed after continuous sintering at 1750°C and 90 min. More porosity and larger β-Si_3N_4 grains are observed than for the unseeded formulation.

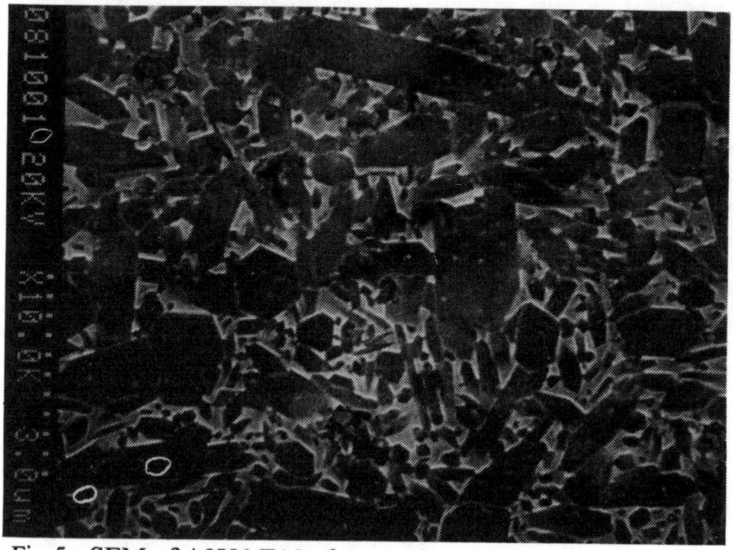

Fig 5. SEM of A2Y6-E10 after continuous sintering at 1750°C and 90 min. The microstructure contains mostly low aspect ratio β-Si_3N_4 grains.

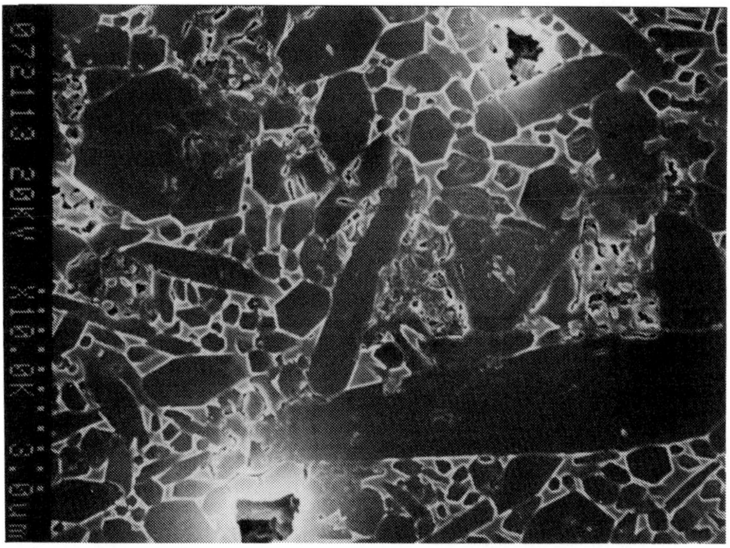

Fig 6. SEM of A2Y6-E10 containing 5 wt% β-Si_3N_4 seed after continuous sintering at 1750°C and 90 min. The microstructure shows much larger β-Si_3N_4 grains than observed in the unseeded formulation.

CHARACTERIZATION OF FAST-PULSE HOT-PRESSED SILICON-CARBIDE COMPOSITES

M.S. Boldrick and M.D. Matthews
Polaris Research, Inc.
P.O. Box 24
Saratoga, CA 95071

ABSTRACT

A method for the non-equilibrium consolidation of silicon-carbide composites containing an electrically-conductive phase was investigated. In this process, a fine powder mixture of silicon carbide and titanium nitride, titanium carbide, or titanium diboride was subjected to mechanical pressure and an intense current pulse with a duration ranging from 0.1 to 10 milliseconds. SEM examination showed that rapid melting occurred in the powder compacts with varying degrees of consolidation, and X-ray diffraction revealed the presence of significant amounts of new phases in some of the compacts.

INTRODUCTION

Superhard refractory materials have important applications in metal-shaping equipment and heat engines due to their high-temperature properties. However, these desirable engineering properties are a major obstacle to the fabrication of useful shapes. In addition, they suffer from low fracture toughness as compared to metals.

Conventional processing methods at or near equilibrium conditions are limited with respect to peak pressure and temperature capability. They are also very limited in their capability to produce heat-treatable ceramic composites. In this work, a non-equilibrium method for consolidation of silicon-carbide composites containing an electrically-conductive phase was studied [1]. A fine powder mixture is subjected

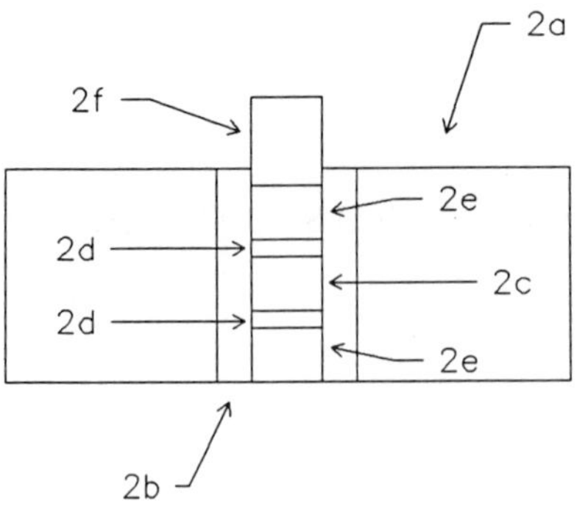

Figure 1. Sample chamber and assembly; 2a - compression ring, 2b- alumina sleeve, 2c - sample, 2d - inner spacer (tungsten), 2e - outer spacer (copper), 2f - piston.

to mechanical pressure and an intense current pulse with a duration that is typically less than 10 milliseconds. Under these conditions, the conductive phase is melted and alloyed with the silicon carbide phase with a minimum of heat loss to the containment structure. The rapid cooling of the compact is expected to result in a heat-treatable solid solution of extremely fine microstructure.

EXPERIMENTAL PROCEDURES

The starting powders for the powder compacts were all obtained from Johnson Matthey/Alfa (one micron beta silicon carbide; 2-5 micron titanium nitride; 2.5-4 micron titanium carbide; -325 mesh titanium diboride). The powders were weighed and mixed using a Glen Mills Turbula mixer/shaker to give a total mass that was five to ten times greater than sample mass. The sample powder charge was then weighed from the powder mixture with the remainder being reserved for characterization.

The sample material was then placed in a sample chamber (Figure 1) and surrounded by copper and tungsten spacers. After mounting in the press, a vacuum chamber was installed and the chamber was pumped to a pressure between 10^{-2} and 10^{-1} torr. At this point, hydraulic pressure was applied to the samples, ranging from 380 MPa (55,000 psi) to 1040 MPa (151,000 psi). Once the working pressure was achieved, a capacitor bank was discharged through the sample and press structure. The extremely large current pulse traveling through the sample results in resistive heating, and ideally, subsequent melting. The quasi-adiabatic heating coupled with the copper and tungsten spacers acting as local heat sinks provides rapid cooling of the sample. The details of the fast-pulse adiabatic heating can be found in reference [1]. After the capacitive discharge, the hydraulic pressure was released and the sample was removed from the chamber. Table I summarizes the composition and consolidation parameters of the six samples discussed in this paper.

The original powder and resulting compacts were examined in the scanning electron microscope to characterize the particle morphology before and after compaction and to look for evidence of melting and/or inter-particle bonding. X-ray diffraction was performed to determine the types and amounts of phases present in the compacts.

Table I. Sample composition and consolidation parameters.

	v/o SiC	v/o TiN	v/o TiC	v/o TiB_2	Mass (g)	$P_{applied}$ (MPa)	E_{input} (J)
Sample 1	90	10	-	-	0.223	100	2160
Sample 2	90	-	10	-	0.224	151	3502
Sample 3	90	-	-	10	0.221	100	2160
Sample 5	50	50	-	-	0.251	55	1575
Sample 6	75	25	-	-	0.243	64	2500
Sample 8	75	25	-	-	0.242	92	2500

The diffraction patterns were collected using Ni-filtered Cu-Kα radiation on a Norelco diffractometer equipped with a proportional counter. The compacts were ground into powder with a mortar and pestle and suspended in a plexiglass tray, and 2θ was scanned from 10 to 80°, corresponding to an S-range of 0.11 - 0.83 $Å^{-1}$. The relative diffracted intensity is plotted as a function of the diffraction vector $S = 2\sin\theta/\lambda = 1/d$, where λ is the wavelength of the radiation, θ is the diffraction angle, and d is the interplanar spacing. The diffraction vector is used in place of the more commonly used diffraction angle, 2θ, because it is independent of the radiation used, making it easy to compare diffraction scans collected with different radiations.

RESULTS AND DISCUSSION

Scanning electron microscopic examination of the samples found localized melting and inter-particle bonding in all six samples. However, there is no correlation between the compaction pressure ($P_{applied}$) or input energy (E_{input}) and the apparent extent of melting in the samples. In all cases, the vacuum chamber surrounding the sample chamber assembly was coated with a fine dust after the pressing and heating operation. The amount of dust observed varied from one run to the next and was greatest for runs in which cracking of the alumina sleeve surrounding the powder occurred. This fine dust appears to be some of the sample material which was blown past the spacers by the gas pressure developed during the rapid melting of the powder. Examples of the post-compaction particle morphology are shown in Figures 2 and 3. Figure 2 shows a large region of complete melting in Sample 6 (75 vol. % SiC - 25 vol. % TiN). The grain structure and some prior-particle boundaries are evident, but most of the voids between particles have been closed. In other regions of the sample, melting was not as complete, but there are sections

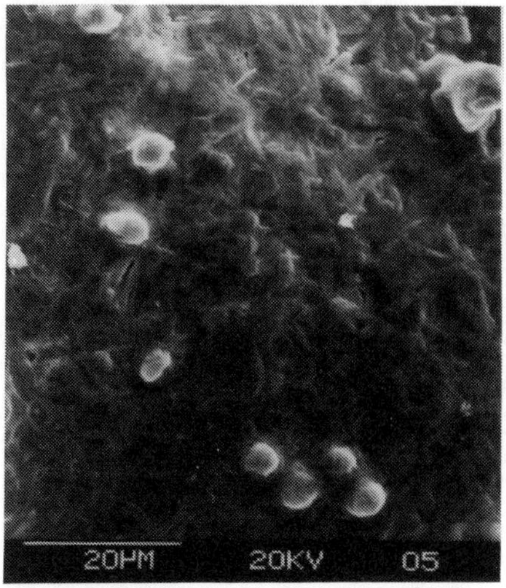

Figure 2. Large area of complete melting in Sample 6; subsurface grain structure and some prior-particle boundaries evident (1000X).

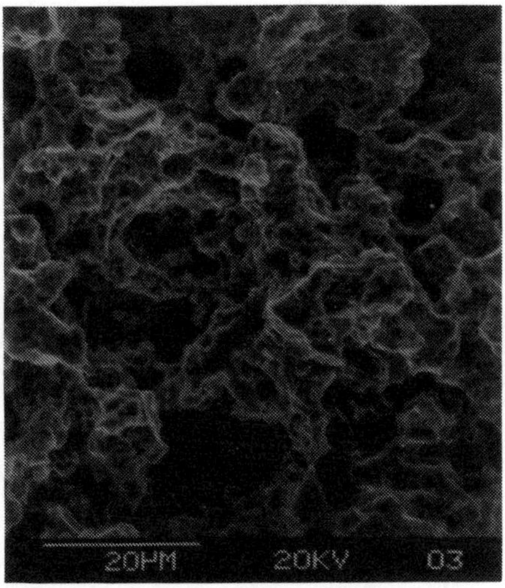

Figure 3. Extensive melting and inter-particle bonding in Sample 3 (10 vol. % TiB_2), although original particles are still largely evident (1000X).

as long as 500 µm in which almost all the inter-particle voids are closed. There is also a region in which numerous spherical particles, with diameters between 100 and 400 µm were formed. The surfaces of the particles appear solid, with only a few voids. The morphology varies significantly throughout the sample, which may indicate preferential current paths through the powder compact.

Samples 1 and 5, with 10 and 50 vol. % TiN, respectively, had a generally similar appearance to Sample 6. Specifically observed in these two samples was that the melting seemed to have been more extensive along the outer diameter of the samples. This would be consistent with current paths (arcing) through the cracked alumina sleeve. Less extensive melting was also observed in the center of Sample 2 (10 vol. % TiC). Inter-particle bonding was incomplete, and the particle morphology was quite similar to that of uncompacted SiC and TiC.

Figure 3 shows the characteristic morphology of Sample 3 (10 vol. % TiB_2). There is extensive melting and inter-particle bonding, but the original particles are still largely evident. The bonding is three-dimensional in appearance, whereas in other areas of the sample, the melting and inter-particle reactions appear to be more planar.

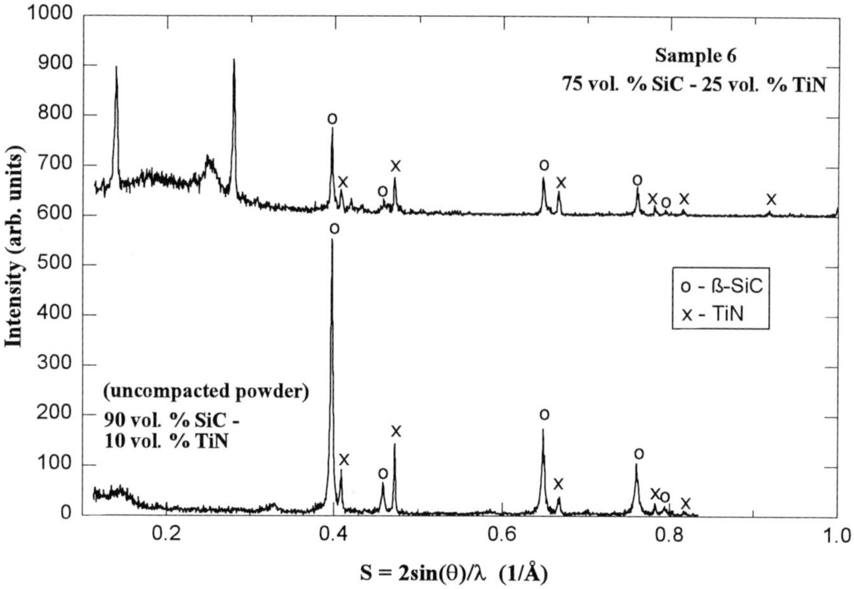

Figure 4. X-ray diffraction profiles of uncompacted 90 vol. % SiC - 10 vol. % TiN and compacted (Sample 6) 75 vol. % SiC - 25 vol. % TiN.

The X-ray diffraction profiles for Sample 6 (25 vol. % TiN) and uncompacted powder with 10 vol. % TiN are shown in Figure 4. Taking into account the different ratio of TiN to SiC, Sample 6 is quite similar in appearance to the uncompacted powder, with the exception of the two strong, sharp reflections at S=0.1394 and 0.2794 $Å^{-1}$ (interplanar spacing d=7.17 and 3.58Å, respectively), and the broad reflection at S=0.2525 $Å^{-1}$ (d=3.96Å). These additional peaks were not identified as to composition, however, based on the extensive melting in the samples, it is believed it is a reaction product and not contamination. This conclusion is further supported by earlier work performed on mixtures of TiC and B_4C, in which large amounts of TiB_2 were formed [2].

The diffraction profiles of uncompacted and compacted (Sample 2) 90 vol. % SiC - 10 vol. % TiC are shown in Figure 5. The relative amount of TiC is slightly decreased in the compact (Sample 2) relative to the starting powder. The strong reflections at S=0.1394 and 0.2794 $Å^{-1}$ observed in the 25 vol. % TiN sample (Sample 6, Figure 4) are present, but they are not as strong or as sharp. The broad reflection at S=0.2525 (d=3.96Å) is quite strong, as is the broad reflection at

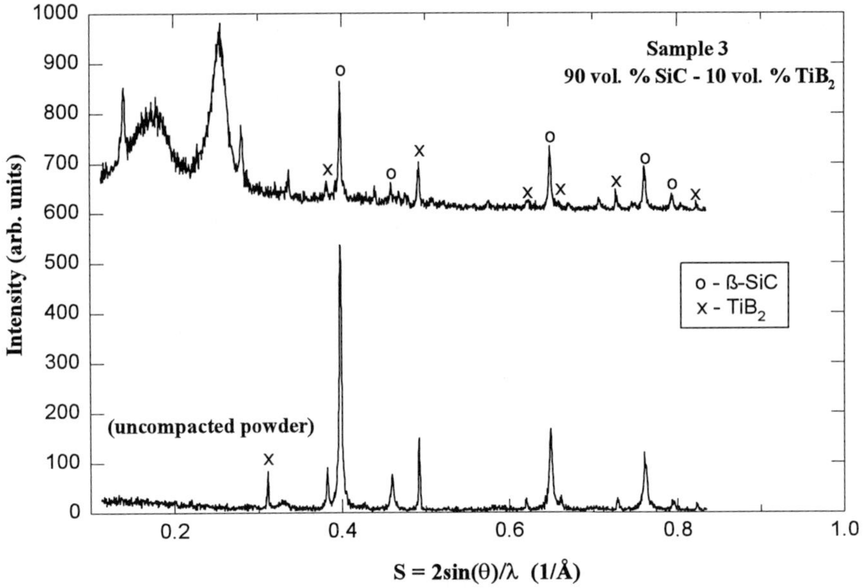

Figure 5. X-ray diffraction profiles of uncompacted and compacted (Sample 2) 90 vol. % SiC - 10 vol. % TiC.

S=0.175 Å^{-1} (d=5.71Å). There are also a number of small, unidentified reflections. The same unidentified major reflections are seen in Sample 3 (10 vol. % TiB_2, Figure 6).

The two unidentified sharp reflections at S=0.1394 and 0.2794 were observed to some extent in all the compacts. Therefore, it is believed that they must be from a reaction between Ti, Si and C. EDX performed on several samples showed only Si and Ti (B, C, and N are not detectable), ruling out reaction with the alumina sleeve or tungsten spacers. As mentioned previously, the formation of a new phase is not unfounded, as previous application of this consolidation process to the TiN/B_4C and TiC/B_4C systems produced large amounts of TiB_2 [2].

The loss of conductive phase relative to the dielectric phase in some of the samples (determined from X-ray diffraction results) can be explained by the 'blow-by' experienced in all experiments, where sample material (primarily the conductive phase) travels between the alumina sleeve and the copper and tungsten spacers. A frozen, silver-colored deposit was observed on the top side of the copper spacer in Sample 8. Although it has not been analyzed, it is expected that it contains titanium

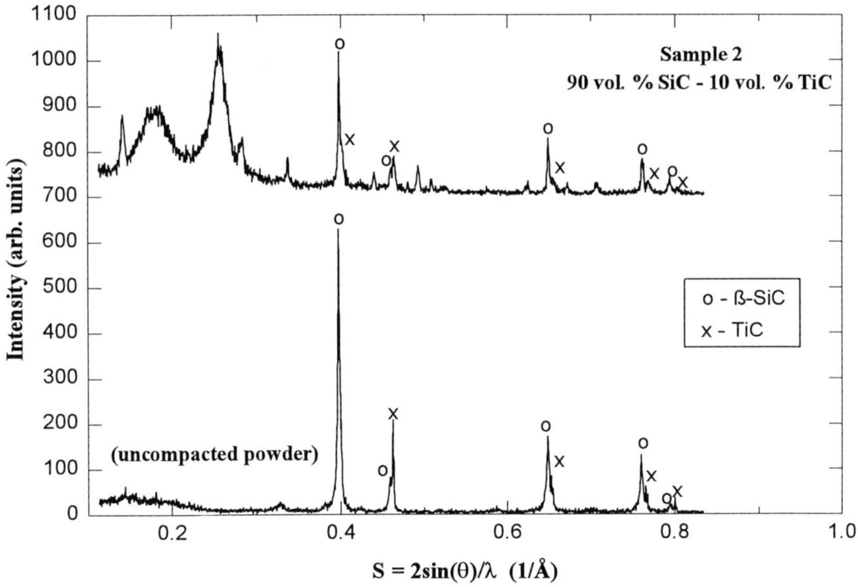

Figure 6. X-ray diffraction profiles of uncompacted and compacted (Sample 3) 90 vol. % SiC - 10 vol. % TiB_2.

nitride. The 'blow-by' of sample material also offers an explanation for the increased melting and density documented near the outer diameters of Samples 1, 5, and 6, as the outer diameter is the exit path. It is interesting to note that blow-by and loss of material even occurred in Sample 2, the sample with the highest applied pressure. In spite of the high applied pressure, the inertial and frictional characteristics of the assembly prevented the rapid follow-up from dynamically maintaining pressure on the sample and the edge seal of the copper spacer when the powder compact melted and collapsed.

CONCLUSIONS

Although fully-dense compacts were not produced, the extensive melting and chemical reactions observed with SEM and X-ray diffraction indicate that extremely fast heating and melting of refractory materials was achieved at pressures of up to 150,000 psi using the fast-pulse hot-pressing system. Based on the extensive melting in the samples (estimated at 30 - 50 percent), full densification is considered technically feasible; however, the mechanical and electrical properties of the fully-dense materials must still be determined. In addition since there is evidence of chemical reaction in the short time span involved in the process, there is a prospect of rapid combustion synthesis, under pressure, of appropriately selected materials. By initiating the combustion throughout the sample with a rapid current pulse and not relying on the propagation of a front, the peak temperature will be higher, allowing for better consolidation.

ACKNOWLEDGMENTS

This material is based upon work supported by the National Science Foundation under award number III-9260345.

REFERENCES

1. M.D. Matthews, "Electric Pulsed Power Vacuum Press," U.S. Patent No. 5,122,043 (6 June 1992).

2. M.S. Boldrick, M.D. Matthews, and I.A. Aksay, "Fast Pulse Hot Pressing of TiC/B_4C Composites," oral presentation at The 45th Pacific Coast Regional Meeting of the American Ceramic Society, San Francisco, CA (November 1992).

SILICON NITRIDE FIBRES SYNTHESISED BY A GAS-PHASE PROCESS

Ulrich Vogt, Jakob Kübler and Georg Engeli
Swiss Federal Laboratories for Materials Testing and Research (EMPA)
Überlandstr. 129, CH-8600 Dübendorf, Switzerland

ABSTRACT

A high temperature CVD-process has been developed to produce Si_3N_4-fibres for high temperature applications. The reaction $3SiO_2 + 4NH_3 \rightarrow Si_3N_4 + 6H_2O$ operates at temperatures above 1400°C by evaporation of SiO_2 to SiO and dissociation of NH_3 into active nitrogen and hydrogen atoms. The active nitrogen atoms react with unbounded Si atoms thus forming Si_3N_4 fibres. The growth process is mainly influenced by the reaction temperature, reaction time and the gas flow rate in the reaction chamber. Depending on the reaction parameters temperature and time, amorphous or crystalline fibres can be grown. The oxygen content of the fibres decreases with an increase of crystallinity. 29-Si-NMR analysis shows that the oxygen in the fibres is not present as silica, but in the composition of non-stoichiometric silicon oxynitride. The commencement of oxidation in air could be established at 1230°C by DTA/TG analysis. Annealing experiments in argon at 1500°C for 16 hours show no crystallisation or recrystallisation of the fibres. Tensile strength measurements of flexible amorphous fibres yield values up to 5 GPa.

1. INTRODUCTION

The development of new materials for high temperature applications requires a new generation of high performance ceramic fibres. Commercially available SiC fibres based on polycarbosilane precursor or vapour deposited SiC have a long service time in air only at temperatures below 1000°C. Above 1200°C they change their structure due to crystallisation or recrystallization. Therefore, a European research group (CEFIR) is calling for the development of a new fibre generation for applications up to 1400°C with a room temperature strength of 2.5-3 GPa. The Si_3N_4 fibre described in this paper should serve for applications up to 1200°C in air and at least 1400°C in non-oxidising atmosphere.
The formation of silicon nitride and silicon oxynitride powder from silica in a stream of ammonia has been described in several publications [1-6], the new developed gas-phase process from EMPA is described in detail in [1].
The decomposition of ammonia above 850°C is an assumption for this process. According to Bartnitskoya [2] at 1000°C-1500°C the gas mixture consists of

atomic and molecular nitrogen, hydrogen, as well as NH- and NH_2-radicals, produced by incomplete dissociation of ammonia. Atomic nitrogen and hydrogen which is formed by dissociation of ammonia has a much higher activity than the molecular species. These active hydrogen atoms dissolves the SiO-bonds in SiO_2 at high temperatures and the active nitrogen atoms react with free silicon atoms to silicon nitride. The dissociated oxygen is eliminated from the surface as a volatile species. It should be noted that the flowing system is a dynamic equilibrium, since some of the products are continuously being removed. The influence of substrate materials on the formation of silicon nitride has been described by Gray and Hendry [4]. They have grown Si_3N_4 whiskers on substrates of carbon and recrystallised alumina to determine their influence on product phase and morphology. Gribkov supposed that whiskers which grow from a vapour-gas phase may involve two deposition schemes: vapour-solid (VS) and vapour-liquid-solid (VLS) [5]. It is generally assumed that in the first case, condensation from the vapour phase directly into the solid phase (VS), whiskers grow by an axial-screw-dislocation mechanism corresponding to the Sears model [6]. In the second model (VLS) no screw dislocation is required.

2. EXPERIMENTAL RESULTS

2.1 Fibre Synthesis

For experimental work a tube furnace with an inner tube diameter of 43 mm and a tube length of 1 m was used. The influences of the reaction parameters temperature, time, gas flow rate, type of silica and metallic additives were tested. As substrate materials we used SiC crucibles, as metallic additives Ti, Fe, V and Cr which were mixed with the silica. In addition, three different silica precursors were tested. In order to reduce the high number of experiments which are necessary for the different parameters, a program for statistical design of experiments and statistical analysis by multiple regression was used (RS1).

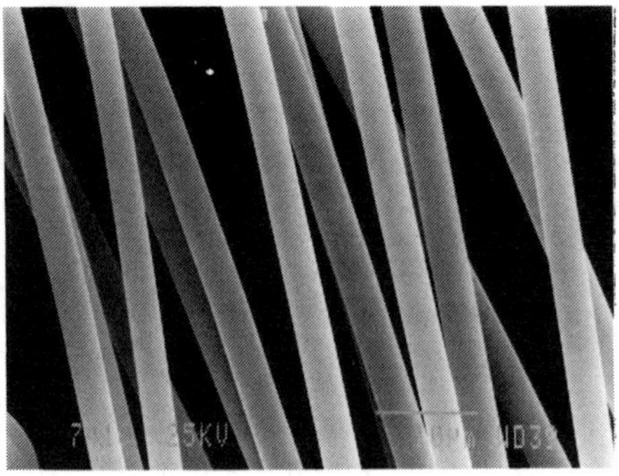

Figure 1
flexible amorphous fibres
—— 10 μm

Figure 2
crystalline fibres
—— 10 μm

At temperatures below 1400°C the reaction rate is very slow, making 1450°C the optimum temperature for the synthesis of amorphous fibres as shown in figure 1. At temperatures above 1500°C crystalline α-Si_3N_4 fibres can be grown (figure 2). The type of silica precursor as well as the type of metallic additive has no significant influence on the reaction. The main effects in response to a low oxygen content of the fibres can be achieved by an increase in reaction temperature, reaction time and gas flow rate. Low reaction temperatures and short reaction times lead to amorphous fibres with a high oxygen content. With an increase in reaction temperature and time the oxygen content diminishes (figure 3). Lower oxygen contents can also be achieved by a higher gas flow rates (figure 4). At high gas flow rates, an oxygen content of 4 wt% can be achieved after a reaction time of 4 h, whereas a low gas flow rate requires nearly 18 h at 1480°C reaction temperature. An increase of the α-Si_3N_4 phase can be accomplished mainly by a rise of the reaction temperature and time and this again coincides with a decrease in the oxygen content of the fibres.

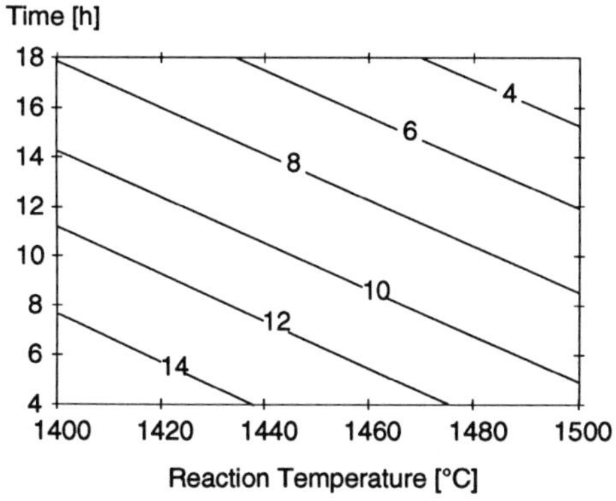

Figure 3:
Iso-lines of the oxygen content [wt%] of Si_3N_4 fibres at different reaction times and temperatures at low gas flow rates

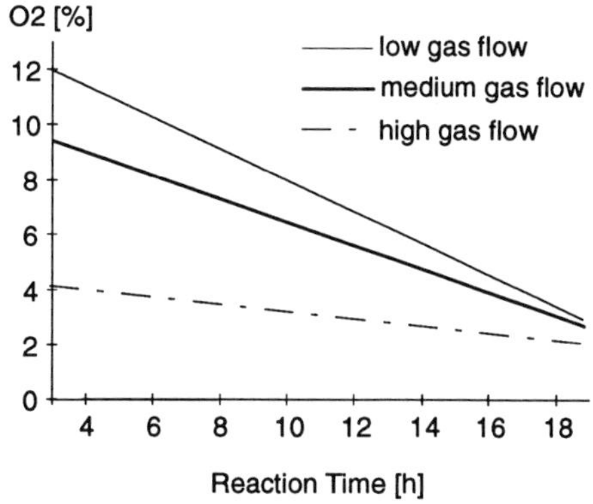

Figure 4:
Oxygen content of Si_3N_4 [wt%] fibres in relation to the reaction time and gas flow rates at 1480°C reaction temperature

2.2 Fibre Characterisation

The fibres were analysed by SEM, XRD and Si-29 MAS-NMR. The fibres were also characterised with respect to their nitrogen- and oxygen content with a LECO TC 463 analyser. The oxygen content of the fibres varies from 0.8 wt% up to 18 wt%, depending on the reaction parameters temperature, time and gas flow rate. Amorphous fibres have oxygen contents in the range of 5-15 wt%, while in crystalline or partially crystalline fibres oxygen contents between 0.8 wt% and 12 wt% were observed (figure 9).

2.2.1 Solid State NMR-Spectroscopy

29-Si Solid State NMR spectra were acquired with the Magic Angle Spinning technique. The instrument was a Bruker ASX 400 spectrometer (29-Si resonance at 79.49 MHz). The pulse sequence was a single pulse of 10°, followed by acquisition and 100 seconds relaxation-recycle delay. The samples were of 100-400 mg, the MAS rate 4000 Hz. Typically 180 pulses were sampled. NMR allows the detection and identification of the crystalline as well as the amorphous phase content of the fibres. But quantification of the phases would require extremely long measuring times because silicon nitride has a relaxation time of 43 minutes. Therefore we did not measure with full relaxation, but used a reference sample of known phase content. So it was possible to do an approximate calculation of the phase content of the Si_3N_4 fibres. The reference sample was made of three different crystalline phases: α–Si_3N_4 (UBE SN E-10), Si_2N_2O (research sample) and SiO_2 (quartz) (figure 5 D). Three different fibre samples were measured, with oxygen contents of 12.6 wt%, 6.5 wt% and 0.8 wt% (figure 5, A-C).
The fibres consist of α–Si_3N_4 with a chemical shift of -46.8 and -48.9 ppm and amorphous Si_3N_4 with a chemical shift of -46.4 ppm. The amount of amorphous Si_3N_4 increases with the oxygen content of the fibres. SiO_2 with a chemical shift of -110 ppm was not detected, even in fibres with an oxygen content of 12.6 wt%. Therefore we presume, that the oxygen is present as a silicon oxynitride phase with a composition of $Si_3N_{4-x}O_{1.5x}$. This phase was be detected at a chemical shift of -56.5 ppm (figure 5 A) and has been described as 'non-stoichiometric silicon oxynitride' by Carduner et al. [7]. Amorphous fibres contain more than 50 wt% of this phase, crystalline fibres only traces.

2.2.2 XRD Analysis

X-ray powder diffractograms were taken of ground fibres on a Siemens D500 diffractometer with $Cu_{K\alpha}$ radiation. XRD confirmed the results from NMR concerning the cristallinity of the fibres. The only crystalline phase that could be detected was α–Si_3N_4 (figure 6). Fibres which had been synthesised at temperatures of 1500°C and higher also contain traces of β–Si_3N_4. The crystallinity of the fibres increases with lower oxygen content. The non-stoichiometric phase which had been identified by NMR was not visible in XRD. This can be explained by the fact that a non-stoichiometric phase cannot have a defined crystal structure. No other phases such as Si_2N_2O, SiO_2 or a continuum of $Si_3N_{4-x}O_{1.5x}$ were visible by XRD, thus confirming the NMR results. The peak intensity distribution of crystalline α–Si_3N_4 fibres was different from UBE α–Si_3N_4 powder (figure 6, C-D). This is due to the morphology of the fibres. From SEM images it was observed that the fibres are of rectangular cross-section and cannot be ground to homogenous particles during XRD sample preparation. The flatness of the fibres prevents them from random orientation but favours orientation in the direction of the sample holder. Therefore all hkl-lines with $l \neq 0$ are of reduced intensity.

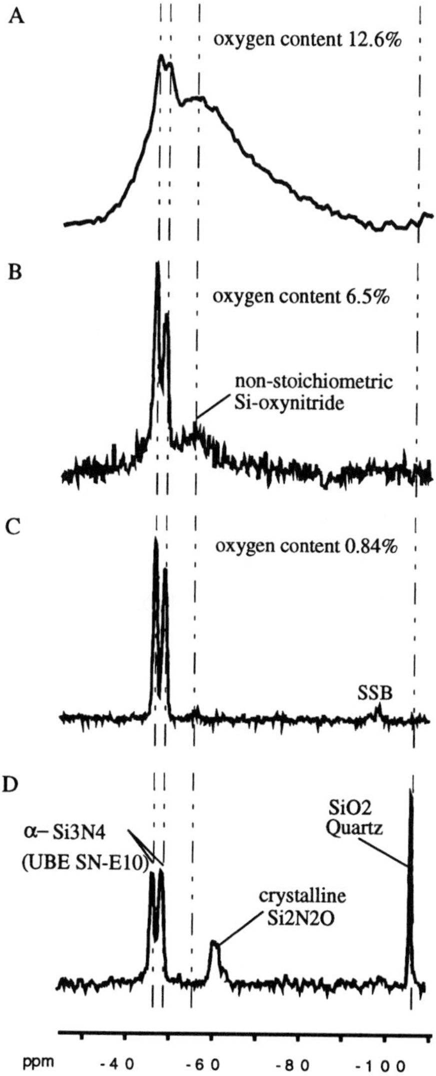

Figure 5: NMR spectroscopy

A: amorphous fibres
B: partially crystalline fibres
C: crystalline fibres
D: reference sample consisting of Si_3N_4 (UBE SN E-10), Si_2N_2O and quartz powder

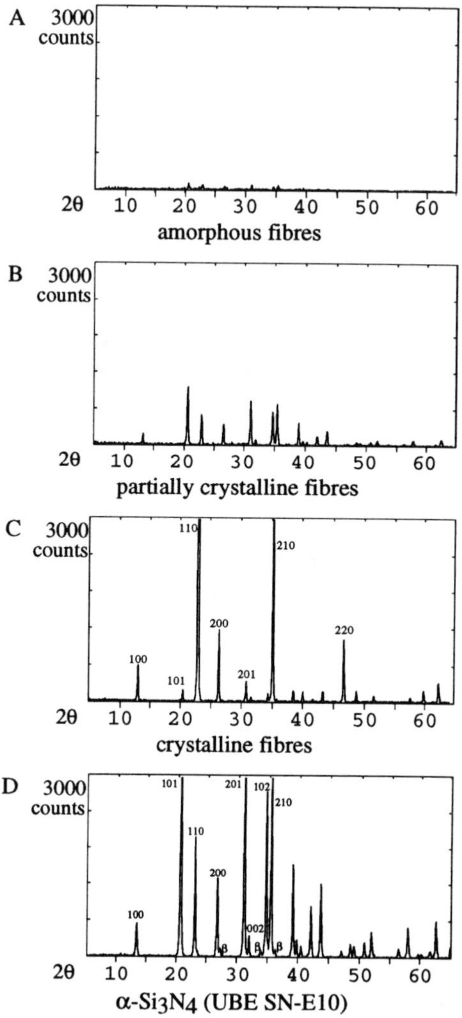

Figure 6: Powder X-ray diffraction

A: amorphous fibres
B: partially crystalline fibres
C: crystalline fibres
D: reference sample Si_3N_4 powder (UBE SN E-10),

2.2.3 Thermal Annealing

Amorphous Si_3N_4 fibres were heated in air and argon for investigations concerning their high temperature behaviour. To determine the start of oxidation, DTA/TG experiments were carried out with amorphous fibres in air up to 1500°C in a Netzsch STA 409. At temperatures up to 1200°C there is only a small increase in weight, active oxidation starts at 1230°C. After annealing in air for 16 hours at 1000°C and 1200°C the fibres show no visible change of the surface structure (figure 7) and XRD proves them to be amorphous. After annealing in argon at 1500°C for 16 hours there is also no visible change of the smooth fibre surface and the fibres are still amorphous (figure 8), no crystallisation occurs.

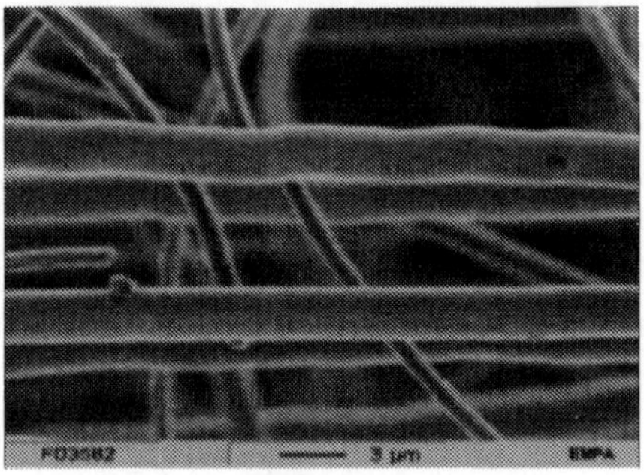

Figure 7:
Fibre surface after annealing in air at 1200°C for 16 h

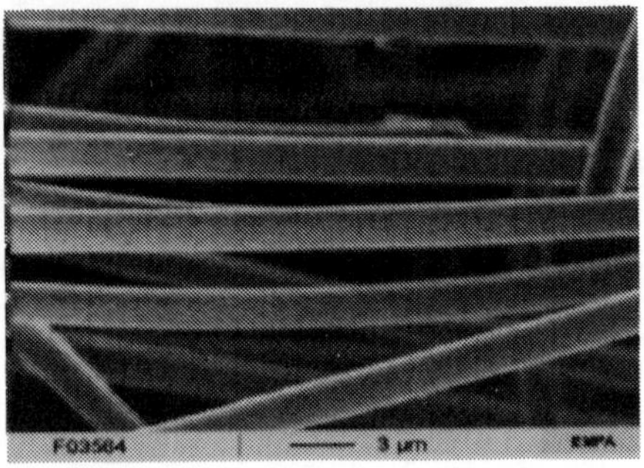

Figure 8:
Fibre surface after annealing in argon at 1500°C for 16 h

2.3 Mechanical Strength Measurements

For room temperature tensile tests, we used a screw-driven universal tensile testing machine with a special measure cell unit. The tests were carried out in accordance with Standard ENV 1007-4, JIS R 7601 and ASTM D 3379. The fibres were glued onto paper holders with a commercially available rubber cement and have a gauge length of 10 mm. For the tensile tests the paper holders were fixed in the test apparatus where the force at failure was measured. The fibre fracture was investigated by SEM and the diameter measured in order to calculate the tensile strength.
The diagram in figure 10 shows a large spread in strength of the measured fibres. It is an average of fibres from different production lots which are different in surface morphologie as well as fibre diameter. Nevertheless, the trend of the tensile strength data for the measured Si_3N_4 fibres can be seen clearly. Fibres with diameters of 60 μm show values of 1 GPa, while fibres with diameters of 15 μm can reach the high tensile strength of 5 GPa. Only amorphous fibres without surface damage as shown in figure 1 reach the high "upper strength limit". Fibres with visible surface defects like bends, kinks or even intergrowth only reach low values.

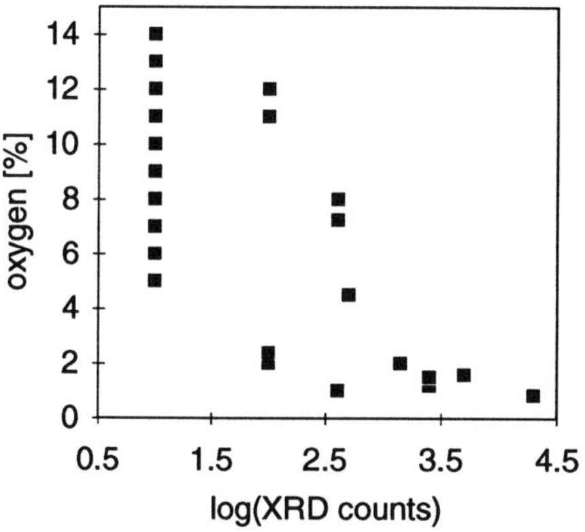

Figure 9
Correlation between oxygen content and crystallinity of Si_3N_4 fibres (background noise 10 counts)

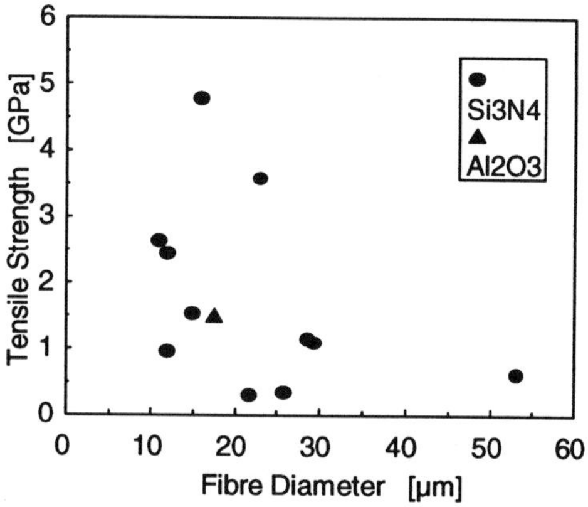

Figure 10
Tensile strength vs. Si_3N_4 fibre diameter at room temperature.
For comparison a Sumitomo Al_2O_3 fibre

3. DISCUSSION

The formation of Si_3N_4 fibres can be controlled mainly by the reaction parameters temperature, time and ammonia gas flow rate. Below 1400°C the reaction is very slow. Up to a temperature of 1500°C the synthesised fibres contain crystalline Si_3N_4, amorphous Si_3N_4 and non-stoichiometric silicon oxynitride of the composition $Si_3N_{4-x}O_{1.5x}$ with x<<1. The amount of crystalline Si_3N_4 increases with the reaction-temperature and -time, above 1500°C the fibres consist mainly of crystalline α-Si_3N_4. The non-stoichiometric oxynitride phase is completely amorphous and could therefore be identified only by NMR and not by XRD. Increasing ammonia gas flow rate coincides with an increase in nitridation rate and a lower oxygen content because of the higher reduction potential due to the hydrogen formed. It should be noted that the flowing system is in dynamic equilibrium since some of the products are continuously being removed. Metallic additives such as Fe, Cr, V or Ti, which are common catalysts for the ammonia synthesis, also have a positive influence on the fibre growth. It is believed that the catalytic effect of metals on the decomposition of NH_3 also enhances the fibre growth rate. The SiC substrates promote the fibre growth. According to Gray and Hendry [4], a nucleation process on a suitable substrate is assumed, with a heterogeneous nucleation from the gas phase on the substrate.
In air the fibres remain intact up to 1200°C with no surface defects or crystallisation, but a slow increase in oxygen content. Active oxidation in air starts at 1230°C. In argon the fibres are stable up to 1500°C without crystallisation during 16 hours annealing time.
Investigations concerning the room temperature strength show the potential of the fibres. A representative average of fibres were tested. Fibres without defects and

diameters of 30 μm achieve the strength goal for the new fibre generation of 2.5 to 3 GPa, fibres with diameters of 15 μm achieve strength values up to 5 GPa. The production of amorphous weavable fibres with diameters of 5-15 μm and room temperature strength of 5-6 GPa will be the goal of the ensuing research activities. Investigations with focus on the growth mechanism and the mechanical behaviour at room- and elevated temperatures will be carried out in the future.

4. CONCLUSION

Si_3N_4 fibres can be produced from silica and ammonia via a gas phase reaction at elevated temperatures. Important parameters are reaction temperature, reaction time, ammonia gas velocity, substrate materials and metallic additives. Depending on the reaction parameters the fibres grow amorphous or crystalline. Amorphous fibres have oxygen contents between 5-15 wt%. They consist of amorphous Si_3N_4 and a non-stoichiometric silicon oxynitride phase. Crystalline fibres can contain less than 1 wt% oxygen and consist of α-Si_3N_4 with traces of β–Si_3N_4. Amorphous fibres with no visible defects can reach strength values up to 5 GPa.

REFERENCES

1) U. VOGT, H. HOFMANN AND V. KRÄMER, Synthesis of Si_3N_4 Fibres by a Gas-Phase Process; Key Engineering Materials Vol. 89-91 (1994) 29-34
2) T. S. BARTNITSKOYA, P. P. PIKUZA, E. S. LUGOVSKAYA and T. YA. KOSOLAPOVA, Formation of Silicon Nitride Fibres from Silicon Oxide in a Stream of Ammonia; Translated from Izvestiya Akademii Nauk, SSSR, Neorganicheskie, Materialy, **18**, No. 10 (1982) 1733-35.
3) M. HOCH and K. M. NAIR, Preparation and Characterization of Ultrafine Powders of Refractory Nitrides: I, AlN and Si_3N_4; Ceram. Bull., 58 (1979) 187-190
4) D. GRAY and A. HENDRY, The Influence of Substrate Material on the Formation of Silicon Nitride; Journal of the European Ceramic Society, **10** (1992) 75-82.
5) V. N. GRIBKOV, V. A. SILAEV, B. V. SHCHETANOV, E. L. UMANTSEV and A. S. ISAIKIN, Growth Mechanism of Silicon Nitride Whiskers; Soviet Physics-Crystallography, **16**, No.5 (1972) 852-854
6) G. W. SEARS, A Growth Mechanism for Mercury Whiskers; Acta Metallurgica, **3** (1955)
7) K. R. CARDUNER, R. O. CARTER, M. E. MILBERG, G. M. CROSBIE, Determination of Phase Composition of Silicon Nitride Powders by 29-Si MAS-NMR, Anal. Chem., **59**, (1987), 2794-2797

NOVEL DIAMOND/CARBON/CARBON COMPOSITES FOR HEAT DISSIPATION IN ELECTRONIC DEVICES

Jyh-Ming Ting, A. G. Lagounov, and Max L. Lake
Applied Sciences, Inc.
Cedarville, OH 45314

ABSTRACT - We have investigated a novel composite material consisting of carbon/carbon composite deposited by polycrystalline diamond useful as an integral dielectric heat sink. C/C composite specimens were deposited with diamond by a microwave plasma enhanced chemical vapor deposition (MPECVD) technique. Physical properties and microstructure of the resulting composite were examined. Effect of temperature on diamond coating morphology and growth rate on C/C composite specimens was investigated, and an activation energy of 34.4 KJ/mol was determined in a temperature range of 860 °C to 1060 °C.

INTRODUCTION

High performance thermal materials are required for future electronic devices. Predictions based on present chip design indicate that heat flux levels in ICs will reach 100 W/cm^2 at the chip level, and 25-50 W/cm^2 at the module level by the end of this decade. [1] [2] A need for a significant level of research and development in thermal management materials for advanced packaging technologies has therefore been identified. In this paper, an innovative integral dielectric heat sink material is described. The novel material, DCC, consists of carbon/carbon composite deposited by polycrystalline diamond. DCC material exhibits a thermal conductivity much greater than any other materials, except free-standing diamond. The diamond deposit serves as a very high thermal conductivity dielectric coating for the electrically conductive C/C composite. Due to its composite nature, the material provides tailorable coefficient of thermal expansion and better mechanical properties than monolithic materials. In this paper, diamond deposition into two types of C/C composites was investigated, and properties of the resulting composites were examined.

EXPERIMENTAL

Two types of carbon/carbon (C/C) composites were used as substrates. The first type is a unidirectional (1D) vapor grown carbon fiber (VGCF) reinforced carbon (VGCF/C) composite prepared in our laboratory. [3] The second type is called AL-C composite, which was provided by BFGoodrich Aerospace (Santa Fe Spring, California, U.S.A.). This type of composite is produced by densifying three-directional (3D) carbon fiber preform using a combination of a pitch infiltration technique and a chemical vapor infiltration technique. For diamond deposition, the C/C composite specimens were subjected to a microwave plasma enhanced chemical vapor deposition (MPECVD) reactor, as described elsewhere. [4] Specimens were characterized for thermal conductivity and coefficient of thermal expansion (CTE), [5] and analyzed by scanning electron microscopy (SEM) and micro-Raman spectroscopy.

RESULTS AND DISCUSSION

C/C Composites

Thermal conductivities and CTE of the 1D VGCF/C and the 3D AL-C composites are given in Table I. VGCF/C composites have very high thermal conductivities in the direction of fiber reinforcement (X-axis) due to the fact that VGCF exhibits the highest thermal conductivity (1950 W/m-K at room temperature) among known carbon fibers. [6] In contrast, the AL-C composites exhibit moderate, but isotropic, thermal conductivity.

DCC Material

Diamond deposition was first performed on VGCF/C composite specimens. The morphology of the diamond deposit is shown in Fig. 1. The diamond deposit is free of pin-hole and crack. The deposition conditions were H_2/CH_4 = 99.5 sccm/0.5 sccm, temperature = 950 °C, and deposition time = 20 hours.

Table I. Characteristics of Substrate C/C Composites. (V_f = Fiber Volume Fraction, κ = Thermal Conductivity.)

Substrate	Fiber	V_f (%)	Density (g/cc)	κ (W/m-K)	CTE (ppm/K)
92.14HT	1D VGCF	58	1.69	824(X)	-1.41
				89(Y)	0.71
				24(Z)	-
AL-C[a]	3D	<30	1.80	234(X)	1.49
				202(Y)	1.41
				170(Z)	1.54
93.01HT	1D VGCF	70	1.88	910(X)	-
				84(Y)	-
				33(Z)	-

As mentioned above, the CTE of DCC material can be engineered. The nature of the composite substrate of DCC allows one to obtain desirable CTE by varying the fiber architecture and densification protocol. Examples are shown in Table I. In addition, diamond deposition on C/C composite further modifies the CTE of resulting material, i.e. DCC. Table II compares the CTE of VGCF/C composite specimens before and after diamond deposition. The CTE measurements were conducted near 100 °C along the fiber direction. It is clear that diamond deposition alters the C/C composite CTE. This is due to a fact that the diamond partially infiltrates into the C/C composite specimens as observed using SEM. The CTE specimens had dimensions of 2" x 1/4" x 1.4", and all the surfaces were deposited by diamond.

Diamond deposition was then performed on AL-C composite specimens. The diamond deposition was conducted at a gas mixture of H_2/CH_4 = 99.5 sccm/0.5

---

[a] Specimens and property data of AL-C C/C composite were provided by BFGoodrich Aerospace, Santa Fe Springs, CA.

sccm and a deposition time of 10 hours under various temperatures, 860±5 °C, 900±5 °C, 940±5 °C, 980±5 °C, 1020±5 °C, and 1060±5 °C. Figure 2 shows the microstructure of the diamond deposits. Faceted diamond is seen in all the specimens. It appears that effect of processing temperature is mainly on the size of diamond crystalline, and to a lower degree, on the crystal orientation. Higher processing temperatures result in larger diamond crystals due to, apparently, higher growth rates. The diamond deposit obtained at 1060±5 °C shows preferred crystal orientation of (111).

Table II. CTE, Near 100 °C, of Uni-directional VGCF/C Composites Before and After Diamond Deposition.

ID	V_f (%)	Before (ppm/K)	After (ppm/K)
1	45%	-0.23	0.17
2	28%	0.95	0.72

To determine the deposition rates, cross sections of these specimens were examined. Typical cross sectional views are shown in Fig. 3. The thickness of diamond grown at each temperature was determined by measuring at least 30 different sites on each specimen. The average thicknesses were then used to generate an Arrhenius plot as shown in Fig. 4. The slope of the line shown in Fig. 4 was determined to be -1.80×10^3. Based on this, the activation energy of diamond growth on C/C or AL-C composite at a temperature range of 860 °C to 1060 °C was calculated to be 34.4 KJ/mol. The value is lower than previously reported data for diamond growth on Si at lower temperature ranges. Maeda et al. reported a value of 43 KJ/mol at a temperature range of 640 °C to 950 °C. [7] Kamo et al. reported 45-47 KJ/mol at a temperature range of 600 °C to 800 °C. [8] Kweon et al. reported 46 KJ/mol below 900 °C. [9]

The quality of diamond deposited on C/C composite was evaluated using micro-Raman microscopy. The characteristic peak, at 1332 cm^{-1}, of diamond is clearly seen in Fig. 5. The high quality diamond deposit on C/C composite serves as a dielectric coating for the electrically conductive C/C composite. [10]

CONCLUSION

A novel integral dielectric heat sink material, i.e. DCC, with ultra high thermal conductivity has been described. The material consists of carbon/carbon composite partially infiltrated with high quality polycrystalline diamond. Effect of temperature on diamond growth and coating morphology on carbon/carbon composite was examined. An activation energy of 34.4 KJ/mol was determined for diamond growth on carbon/carbon composite in a temperature range of 860 °C to 1060 °C.

ACKNOWLEDGEMENT

The authors express their appreciation to W. Shih of BFGoodrich Aerospace, Santa Fe Spring, CA for the supply of AL-C composite. This work was supported by the Aero Propulsion and Power Directorate of Wright Laboratory, United States Air Force Systems Command, Wright-Patterson AFB OH 455433-6563 under Contract Number F33615-92-C-2240.

REFERENCES

1. M. Kokado, M. Yoshida, N. Miyoshi, K. Suzuki, M. Takaika, N. Tsuzuki, and H. Harada, IEEE J. of Solid State Circuits, **24** [5] 1271 (1989).
2. A. Goyal, R.C. Jaeger, S.H. Bhavnani, C.D. Ellis, N.K. Phadke, M. Azimi-Rashti, and J.S. Gooding, p.25, Proc. Semi-Therm 1992, Austin, TX, February, 1992.
3. J.-M. Ting and M.L. Lake, J. Nuclear Mat., 212-215 (1994) 1141-1145.
4. J.-M. Ting, M.L. Lake, and D.C. Ingram, Diamond and Related Materials, **2** [5-7], p. 1069 (1993).
5. R.E. Taylor, Thermophysical Properties Research Laboratory Report # 181A, School of Mechanical Engineering, Purdue University, July, 1985.
6. Heremann and Beetz, Phys. Rev. B. **32** [4] 1981 (1985).
7. H. Maeda, M. Irie, T. Hino, K. Kusakabe, and S. Morooka, Diamond & Relat. Mat., **3** [7] 1072 (1994).
8. M. Kamo, T. Ando, Y. Sato, K. Bando, and J. Ishikawa, Diamond & Relat. Mat., **1** [2-4] 104 (1992).
9. D.W. Kweon, J.Y. Lee, and D. Kim, J. Appl. Phys. Lett., 59 (1991) 488.
10. J.-M. Ting and M.L. Lake, Diamond & Relat. Mat., **3** [10] 1243 (1994).

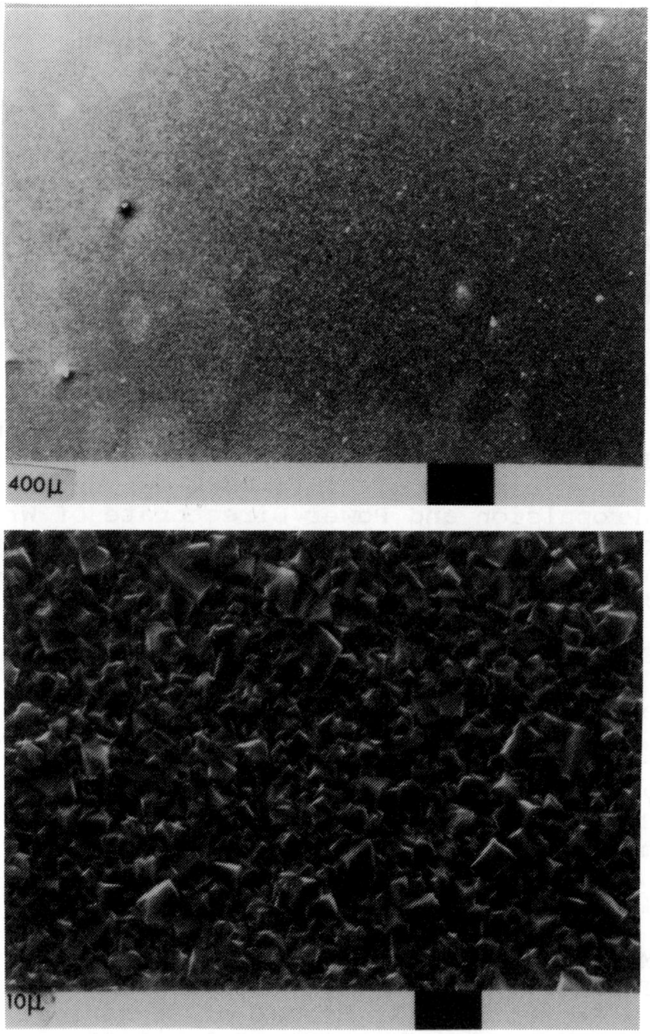

(A)

(B)

Fig. 1. Morphology of the diamond deposit on VGCF/C composite. The deposition conditions were H_2/CH_4 = 99.5 sccm/0.5 sccm, temperature = 950 °C, and deposition time = 20 hours. (A) Crack-free deposit and (B) pin-hole-free deposit.

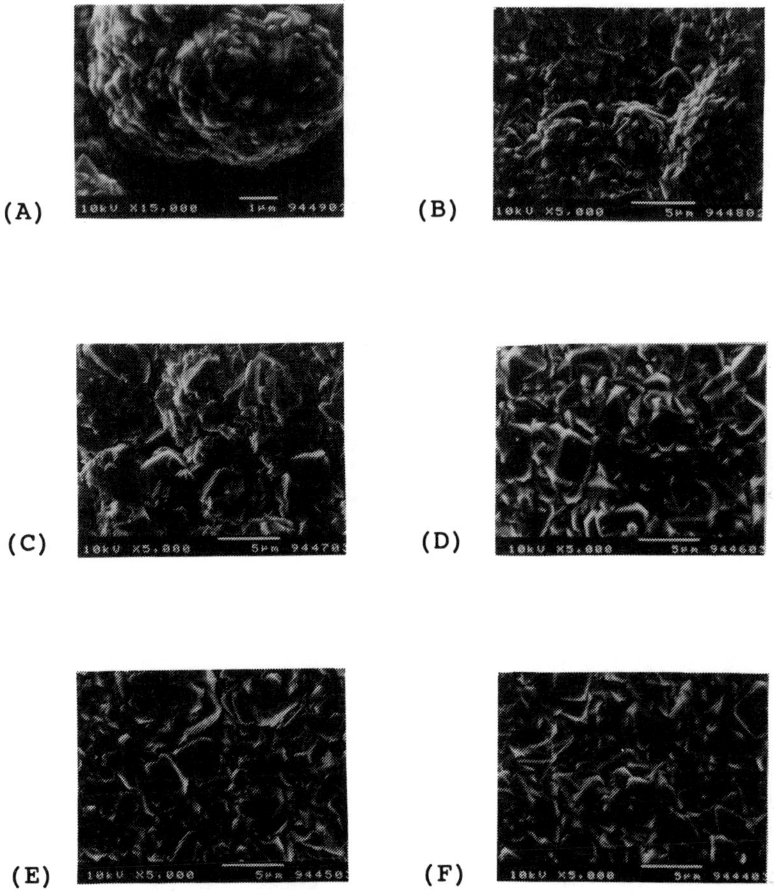

Fig. 2. Morphologies of diamond deposits on AL-C composite specimens. The deposition conditions were H_2/CH_4 = 99.5 sccm/0.5 sccm, deposition time = 10 hours, and temperature = (A) 860±5 °C, (B) 900±5 °C, (C) 940±5 °C, (D) 980±5 °C, (E) 1020±5 °C, and (F) 1060±5 °C.

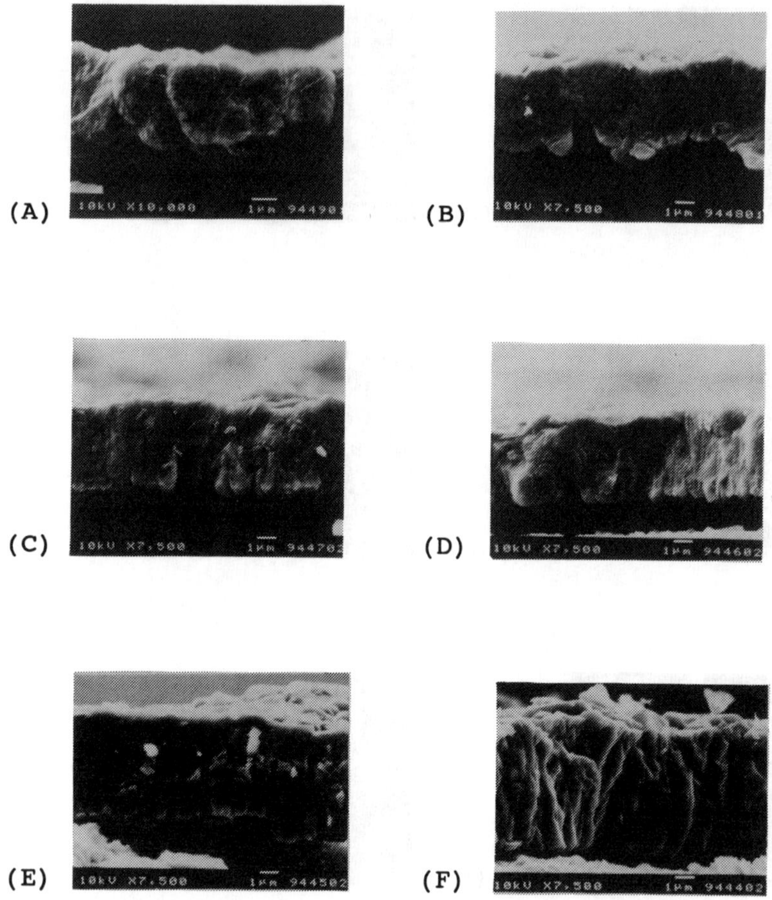

Fig. 3. Cross sections of diamond deposits on AL-C composite specimens. The deposition conditions were H_2/CH_4 = 99.5 sccm/0.5 sccm, deposition time = 10 hours, and temperature = (A) 860±5 °C, (B) 900±5 °C, (C) 940±5 °C, (D) 980±5 °C, (E) 1020±5 °C, and (F) 1060±5 °C.

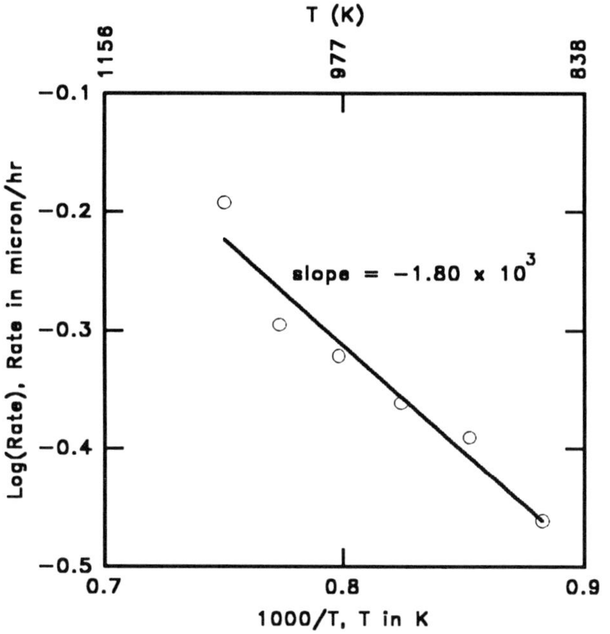

Fig. 4. An Arrhenius plot of diamond growth on C/C composite substrates.

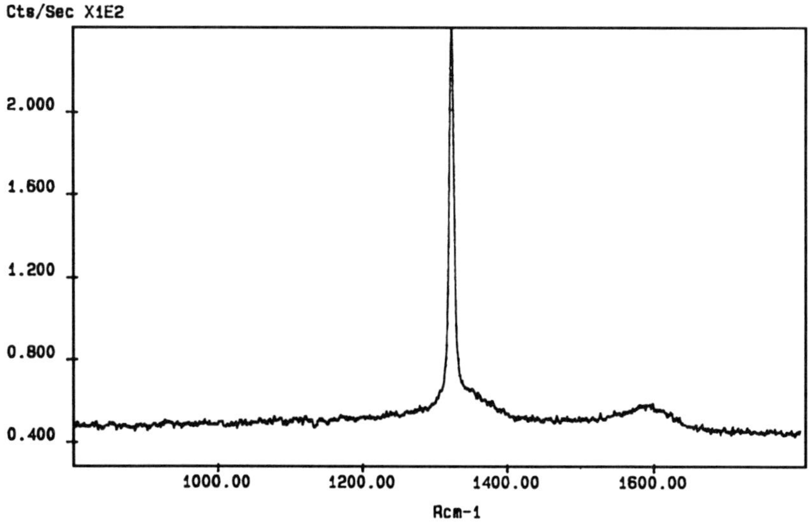

Fig. 5. Micro-Raman spectra of diamond deposited on C/C composite substrate.

A COMPARISON OF THE PROCESSING CONDITIONS FOR THE FABRICATION OF SiC/SiO_2 AND Al_2O_3/Al_2O_3 WOVEN FIBRE REINFORCED COMPOSITES

P. A. Trusty*, T. J. Illston, E. G. Butler, P. M. Marquis and C. B. Ponton
Interdisciplinary Research Centre in Materials for High Performance Applications, The University of Birmingham, Edgbaston, Birmingham, U.K.

ABSTRACT

The fabrication routes for a woven SiC fibre reinforced SiO_2 matrix composite and a woven alumina fibre reinforced alumina matrix composite have been investigated. The woven fibre preforms were infiltrated with the appropriate precursor sol systems using a novel electrophoretic deposition technique. The formation of green body compacts from these infiltrated woven fibre mats and the subsequent high temperature consolidation treatments will be described for the two systems. Microstructural examination, density measurements and mechanical testing will highlight the importance of the green body infiltration stages in the processing of woven fibre reinforced ceramic matrix composites.

INTRODUCTION

The need for ceramic materials in engineering applications has led to much work on the fabrication of ceramic matrix composites (CMCs). The majority of the fibre reinforced ceramic systems for high performance applications, however, have contained unidirectional fibres. These fibres may impart improved toughness or strength to the composite over that of the matrix material, but the resultant composite properties are invariably anisotropic. The use of a woven fibre mat as the reinforcement is more applicable to the fabrication of shaped components and will produce composites which have more isotropic properties. The problem of infiltrating the woven fibre inter/intra tow regions with matrix material, however, becomes more accentuated also. Matrix infiltration techniques such as chemical vapour infiltration, have been used widely, but invariably require long processing times and are costly. The novel technique of electrophoretic deposition (EPD), however, has been found to be a rapid, less costly, technique for the infiltration of woven ceramic fibre preforms(1).

The processing route used to fabricate woven fibre reinforced CMCs using EPD will depend on several factors including the conductivity of the fibre surface, the fibre weave architecture and the high temperature behaviour of the matrix material(2). This study aims to highlight the differences in the processing routes for a plain woven SiC fibre (Nicalon 607, Nippon Carbon Co., Japan) in a silica matrix and a plain woven alumina fibre (Almax, Mitsui Mining Co., Japan) in an alumina matrix when using EPD as the initial infiltration technique.

ELECTROPHORETIC DEPOSITION

If a particle in suspension is considered, the particle/suspension interface usually becomes charged. The sign and magnitude of this charge is controlled by the properties of the two materials. The charge present on the surface of the particle will relax to neutrality at a certain distance away from the particle into the suspending liquid. If this distance is large compared to the radius of the particle, then the particle, when placed in an electric field, will move with respect to the suspending liquid. Particles will move in a direction opposite or parallel to the external current, depending on their charge, and deposit on either the anode or the cathode. This phenomenon is known as electrophoresis and is a process which dates back to as early as the 17th century. It is observed generally in colloidal materials with sub-micron particles.

Considering a ceramic sol particle in aqueous suspension, the movement of the particle in an external field will be governed mainly by such factors as the magnitude of the applied field and the pH of the solution. At higher voltages than about 2 V, oxygen and hydrogen are evolved at the electrodes due to the electrolytic decomposition of the aqueous medium. During EPD, these gas bubbles may become entrapped between the depositing sol particles leading to porosity in the final compact (Figure 1). The problem of gas evolution can be addressed, however, if an organic suspending medium is employed. Voltages in excess of 100 V have been used successfully in the EPD of alumina sols in organic media[3,4] but with reduced deposition rates and poor green strengths of resultant compacts.

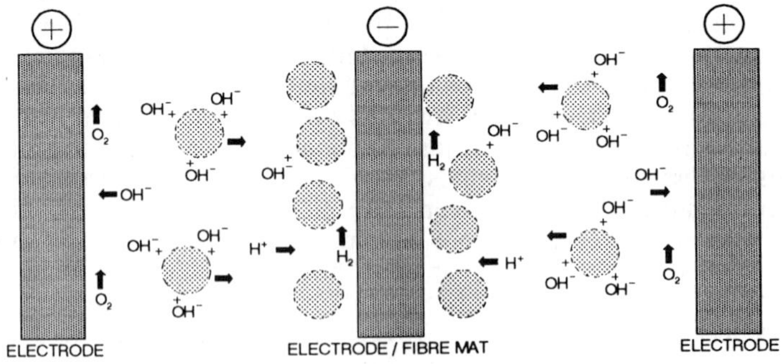

Figure 1: Electrophoretic Deposition Cell for an Alumina Sol in Aqueous

Higher voltages have been employed in the EPD of aqueous media by Clasen[5]. He formed dense silica compacts with an applied voltage of 45 V by using a membrane filter method. This membrane physically separated the sol particles from the oxygen bubbles evolved at the cathode. A dense compact was deposited onto the filter, therefore, which did not contain entrapped oxygen.

INFILTRATION OF WOVEN FIBRE PREFORMS

The particular grade of Nicalon fibres used in this work possess a surface layer of carbon (≈100 nm thick) which enables them to carry a charge. It is possible, therefore, for these mats to be used as one of the electrodes in an EPD cell. Infiltration of the Nicalon mats was performed using a commercially available silica sol (Syton HT50, Monsanto plc, particle size 20 nm) with additions of alumina and boria to suppress crystallisation and decrease viscosity respectively. The fibre mat was used as the positive electrode and a stainless steel plate as the negative electrode. A voltage of 4 V was then applied to the electrodes which caused the silica particles to migrate towards, and deposit onto the Nicalon fibre mat. A time period of three minutes was found to be sufficient to infiltrate fully the intra tow regions with the sol. The infiltrated tows were then dried slowly in a humid atmosphere to minimise drying cracks. Six dried compacts were then stacked and bonded with interlayers of a Syton-based slurry containing a small amount of binder. This configuration, which contained a volume fraction of 33% fibres, was then left to dry under load prior to high temperature consolidation.

The infiltration of Almax mats was performed using an alumina sol (Remal A20, Remet Corporation, particle size 50 nm). When Remal is used in an electrophoretic cell, the positive charge on the alumina particles causes them to migrate towards, and deposit onto the anode. In order to utilise this particle migration for a non conducting fibre such as Almax, a modification of the method used by Clasen[5], known as electrophoretic filtration deposition (EFD), was employed. The stages in EFD are shown in Figure 2. Both electrodes in the EFD cell were made from stainless steel and immersed in the Remal sol. A filter membrane (Whatman, pore size 0.1 μm) was placed adjacent to the anode prior to the application of a voltage across the system. On applying a voltage of 60 V,

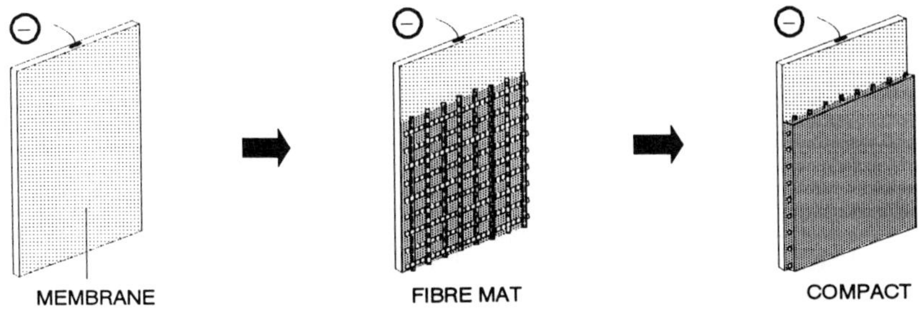

Figure 2: The Sequence of Events for Electrophoretic Filtration Deposition

alumina sol particles migrated and deposited onto the membrane from one direction only, through the fibre mat until a sufficient matrix thickness which envelops the mat was achieved. The high voltage involved caused hydrogen evolution at the anode but the gas was prevented from becoming part of the compact by the presence of the filter membrane.

Even with the higher voltages employed in EFD, the optimum infiltration time was found to be similar to that of Nicalon. The infiltrated mats were dried in the same way as the Nicalon mats and were then stacked and infiltrated with an alumina slurry (RA207, Alcan UK, particle size 0.4 μm) using pressure infiltration. It was envisaged that any cracks still present in the dried infiltrated fibre mats would be filled forcibly with slurry due to the applied pressure in this process. The resultant multi fibre layered composite compact, which contained a volume fraction of 10% fibres, was then dried slowly prior to high temperature consolidation.

COMPOSITE FORMATION

The Nicalon/silica compacts were hot pressed using a radio frequency induction coil heated hot press. At a temperature of 1300 ^{o}C, a pressure of 25 MPa was applied to aid densification and held for 60 minutes. As the silica was in a viscous state at this temperature, load could be applied to the compact without damaging the Nicalon fibres. It was envisaged also that any drying cracks present in the compact at this stage would heal during this viscous flow stage of hot pressing. The applied pressure was removed prior to the cooling down of the specimen. Bend test specimens were made from the resultant composites (50 mm x 10 mm x 3 mm) which were tested in the as-cut condition under 3-point loading. Specimens were prepared for microscopic examination also.

Hot pressing of the alumina/alumina composites could not be performed due to the lack of viscous flow on high temperature consolidation of alumina. Any load applied at sintering temperature would be transferred directly to the fibres causing fibre damage. It is for this reason that the alumina/alumina compacts were pressureless sintered in air at a temperature of 1500°C for two hours, and then allowed to furnace cool. Specimens were then sectioned, and polished to a one micrometre finish prior to microscopic examination.

RESULTS AND DISCUSSION

Figure 3(a) shows a reflected light micrograph of the Nicalon/silica composite. At this low magnification it can be seen that hot pressing has enabled good matrix consolidation around the fibres without fibre damage. The density of this compact was measured and found to be 98.5% theoretical. The discrepancy has been attributed to porosity. The scanning electron photomicrograph in Figure 3(b) shows that by using EPD, the silica sol particles are able to infiltrate in between fibres which are in extremely close proximity (on the nanometre scale).

One of the most important factors which influences the infiltration of the fibre preforms is the tightness of the fibre mat weave. Although the plain weave architecture of the Nicalon fibre mats is one of the most tightly woven architectures, it could still be infiltrated fully. This implies, therefore, that EPD could be utilised to infiltrate other, less tightly woven, Nicalon fibre mats (e.g. Nicalon 607; 5 harness satin).

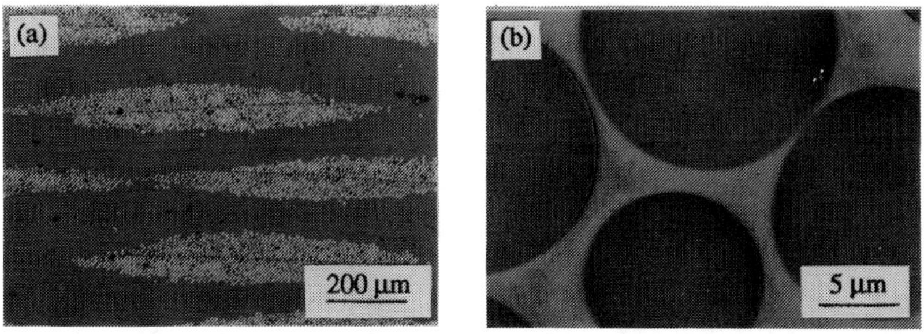

Figure 3: Photomicrographs of Nicalon/Silica Composite Material

Table 1 shows the mechanical properties of the Nicalon/silica composite. Four specimens were tested, each containing six layers of Nicalon fibre mats. The stress strain curve for each specimen contained regions corresponding to matrix microcracking and fibre pull out. The average flexural strength value measured for these composites indicates an increase in the strength by over a factor of three when compared to the monolithic silica matrix .

Table 1: Mechanical Properties of the Nicalon/Silica Composite System

Crosshead Speed (mm/min)	Onset of Matrix Cracking (MPa)	Flexural Strength (MPa)	Work of Fracture ($kJ\ m^{-2}$)
0.5	166.4	290.9 (4 Specimens)	8

Figure 4(a) is a scanning electron photomicrograph of the fibre tow region of a sintered alumina/alumina composite. As can be seen, the Remal sol particles have infiltrated fully the fibre tows, including those fibres which are in close proximity. This result shows that the filtration technique is just as effective as the EPD technique used with Nicalon. Hence, by using either EPD or EFD as the infiltration method, compacts can be fabricated from both conducting and non-conducting fibre mats.

Green body alumina/alumina compacts could be fabricated by the pressure infiltration of a ceramic slurry around the infiltrated alumina fibre mats. Pressure infiltrated green compacts were 64% dense and contained no cracks in the as-infiltrated, wet state. On slow drying, however, the shrinkage of the matrix was retarded by the rigid fibres resulting in matrix microcracking. These cracks became more accentuated on sintering due to the increased shrinkage and produced a composite which contained gross matrix microcracking. A scanning electron photomicrograph of the alumina composite system is shown in Figure 4(b). As can be seen, the region adjacent to an infiltrated fibre tow region contains cracks in the order of 10 μm in width. These cracks propagated throughout the inter fibre mat regions. This shrinkage of the matrix was to be expected due to the small particle size (0.4 μm) of the alumina. In order to address this problem, it would be necessary to use a matrix material which exhibits very little shrinkage on sintering. Candidate materials are the reaction bonded aluminas which have been shown to exhibit less than 1% shrinkage on sintering(6).

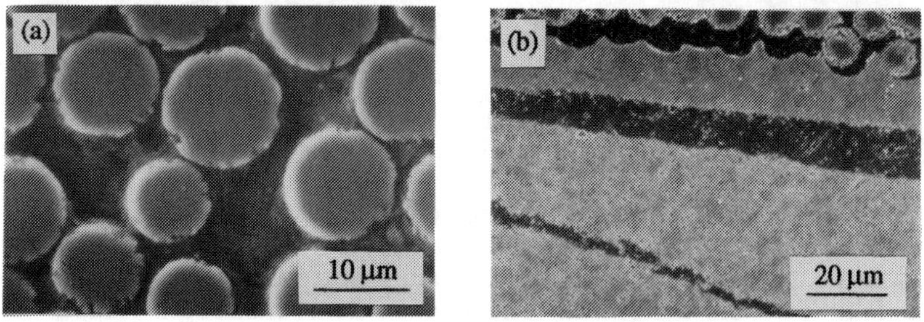

Figure 4: Photomicrographs of Al_2O_3/Al_2O_3 Composite;
(a) Intra-tow region, (b) Matrix region.

CONCLUSIONS

It has been shown that the fabrication of Nicalon/silica composites is possible using electrophoretic deposition. Infiltration of closely spaced fibres and the subsequent stacking/hot pressing of fibre mats produces dense glass matrix composites. These composites have a greatly improved flexural strength over that of the glass matrix.

An electrophoretic filtration deposition approach has been shown to be effective with the non-conducting alumina fibre system producing infiltration on the same level of efficiency as with electrophoretic deposition. Pressure filtration has been shown to be a promising technique for the formation of multi fibre layered green body compacts.

The gross matrix cracking which occurred as a result of sintering shrinkage, however, has not enabled the formation of dense alumina/alumina composites. The type of material used as the matrix material in pressureless sintering, therefore, must be such that shrinkage on sintering is at a minimum. It is envisaged that the use of reaction bonded aluminas in conjunction with EFD will enable the fabrication of high performance alumina/alumina composites.

ACKNOWLEDGEMENTS

This work was performed with the support of the Defence Research Agency U.K., High Temperature Materials Program.

REFERENCES

1. Illston, T. J. , "The Manufacture of Ceramic Matrix Composites Using Electrophoretic Deposition," *MPhil Report*, IRC in Materials for High Performance Applications, University of Birmingham, UK, 1993.
2. Trusty, P. A. "Development, Processing and Characterisation of Oxide Matrix Ceramic Fibre Composites," *Progress Report I,* IRC in Materials for High Performance Applications,University of Birmingham, UK, 1994.
3. Andrews, J. M., Collins, A. H., Cornish, D. C., Dracass, J., "The Forming of Ceramic Bodies by Electrophoretic Deposition," *Proceedings of the British Ceramic Society,* Fabrication Science, 2, **12**, 211, 1969.
4. Foissey, A., Robert, G., "Electrophoretic Forming of Beta-Alumina from Dichloromethane Suspensions," *Am. Ceram. Soc. Bull.,* **61**, 251, 1982.
5. Clasen, R. "Forming Compacts of Submicron Silica Particles By Electrophoretic Deposition," *Powder Processing Science,* 2nd International Conference, Ed. Hauser, 1989.
6. Claussen, N., Le, T., Wu, S.,"Low-shrinkage Reaction-bonded Alumina," *J. Euro. Ceram. Soc.* , **5**, 29, 1989.

PLASMA ENHANCED MICROWAVE JOINING

T. YIIN and M. BARMATZ, Jet Propulsion Laboratory, California Institute of Technology, Pasadena, CA 91109; and A. SAYIR, NASA Lewis Research Center, Cleveland, OH 44135.

ABSTRACT

A new method for plasma enhanced microwave joining of high purity (99.8%) alumina has been developed. The controlled application of a plasma between the adjoining surfaces of two rods initially heats the microwave-low-absorbing alumina rods to temperatures high enough for them to absorb microwave energy efficiently. With this technology, the adjacent surfaces of alumina rods can be melted and welded together in less than three minutes using approximately 400 watts of microwave energy. Four point bending tests measured fracture strengths of up to 130 MPa at the joined interface. Optical and SEM micrographs indicated that exaggerated grain growth prevailed for all joints studied.

INTRODUCTION

The use of microwaves to join a variety of materials has been successfully achieved;[1-4] however, joining very low absorbing ceramics is still a challenge. To significantly absorb microwaves, a material must have a large enough imaginary dielectric constant, ε'' (≥ 0.05). In general, ε'' increases with temperature and thus most materials are reasonably good absorbers of microwave energy at higher temperatures ($\geq 800°C$). However, materials must be efficiently heated from room temperature for commercial microwave processing to be economical. Conventional microwave heating techniques attempt to heat a material by placing it at a high electric or magnetic field position in a microwave cavity depending on whether the material is insulating or conducting. However, the electromagnetic field strengths attained, even for optimum power transfer conditions, are usually not sufficient to heat materials with low ε'' (≤ 0.01). Furthermore, when many materials are heated to high temperature, thermal runaway effects may cause hot spots or internal melting that are detrimental to the microwave heating process. Although high power magnetron and hybrid heating techniques have been used,[5,6] these techniques are expensive, slow, and still are not able to join very low absorbing ceramics starting from room temperature. We have developed and tested a new plasma enhanced joining method that can supplement microwave heating starting at room temperature and also continue to heat the rod interfaces to high temperature.

EXPERIMENTAL PROCEDURE

Commercially available 99.8% purity alumina rods from Coors Ceramics Co. were used in this work. The samples were 7.6 cm long and 0.5 cm in diameter. No further surface preparation was needed. In this study, we used a magnetron source to excite a rectangular waveguide cavity in the TE_{102} mode at 2.45 GHz. This mode has electric field maxima along the lines defined by $z/L_z = 0.25$ and 0.75 and $x/L_x = 0.5$ (see Fig. 1). An alumina rod was placed along one of these maximum E field lines ($z/L_z = 0.75$ and $x/L_x = 0.5$) with its end inserted approximately half way down the y axis ($y/L_y = 0.5$). This experimental arrangement was chosen since a plasma is generally ignited at a high electric field position in a microwave cavity. Helium gas flowed through the cavity to ignite the plasma around the tip of the rod. The temperature near the tip was measured using a non-contact Everest Interscience IR thermometer (range -30 - 1100°C) and an Accufiber non-contact pyrometer (range 500 - 3000°C). After the Helium plasma was ignited, a second rod was inserted through the opposite hole below the first rod as shown in Fig. 2. The new rod was positioned with ≈ 1.3 cm gap between the rods. The plasma surrounded the gap due to the concentrated microwave field in the gap. After a short time, both adjacent rod tips were preheated by the plasma to temperatures ≈ 1000°C. The Helium gas was then turned off, extinguishing the plasma and allowing the heated rod tips to directly absorb microwave energy. After the rods quickly reached their melting temperature, they were pushed together with ≈ 10 lbs of force using a force-controlled pneumatic unit to form a joint. The joined rods were then cooled at a controlled rate by gradually reducing the microwave power. All samples were tested to failure in a four-point bend fixture. The fixture design was taken from ASTM standard C 158-72. The strength tests were performed at a constant crosshead speed of 0.002 cm/s in a laboratory environment on an Instron testing machine (Instron Universal Testing Machine: Canton, MA) with a calibrated load cell. Requirements for accurate bend-testing of four-point circular beams were followed as described by Baratta et al.[7]

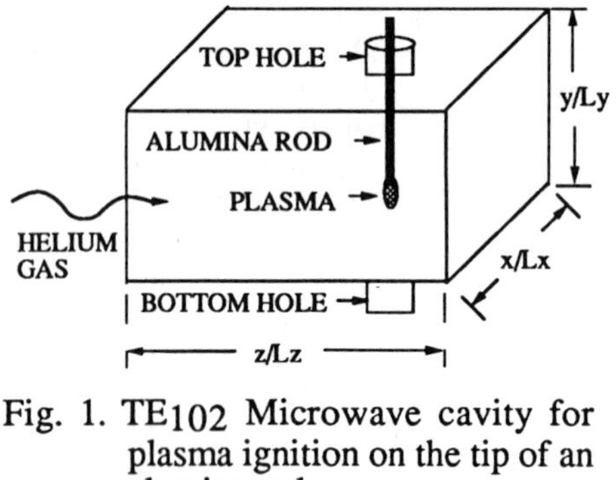

Fig. 1. TE_{102} Microwave cavity for plasma ignition on the tip of an alumina rod.

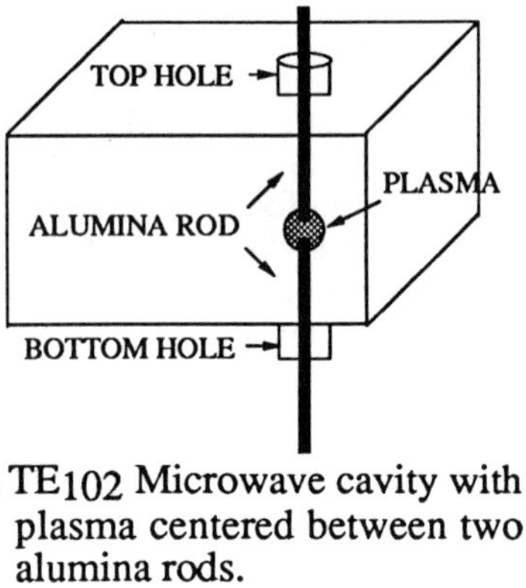

Fig. 2. TE_{102} Microwave cavity with plasma centered between two alumina rods.

RESULTS AND DISCUSSION

The enhancement in heating a low absorbing alumina rod using a plasma is illustrated in Figs. 3 - 6. In each case, the experimental arrangement corresponds to that shown in Fig. 1. Typical temperature and transmitted power heating curves for the rod tip are shown in Fig. 3 with no Helium plasma gas in the cavity. We see that the sample temperature rose very slowly as the transmitted power increased in steps up to approximately 500 watts. The maximum temperature attained with no plasma was ≈ 65°C. When Helium gas flowed through the cavity (≈ 100 ml/min at STP) and the microwave power was slowly increased, we obtained the temperature and transmitted power heating curves shown in Fig. 4 (the temperature was not measured below 500°C). The Helium plasma was excited suddenly using only ≈ 60 watts of transmitted power. The initially excited plasma quickly heated the tip of the rod to a temperature of ≈ 900°C. Once the plasma was initially excited, lower rod temperatures could be attained by decreasing the transmitted power. The Helium plasma was extinguished when the transmitted power was reduced below ≈ 15 watts which corresponded to a tip temperature of ≈ 700°C.

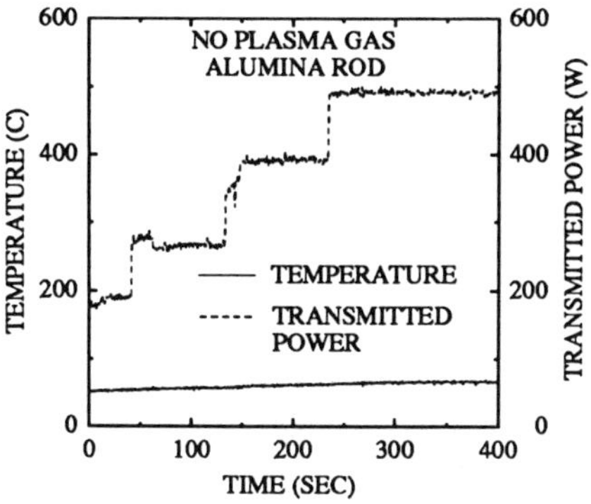

Fig. 3. Temperature and transmitted power curves with no plasma.

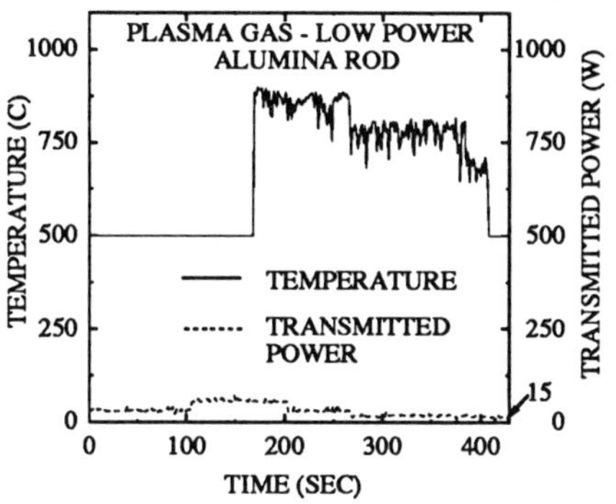

Fig. 4. Temperature and transmitted power curves with Helium plasma (low power).

High purity alumina has a room temperature imaginary dielectric constant of ε" ≈ 0.004. This is a very low value and very high microwave power levels would be required to heat this material initially as indicated in Fig. 3. Figure 5 shows how easily the rod tip can be heated to high temperatures in the presence of the Helium plasma. The initially ignited plasma again heated the rod tip to ≈ 900°C (ε" ≈ 0.09). The temperature of the rod tip could be increased further to ≈ 1550°C by increasing the transmitted power to ≈ 430 watts. When the microwave power was turned off (at the end of the 1550°C plateau) the rod tip temperature cooled very quickly. As the transmitted power was increased, the rod began to absorb more microwave energy. However, the plasma surrounding the rod tip absorbed most of the microwave energy in the cavity and also partially shielded the rod from the remaining microwave energy. While the Helium plasma could quickly heat the alumina rod tip to relatively high temperatures, it could not heat the tip to melting (≈ 2000°C) without using significantly higher transmitted power (> 500 watts). In order to heat the alumina rod tip to melting, we extinguished the plasma by turning off the Helium gas after the rod was initially heated to ≈ 900°C. Figure 6 shows how this plasma enhanced technique was applied. For this demonstration, the transmitted power was initially set to 200 watts. The plasma was excited after ≈ 45 seconds and the sample reached a steady state plateau of ≈ 900°C after 120 seconds. The Helium gas was then turned off at ≈ 200 seconds and the rod tip quickly heated to above 1600°C by directly absorbing microwave energy. Further increases in the transmitted microwave power to ≈ 300 watts heated the tip of the alumina rod to melting.

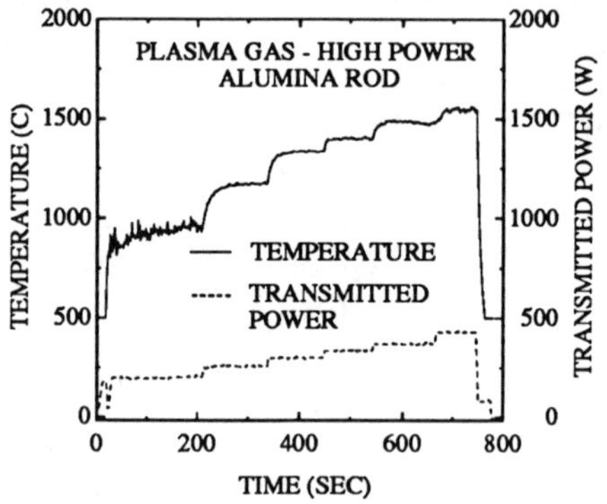

Fig. 5. Temperature and transmitted power curves with plasma gas (high power).

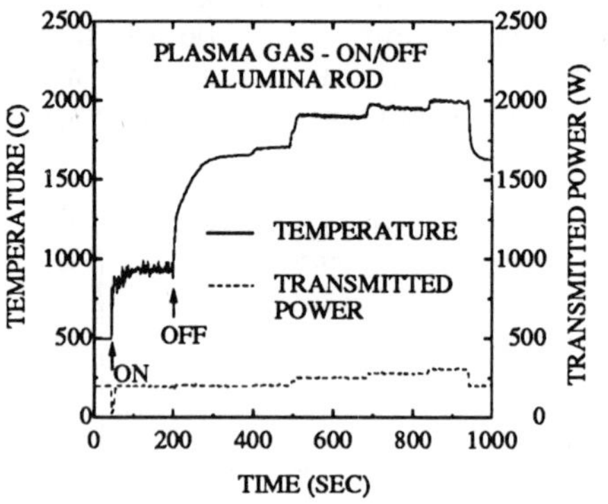

Fig. 6. Temperature and transmitted power curves with gas on and off.

Figure 7 shows the typical temperature and transmitted power heating curves for joining two alumina rods. The temperature was measured near the tip of the top alumina rod. The plasma was ignited immediately after the power (200 watts) was turned on. After ≈ 30 seconds, a second rod was inserted through the bottom hole in the cavity (see Fig. 2). The plasma, positioned in the gap (≈ 1.3 cm) between the two rods, quickly heated the tips of the rods. To improve the preheating of the rod tips, the transmitted power was further increased after 50 seconds until the temperature reached ≈ 1000°C. The plasma was turned off after 80 seconds and the temperature of the rod tips then increased rapidly to the melting point of alumina due to the enhanced absorption of microwave energy. After molten alumina droplets formed at the tips, the two rods were pushed together (at ≈ 150 seconds). The rod interface temperature then dropped quickly due to the sudden impedance mismatch.

With this plasma method, joint strengths in the range 80 - 130 MPa at room

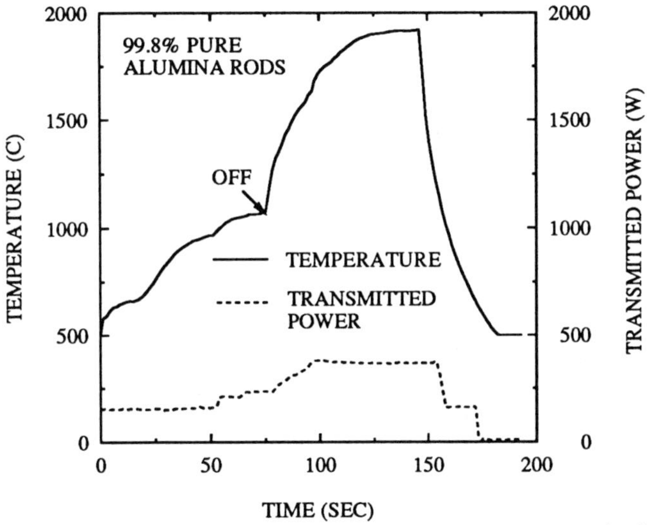

Fig. 7. Temperature and transmitted power curves during plasma enhanced microwave joining of 99.8% pure alumina.

temperature were measured using a four-point bending test. The wedging stress and friction at rod contact points were negligible since the failure occurred at the joint and the required load was small. After the joined rods were tested to failure, the fracture origins were examined with a scanning electron microscope (Joel 840, UTW Kevex EDS) to characterize rod microstructures and identify various phases. A typical optical micrograph of the interface between the joined samples and a representative fracture surface at the joint interface are shown in Figs. 8 and 9 respectively. The SEM characterization indicated that exaggerated grain growth prevailed for all joints studied. When the rods were joined, we assume that most of the molten material at the center of the rods was squeezed out to the surface. This process leads to a temperature gradient with the center of the interface being cooler than the surface. The microstructure in the rod interior consisted of exaggerated grain growth and fine equiaxed microstructure. We attribute this microstructure to the faster cooling rate in the center as shown in Fig. 10. This rod also had a skin layer of ≈ 1 to 1.5 mm with a well developed dendrite microstructure presumably due to the slower cooling at the surface.

CONCLUSION

In conclusion, we have demonstrated that a novel plasma enhanced microwave joining technique can lead to more efficient joining of low absorbing alumina. This approach should also permit the joining of many more low absorbing ceramics that have high commercial value. The possibility of obtaining even higher strength joints with better microstructural homogeneity will depend on the optimization of processing conditions. We plan to perform additional joining investigations under more controlled cooling conditions.

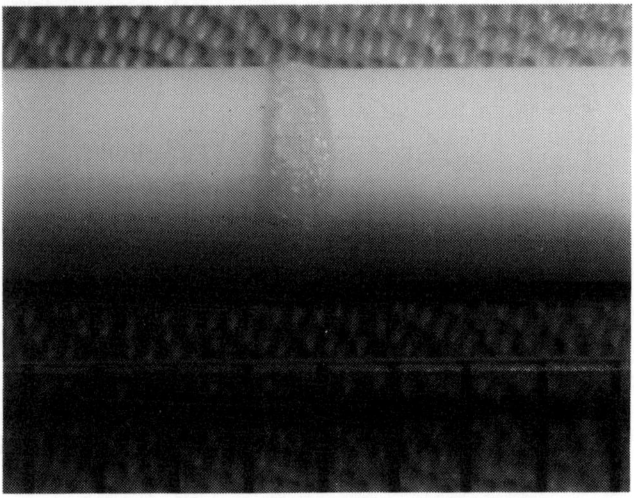

Fig. 8. Optical micrograph of the joined region (scale divisions equal 1.6 mm).

Fig. 9. Fracture surface of the interface of the joined rods (fracture occurred at the joint).

Fig. 10. Skin layer and dendrite microstructure of the fracture joint.

ACKNOWLEDGMENT

The authors wish to acknowledge Mr. O. Iny and Mr. R. Zanteson for the technical support. The research described in this article was carried out at the Jet Propulsion Laboratory, California Institute of Technology, under contract with the National Aeronautics and Space Administration.

REFERENCES

1. D. Palaith, R. Silberglitt, C. C. M. Wu, R. Kleiner, and E. L. Libelo "Microwave Joining of ceramics," **Microwave Processing of Materials**, Edited by W. Sutton, M. Brooks and I. Chabinsky, *Mat. Res. Soc. Symp. Proc.*, 124, 255-266, (1988).

2. H. Fukushima, T. Yamanaka, and M. Matsui, "Microwave Heating of Ceramics and Its Application to Joining", **Microwave Processing of Materials**, Edited by W. Sutton, M. Brooks and I. Chabinsky, *Mat. Res. Soc. Symp. Proc.*, 124, 267-272, (1988).

3. T.Y. Yiin, V.V. Varadan, V.K. Varadan, and J.C. Conway, "Microwave Joining of Si-SiC/Al/Si-SiC," **Microwave: Theory & Application in Materials Processing**, Edited by D. Clark, F. Gac and W. Sutton, Ceramic Transactions, 21, 507-514, (1991).

4. S. Arunajatesan, A. H. Carim, T.Y. Yiin, V. K. Varadan, "TEM Investigation of Microwave Joined Si-SiC /Al /Si-SiC and α-SiC /Al / α-SiC," **Joining and Adhesion of Advanced Inorganic Materials**, Symposium Proceeding, 314, 131-136, (1993).

5. S. Al-Assafi, I. Ahmad, Z. Fathi and D. E. Clark, "Microwave Joining of Ceramics," **Microwave: Theory & Application in Materials Processing**, Edited by D. Clark, F. Gac and W. Sutton, Ceramic Transactions, 21, 515-521, (1991).

6. M. Samandi and M. Doroudian, "Plasma Assisted Microwave Sintering and Joining of Ceramics," **Microwave Processing of Materials**, Edited by M. Iskander, R. Lauf and W. Sutton, *Mat. Res. Soc. Symp. Proc.*, 347, 605-615, (1994).

7. F. I. Baratta, W. T. Matthews, and G. D. Quinn, "Errors Associated with Flexure Testing of Brittle Materials," U. S. Army Mat. Tech. Lab., MTL TR, 87-35, (1987).

Nanotechnology and Advanced Characterization

SYNTHESIS AND PROCESSING OF ZnS/GaP NANO COMPOSITES

Yong Han and Mufit Akinc
Department of Materials Science & Engineering
Iowa State University
Ames, IA 50011

ABSTRACT

The zinc sulfide/gallium phosphide system was chosen as a model for studying nano composite synthesis and processing. Nano composites were synthesized by a novel impregnation method. Zinc sulfide was synthesized by homogeneous precipitation from aqueous solutions whereas gallium phosphide from organometallic routes. Nano-porous ZnS particles were impregnated with phosphinogallane liquid precursor followed by flash pyrolysis to precipitate GaP in the pore network. Composite powders were characterized with respect to chemistry and structure. Compacts were hot isostatically pressed to dense parts. Compacts maintained nanocrystallinity even after HIPping at 900°C exhibiting crystallite size less than 85 nm. Nano composites exhibited hardness values that were double that of ZnS.

INTRODUCTION

Infrared transmitting windows and domes aboard high speed aircraft are expected to perform under harsh environments. The primary function of these components is to protect the thermal imaging devices. To serve this purpose, the window material is required to be transparent to the IR region of the spectrum, strong, tough and have high thermal shock resistance. A variety of oxide ceramics such as magnesium aluminate spinel, sapphire, yttria, aluminum oxynitride are being used in the 3-5 μm region. However only a limited number of materials are known to show a tolerable combination of thermomechanical properties and transparency in the 8-12 μm region. Despite its poor mechanical properties, zinc sulfide is currently being used for 8-12 μm transmitting window and dome applications. Therefore, some attempts have been made to improve the properties of ZnS by applying a protective coating[1], by solid solution strengthening with Ga_2S_3[2], or by forming a composite with diamond.

Recent studies[4] indicated that the nanostructured materials show improved mechanical properties. Traditional composite processing is mainly based on either mechanical mixing or precipitation of a secondary phase from a melt. It is, however, difficult to control the size and the distribution of the secondary phase within the matrix. Uniform and ultrafine microstructures in composite materials may be achieved by controlling the chemistry and morphology of the constituents. In this study, we demonstrate that the scale of the microstructure can be reduced well below 100 nm using chemically designed materials. The basic concept is depicted in Figure 1.

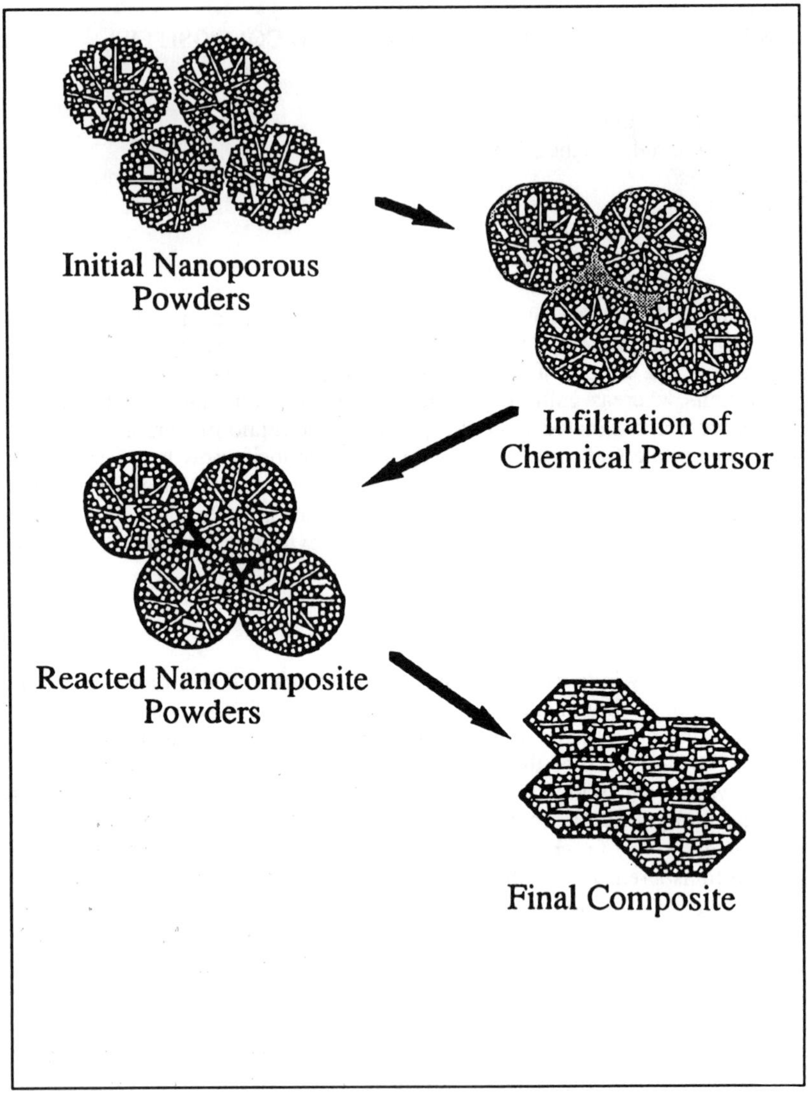

Figure 1. Schematic illustration of nanocomposite synthesis

EXPERIMENTAL

Zinc Sulfide Powder Precipitation

Spherical nanoporous zinc sulfide powder was homogeneously precipitated by thermal decomposition of thioacetamide[5]. A 0.05 M solution of zinc nitrate was acidified to a pH of 2.0 with nitric acid and heated to 70°C in a water bath. Thioacetamide was added to the solution to give a concentration of 0.8 M. The precipitation began in 4-6 minutes. After aging for 440 minutes, the solution was quenched in an ice bath and the precipitate was collected by centrifugation. The precipitate was washed with dilute ammonium hydroxide solution (pH=10.0) followed by centrifugation. Washing step was repeated until the pH of supernatant reached 10. Final wash was carried out with acetone, after which the powders were dried at 120°C.

Phosphinogallane Synthesis

Detailed procedure for the phosphinogallane synthesis is given elsewhere[6]. Monomeric tert-butyl phosphinogallane, $(t\text{-Bu})_2Ga\text{-}P(t\text{-Bu})_2$, was synthesized by a stoichiometric reaction between $LiP(t\text{-Bu})_2$ and $ClGa(t\text{-Bu})_2$. Reaction was carried out in a double manifold vacuum line following the standard inert atmosphere and Schlenk techniques[7].

ZnS/GaP Composite Powder Synthesis

Synthesis of ZnS/GaP composite powders[8] was carried out in the aforementioned vacuum line. Zinc sulfide powder (20 g) was placed in a pyrex tube and outgassed at 400°C until the pressure stabilized at 0.02 Torr. While cooling the tube to room temperature, phosphinogallane stored in a glove box was weighed out to provide various GaP contents (0.5-10.0 weight % in the ZnS/GaP composite), then it was transferred into a 50 mL flask with a side arm. A septum was inserted into the flask joint and the flask was taken out of the glove box. Diethyl ether (10-15 mL), which was dried over sodium metal and outgassed just before use at liquid nitrogen temperature to remove oxygen, was added to the flask through the septum to dissolve phosphinogallane. The solution was injected to the Pyrex tube which contained ZnS powder and the solvent was vacuum evaporated at -15°C. To convert the phosphinogallane to GaP, the mixture was flash pyrolyzed at 600°C for 30 minutes in a flowing argon atmosphere.

Densification of ZnS/GaP Composite Powders

The powders were sieved through a 200-mesh nylon screen to break large agglomerates and compacted into pellets by dry pressing uniaxially with a 22 MPa pressure. The pellets were then cold isostatically pressed at 207 MPa. After spray coated with boron nitride to a minimum thickness of 0.1 mm, the pellets were encapsulated in Pyrex glass ampoules under vacuum. The pellets were hot isostatically pressed at 900°C and 207-276 MPa for 30-60 minutes.

Characterization

Structural and morphological changes brought about by heat treatment of ZnS and ZnS/GaP composite powders were studied. Powders were sealed in quartz ampoules at 0.05 Torr and heat treated for 24 hours at temperatures ranging from 500° to 900°C with 100°C intervals. The powders were characterized with XRD, SEM, and TEM. Using x-ray diffraction line broadening, average crystallite size was calculated. From the TEM image of the densified pellets, the average grain size was obtained using the intercept method to compare with that from x-ray line broadening. Young's moduli of the dense composites were measured from ultrasonic technique using longitudinal and shear velocities. Room temperature hardness and fracture toughness were evaluated by indentation method using 1.0 kg and 0.5 kg loads for 15 seconds.

RESULTS AND DISCUSSION

ZnS and ZnS/GaP Composite Powders

SEM micrographs and XRD traces of the as-prepared and heat treated powders are shown in Figure 2 and 3, respectively. As-precipitated ZnS powders were spherical agglomerates of nano-crystallites. From the x-ray line broadening, the average crystallite sizes were estimated to be 8 nm for the as-precipitated ZnS and 12 nm for the as-prepared composite powders. Coarsening of the ZnS powders was observed at temperatures as low as 500°C. At 700°C, the initial powder morphology was lost and grains as large as few micrometers were observed. The average crystallite sizes were estimated to be 24 nm at 500°C and 41 nm at 600°C. Composite powders of ZnS/GaP, however, did not undergo any noticeable changes up to 600°C. Average sizes were 17, 29, and 86 nm at 700°, 800°, and 900°C, respectively. The effect of GaP impregnation on the grain coarsening behavior is quite significant. *In-situ* heating experiments using TEM equipped with a hot stage suggested that grain growth of nanocrystalline ZnS powder is due to material transport via fluid phase[9]. Since ZnS crystallites are covered with GaP in the composite powders, material transport via vapor phase in composite is less significant than it is in pure ZnS powders.

Following hot isostatic pressing (HIPping), all samples showed relative densities greater than 95% with an average value of 96.3 % as determined by the liquid immersion method. Figure 4 is a TEM micrograph of HIPped composite with 3.0 wt.% GaP, where the nano-sized microstructure is evident. The average grain size was estimated to be 34 nm by the interception method[10], while grain size using x-ray line broadening was 20 nm. For the remainder of the work reported here, grain sizes obtained from XRD line broadening are used.

The room-temperature values for hardness, Young's modulus, and indentation fracture toughness along with average grain size are tabulated in Table I. The average grain size gradually decreased with an increasing GaP content.

It can be seen that as GaP content increases, the Vicker's hardness value increases, which is accompanied by a corresponding decrease in grain size, while fracture toughness remains fairly constant. At ≈ 3 wt.% GaP, the hardness values are almost twice as high as that of

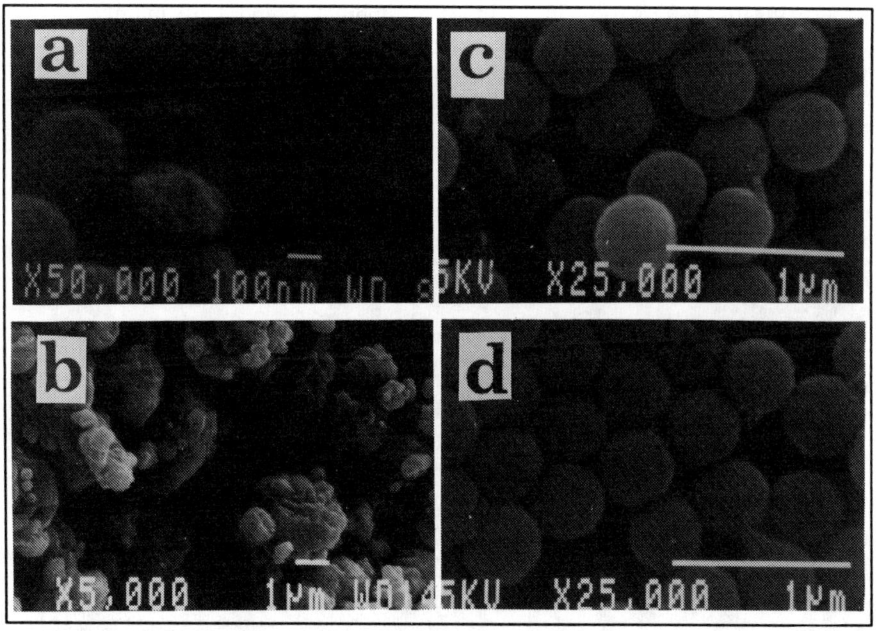

Figure 2. SEM micrographs of (A) ZnS (B) ZnS at 700°C
(C) ZnS/GaP (D) ZnS/GaP at 700°C

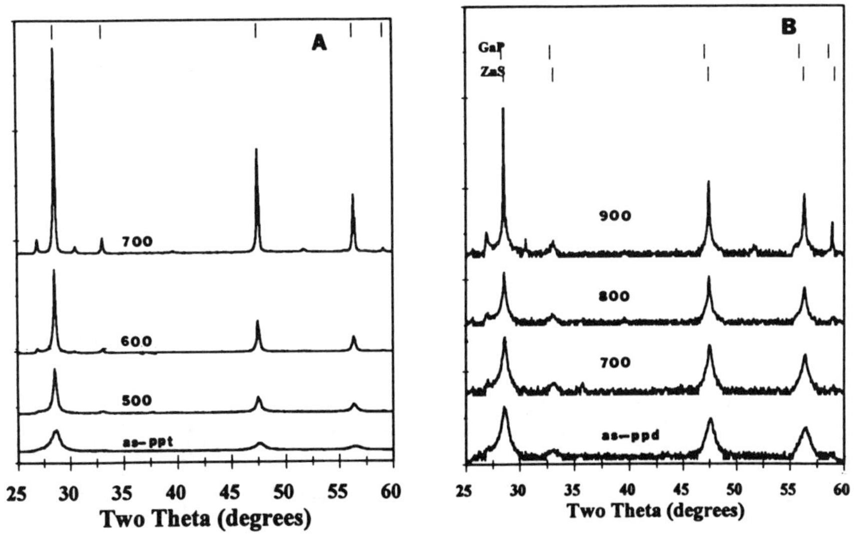

Figure 3. XRD traces of (A) ZnS (B) ZnS/GaP powders

pure ZnS. Plastic deformation in ceramics at ambient temperature mainly occurs via dislocation generation and its movement. As grain size decreases, the chance for dislocation pileup or pinning at the grain boundaries increases, thus increasing the hardness. This is the basis for the well known Hall-Petch relationship. However, Siegel[11] argued that as the

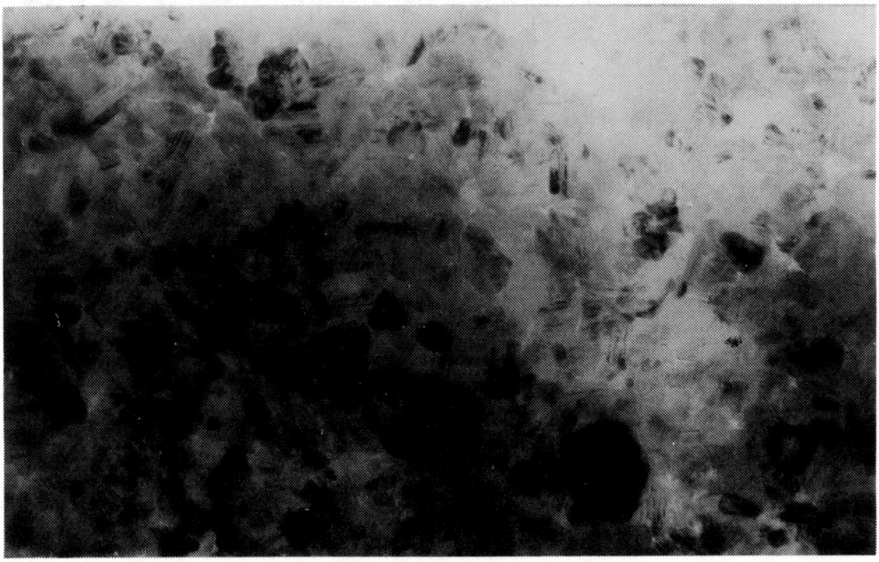

Figure 4. TEM bright field image of ZnS/GaP nanocomposite with 3.0 wt.% GaP HIPped at 900°C and 276 MPa for 30 minutes.

Table I. Room-temperature Mechanical Properties of ZnS and ZnS/GaP Nanocomposites.

GaP Content (wt.%)	density (%)	E (GPa)	d (nm)	H (GPa)	K_{IC} ($MPa \cdot m^{1/2}$)
CVD ZnS[A]	> 99	79	5-60 μm	1.38 ±0.49	-
HP ZnS[B]	99.5	78	>500	2.41±0.23	0.57±0.09
0.51	94.8	67	30±3	3.40±0.04	0.39±0.09
0.64	95.1	70	31±2	3.32±0.04	0.48±0.05
1.20	95.3	70	24±2	3.63±0.07	0.57±0.11
3.00	96.8	73	20±1	4.23±0.05	0.51±0.10
3.20	96.9	74	16±2	4.31±0.08	0.46±0.04
GaP[C]	-	103	single xtal.	8.28	0.80

A. ClearTran: a commercial CVD grown ZnS disc (Wilmad Glass, Buena, NJ)
B. Hot pressed ZnS using homogeneously precipitated powder, 900°C 105 MPa.
C. Data taken in the 100 plane[12].

grain size further decreases into the nano size regime, conventional dislocation generation and its movement become very difficult since the grain size is smaller not only than the critical bowing lengths necessary for a grain to serve as a dislocation source, but also than the typical spacing between dislocations in a pileup. Therefore, it appears that the hardness increase in nanophase material may be due to lack of an easier medium for plastic deformation, i.e. dislocation.

SUMMARY

The preparation, characterization, and hot isostatic pressing of zinc sulfide/gallium phosphide composite powders wre investigated. Monodisperse, spherical, submicron, nanoporous ZnS powder was used as a matrix for the nano composite synthesis. GaP was introduced into the porosity utilizing phosphinogallane solution. Composite powders showed significant grain growth retardation compared to pure ZnS. After HIPping, dense, ultrafine microstructures (as low as 16 nm) were obtained. Hardness and fracture toughness were measured by an indentation technique. While fracture toughness remained comparable to ZnS, hardness was almost doubled by addition of 3.2 wt.% of GaP. Although the exact nature of hardening mechanism is not understood, the lack of an easy medium for plastic deformation, i.e. dislocation, appears to be responsible for the hardness increase.

ACKNOWLEDGMENT

This work was supported in part by the Office of Naval Research and Iowa State University.

REFERENCES

1. K.J. Saunders, T.Y. Wong. T.M. Harnett, R.W. Tustison, and R.L. Gentilman, "Current and Future Development of Calcium lanthanum Sulfide," pp. 72-78 in Infrared and Optical Transmitting Materials, Proceedings of SPIE Vol. 683 (San Diego, CA, August, 1986). Edited by R.W. Schwartz. SPIE-The International Society for Optical Engineering, Bellingham, WA. 1986.

2. J.Zhang, W.W . Chen, and A.J. Ardell, "Solid Solution Strengthening of ZnS," pp. 93-98 in Windows and Dome Technologies and Materials II, Proceedings of SPIE Vol. 1326 (San Diego, CA, July 1990). Edited by P. Klocek. SPIE-The International Society for Optical Engineering, Bellingham, WA, 1990.

3. L.A. Xue and R.Raj, "Grain Growth in Superplastically Deformed ZincSulfide/Diamond Composites," J. Am. Ceram. Soc., **74**[7], 1729-31,1991.

4. K. Niihara, A. Nakahira, and T. Sekino, "New Nanocomposite Structural Ceramics," pp. 405-12 in Nanophase and Nanocomposite Materials, MRS Symposium Proceedings, Vol.286 (Boston, MA, December 1992). Edited by S. Komarneni, J.C. Parker, and G.J. Thomas. MRS, Pittsburgh, PA, 1993.

5. A.Celikkaya, and M. Akinc, "Preparation and Mechanism of Formation of Submicrometer Zinc Sulfide Powders," J. Am. Ceram. Soc., **73**[8], 2360-65, 1990.

6. A.A. Naiini, Y. Han, M.Akinc, and J.G. Verkade, "Synthesis and Pyrolysis Studies of $[(t\text{-}Bu)_2GaPR_2]_x$," Inorg. Chem., **32**[23], 5394-95, 1993.

7. D.F. Shriver and M.A. Drezdon, The Manipulation of Air Sensitive Compounds, pp.7-167. Wiley and Sons, N.Y. , N.Y., 1986.

8. Y.Han, E.T. Voiles, A.A. Naiini, L.S. Chumbley, J.G. Verkade, and M.Akinc, "Synthesis and Characterization of Zinc Sulfide/Gallium Phosphide Nano-composite Powders," J. Am. Ceram. Soc., **77**[12], 3153-60, 1994.

9. E.T. Voiles, L.S. Chumbley, Y.Han, and M.Akinc, "TEM Examination of Heat-Treated Nanocrystalline ZnS," J. Am. Ceram. Soc., **77**[12], 3173-756, 1994.

10. ASTM Standard Test Methods for Determining Average Grain Size (E112), 1989, Annual Book of ASTM Standards, Vol.03.011, American Society for testing and Materials, Philadelphia, PA, 1989.

11. P. Klocek, and D.F. Weirauch, "Pressure SIntering of III-V Compounds," pp. 105-18, in Windows and Dome Technologies and Materials, Proceedings of SPIE Vol. 1112 (Orlando, FL, March 1989), Edited by P.Klocek, SPIE-The International Society for Optical Engineering, Bellingham, WA, 1990.

12. R.W. Siegel, "Synthesis and Properties of Nanophase Materials," pp.189-97 in Materials under Extreme Conditions and Nanophase Materials, Proceedings of Symposium C of E-MRS (Strasbourg, France, November 1992), edited by G. Kostorz, I. Gorynin, V. Trefilov, H.D. Kunze, B.Gunther, B. Skorokhod, and V. Troitsky, North-Holland, N.Y. , N.Y., 1993.

ELECTROACOUSTIC CHARACTERIZATION OF PARTICLE SIZE AND ZETA POTENTIAL IN MODERATELY CONCENTRATED SUSPENSIONS

Vincent A. Hackley and Subhas G. Malghan
National Institute of Standards and Technology, Ceramics Division, Gaithersburg, MD, 20899

ABSTRACT

A novel electroacoustic technique for the measurement of particle size and charge in opaque suspensions is reviewed, and a new commercial instrument based on this method is evaluated for applications to ceramic powder processing.

INTRODUCTION

Preparation and dispersion control of ceramic powder slips during wet processing requires knowledge of the particle size distribution and particle charge properties under various conditions. Determination of size and charge usually necessitates two separate measurements performed under conditions which differ from those under which the slip was prepared. For example, significant dilution is required prior to analysis by light scattering techniques. Rheological measurements on the other hand, provide information on the macroscopic flow properties of the concentrated slip, but yield only limited and indirect information concerning the underlying mechanisms responsible for dispersion.

Recent advances in theory and instrumentation have led to the development of techniques for suspension characterization based on electroacoustic effects [1,2]. These effects, the colloid vibration potential (CVP) and the electrokinetic sonic amplitude (ESA), are reciprocal in nature and result from the periodic displacement of particle and surrounding ion atmosphere in a high frequency (MHz range) acoustic or electric field, respectively. With acoustic-based techniques, the measurement of suspensions containing up to 50 vol % solids is possible without dilution. Measurement conditions, therefore, can more closely resemble a real processing situation, and a potential exists for application to on-line analysis [3]. Single-frequency commercial instruments have been used extensively in recent years to study a variety of particulate and chemical systems related to ceramic processing. Some applications include the study of surface oxide layers on Si_3N_4 powders [4], the effects of hydrolyzable ions [5], and dispersant interaction mechanisms [6].

In this paper we review a relatively recent extension of electroacoustic measurements based on frequency spectrum analysis. In this technique both size and charge data are obtained from ESA measurements over a range of frequencies. We will discuss the basis for these measurements and show some initial results obtained on several materials with a comparison to other measurement techniques.

TECHNIQUE AND INSTRUMENTATION

In the case of ESA, an a.c. electric field induces vibratory electrophoretic motion of the charged particles at the applied frequency. Periodic displacement of the particle and surrounding fluid generates a sound wave that is detected at the electrode surface and converted into a voltage by a piezoelectric transducer attached to an acoustic delay rod. The amplitude of this signal is directly proportional to the frequency-dependent dynamic electrophoretic mobility of the particles, μ_ω, according to O'Brien [7,8]:

$$ESA - c\,\phi\,\Delta\rho\,\mu_\omega\,G_f \qquad (1)$$

where c is the speed of sound in the fluid medium, ϕ is the volume fraction of solids, $\Delta\rho$ is the density difference between the solid and liquid phases, and G_f is a form factor related to the geometric and acoustic characteristics of the electrode system. Additionally, the phase angle of the measured signal, after appropriate calibration against a known material, can be used to determine the polarity of the particles. The dynamic mobility is reduced from the static mobility, μ_0, determined in a conventional d.c. electrophoresis measurement, due to the inertial effects of particle mass [7]:

$$\mu_\omega - \mu_0\,G(\omega a^2/\nu) \qquad (2)$$

where the inertial term $G(\alpha)$ describes the continual decrease in mobility with increasing frequency and particle size, ω ($=2\pi h$) is the angular frequency, h is in Hertz, a is the mass-averaged particle radius, and ν is the kinematic viscosity. This relationship was derived for the thin double-layer case ($\kappa a >> 1$). Figure 1 shows $G(\alpha)$ as a function of frequency, calculated using the parameters for silica (ρ=2 g cm^{-1}) in water at 20 °C.

In order to correctly determine zeta potential from μ_ω, it is first necessary to know the correct particle size so that the inertial term can be calculated properly. At a frequency of 1 Mhz, the nominal operational frequency of commercial instruments, this correction becomes significant for particles much larger than 0.2μm.

As a consequence of O'Brien's theory, $G(\alpha)$ can be used to obtain particle size information from the frequency dependence of the dynamic mobility and phase angle [9]. Frequency spectrum analysis provides several benefits over the previous single-frequency measurement system. First, one can obtain size data *in situ* and simultaneously with particle charge. Second, the properly weighted size can be fed back into the zeta potential calculation to correct for the inertial term, thereby avoiding previous assumptions as to what constitutes the *proper* size for calculating G(α). Matec Applied Science* has recently developed the Acoustosizer (AZR) instrument which measures the ESA response over a frequency range from 0.3 to 11.1 Mhz. The relevant particle size range for this technique is limited primarily by the applied frequency. As size increases, G(α) approaches immeasurably small values due to the large inertial response, and the measured signal is correspondingly reduced. The upper size limit is about 10μm, depending somewhat on particle density and mobility. At the lower size range, particles smaller than 0.1μm do not exhibit sufficient inertial response at the applied frequencies for a measurable phase lag to occur.

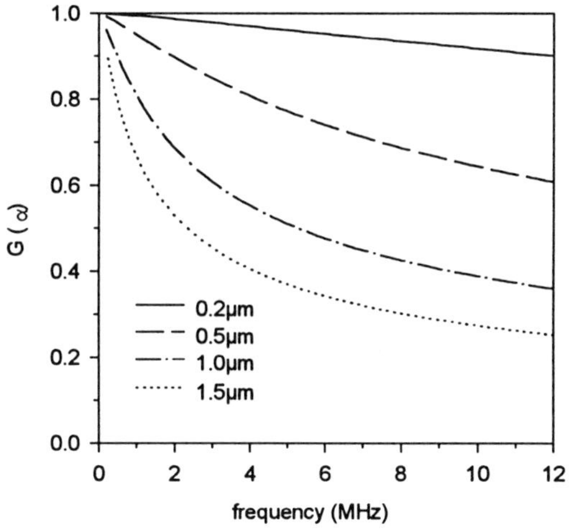

Figure 1 Variation of inertial term with frequency and particle diameter. Calculated using the parameters for silica in water at 20 °C.

The zeta potential, ζ, is calculated from the dynamic mobility according to the relationship of O'Brien [7]:

$$\mu_\omega = \frac{2\epsilon\zeta}{3\eta} G(\alpha)\,(1+f) \qquad (3)$$

where ε is the dielectric constant, η is the fluid viscosity, and *f* is a complex function describing

---

* Certain trade names and company products are mentioned in the text or identified in illustrations in order to adequately specify the experimental procedure and equipment used. In no case does such identification imply recommendation or endorsement by National Institute of Standards and Technology, nor does it imply that the products are necessarily the best available for the purpose.

the effects of surface conduction and particle permitivity. This latter term reduces to ½ in cases where the particle permitivity is small relative to the medium (80 for water at 20 °C) and surface conduction is insignificant (assumed for most inorganic colloids). Under these conditions equation (3) reduces to equation (2).

Since real particulate systems are not monodisperse, but exhibit some range in sizes, the AZR calculations assume a distribution function (log-normal) which defines the mass-averaged mobility. A least-squares fitting routine is used during which the median size, log-normal size spread, zeta potential, and *f* are varied to find the best fit to the measured dynamic mobility spectrum, including both magnitude and phase data.

The AZR instrument is designed to operate in a batch mode. The sample cell contains probes to monitor pH, conductivity, and temperature during analysis. The sample is stirred by an overhead mixer, and temperature control is maintained using a Peltier heat pump. Two electrodes are mounted on opposite sides of the sample cell, acoustically coupled to the transducers by solid delay rods (see Figure 2).

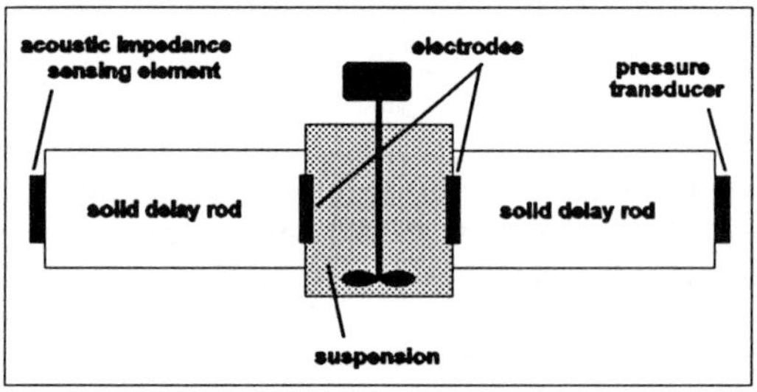

Figure 2 Schematic of Acoustosizer cell and electrode design (after reference 9).

In order to convert the ESA signal to dynamic mobility, the material density and solids volume fraction must be known, and a calibration standard with known signal amplitude is required to determine the instrument constant. Additionally, a correction for the background electrolyte signal may be required under certain conditions. Generally, the electrolyte signal is small relative to the sample signal except near the isoelectric point and at low solids concentrations. If necessary, a background measurement can be run and automatically subtracted from the sample signal.

EXPERIMENTAL

Silica and silicon powders were obtained from Geltech Inc. and Elkem Materials Inc., respectively. Suspensions were prepared at 5 vol % in deionized water using high-intensity ultrasonic treatment. Silicon suspensions were aged at pH 10 for a period of 20 hours. Electroacoustic measurements were made on a Matec Applied Sciences Acoustosizer. Electrophoresis and photon correlation spectroscopy (PCS) measurements were performed on a Malvern Zetasizer III laser light scattering instrument. Suspensions for electrophoresis and PCS were prepared by dilution from a 5 vol % suspension into its own supernatant. Particle size measurements were made by x-ray gravitational sedimentation using a Micrometrics Sedigraph 5100, and by Mie scattering using a Horiba LA900 Laser Light Scattering Analyzer. Sedigraph measurements were made on 2.5 and 5 weight % suspensions for silicon and silica, respectively. Samples for the Horiba system were prepared by diluting 5 wt % samples into deionized water to obtain the appropriate transmittance level.

RESULTS AND DISCUSSION

Analysis of Silica Powders.

Comparative measurements were made on silica powders with nominal diameters of 0.5 and 1.0μm. Scanning electron microscopic analysis of the primary particle size yielded values of 0.58 ± 0.02 μm and 1.08 ± 0.03 μm, respectively, where the indicated uncertainty is the standard deviation of the sample population. The particles in both cases were spherical. The electroacoustic technique is sensitive to the effective mass-weighted particle size, and not necessarily the primary size as represented in the SEM. This is particularly important for making the inertial correction to the zeta potential, as the primary size may not truly reflect the agglomerated size in suspension and as a result greatly underestimate the effect of inertia. Furthermore, changes occurring in the size distribution as a result of varying suspension conditions cannot easily be determined without *in situ* analysis techniques such as ESA.

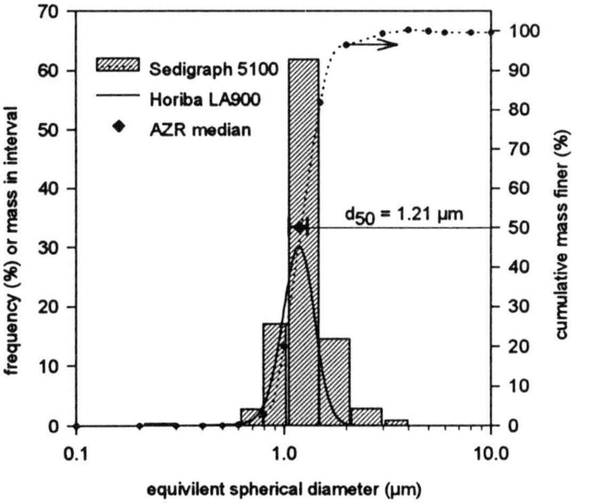

Figure 3 Comparison of size distribution results for a nominal 1 μm silica powder.

Figure 3 is a comparison of size analysis results obtained for 1μm silica near pH 9 where dispersion is generally promoted . Since both AZR and Sedigraph measurements are sensitive to particle mass, and can be run at similar solids concentrations, one would expect to find comparable results. In Figure 3 the cumulative mass determined by sedimentation yields a median (50% mass finer) diameter of 1.21 μm. This compares well with the AZR median diameter of 1.18 μm, and the Horiba mass-weighted median diameter of 1.09 μm.

These values are close to the 1.08 μm SEM value for the primary size, indicating this sample is easily dispersed and contains few agglomerates. The horizontal error bars shown for the AZR data represents the 15 to 85% mass interval size range. A comparison of the measured 15 to 85% size spread for each technique shows that the AZR (1.07 to 1.31) and Horiba (0.9 to 1.31) somewhat underestimate the spread relative to the Sedigraph (0.93 to 1.51). Overall, the agreement between methods is excellent for this material.

Results for 0.5 μm silica (not shown) indicate a greater degree of disparity between the three methods probably as a result of poor dispersion of primary particles. The sedimentation data shows a long tail extending out to 10 μm, indicating that some large agglomerates are present even after aggressive treatment. The median size for AZR and Sedigraph are 0.72 and 0.64 μm, respectively. Again, these methods are in reasonable agreement. The Horiba, on the other hand, significantly overestimates the median size in this case (0.83μm). This is probably due to the presence of a few large agglomerates. Since light scattering has a strong power-law dependence on particle size, it is more sensitive to their presence. The 15 to 85% mass interval range for the AZR (0.65 to 0.8 μm) is less than half that measured by sedimentation (0.5 to 0.95 μm) and light scattering (0.69 to 1.0 μm). This tendency for the AZR to underestimate the distribution width requires further study.

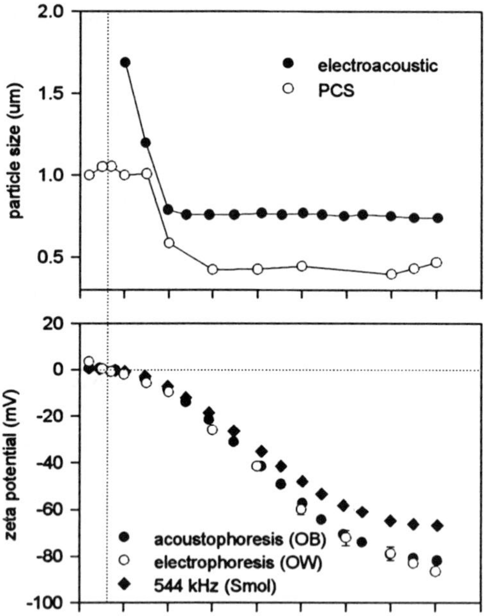

Figure 4 Comparison of electroacoustic, electrophoretic and PCS techniques for nominal 0.5 μm silica.

The next step in this evaluation was to compare zeta potential values derived from dynamic mobility measurements using the AZR, with those determined by the standard d.c. electrophoresis method on dilute suspensions. Measurements were made as a function of pH on 0.5 μm silica in 0.01M $NaNO_3$ solution, starting at pH 10 and titrating to about pH 2. Zeta potential was calculated from electrophoretic mobility using the O'Brian and White algorithm [10]. PCS size measurements were performed subsequent to each electrophoresis measurement using the same Malvern instrument. Figure 4 compares the results of these experiments, where acoustophoresis and electrophoresis refer to the AZR and Malvern results, respectively.

Zeta potential values determined by electroacoustic measurements are in excellent agreement with d.c. results over the entire pH range, apparently validating the O'Brien theory in this case. Additionally, these results offer further evidence that the particle size measured by the AZR instrument is appropriate for correcting the inertial term. The particle size determined by PCS is significantly lower than the electroacoustic median diameter, but both show a similar trend relative to pH. Aggregation rates will be much higher in the concentrated suspension used in electroacoustic measurements, compared to the much more dilute suspension used in PCS. This would account for the differences below pH 4 where the particles become unstable due to low zeta potentials. Both sets of data indicate an isoelectric point at about pH 2.3-2.5.

Anomalous Electroacoustic Response of Silicon

In the preceding section we described the evaluation of a new electroacoustic technique using silica spheres as a model system. This work is important in validating the basic theory and method of multiple-frequency electroacoustic analysis. We will now describe some preliminary results obtained for a less than ideal system. This powder consists of non-spherical silicon particles with a broad size distribution.

Previous dynamic mobility measurements performed on a single-frequency Matec ESA8000 instrument indicated unusual electroacoustic behavior for aqueous silicon suspensions. A more recent analysis of silicon using the AZR instrument showed particle size *decreasing* with *decreasing* zeta potential as the pH was lowered from 10 to 2. Since this is counter to expected behavior, Sedigraph measurements were also made on a similar suspension as a function of pH. Sedigraph results demonstrated the size dependence observed on the AZR instrument is not a real physical effect in terms of particle size. Additionally, it was observed that the AZR median diameter at low pH approached the "real" value determined by sedimentation analysis (about 3 μm). Unusually high zeta potential values were also suspect. Further studies showed that the effect depended not on pH, but on solution conductivity which varied with pH in low electrolyte systems. The apparent median size varied a factor of 5 with conductivity from O0 to 1400 μS/cm. Aged silicon powders have a silica-rich surface layer resulting from oxidation/hydrolysis of the unstable Si in water. Consequently, silica powder was tested for the size effect, but showed no significant size dependence on electrolyte concentration over the same conductivity range. The effect was not apparent for TiO_2 or Si_3N_4 powders either, but did occur to a lesser degree in the case of SiC. It is currently thought that this effect is related to the semiconductor (electron/hole pair) properties of Si and SiC, and will have to be studied further experimentally and with respect to O'Brien's current theory.

CONCLUSIONS

A new multiple-frequency electroacoustic instrument (AZR) for measurement of particle size and zeta potential has been evaluated using model silica powders. The AZR was shown to produce results comparable to established methods for size and potential analysis. Although some *a priori* knowledge of the material properties is required to provide accurate and reliable data, this requirement is reasonable relative to other techniques. Particle size values generated by the AZR correspond closest with x-ray sedimentation, both being sensitive to particle mass. The size spread is consistently underestimated by the electroacoustic technique, when compared to sedimentation or light scattering. Zeta potential measured by the AZR showed excellent agreement with d.c. electrophoresis results for a corresponding dilute suspension over a broad pH range. Anomalous results were obtained for Si and SiC, indicating that further studies are needed to fully apply this technique. Overall, the AZR offers the capability to characterize both size and charge *in situ* in particle suspensions at moderate to high concentrations. This is an improvement over existing methods, and has potential for on-line applications in the future.

REFERENCES

1. B.J. Marlow, D. Fairhurst and H.P. Pendse, "Colloid vibration potential and the electrokinetic characterization of concentrated colloids", *Langmuir*, **4**, 611-626 (1988).

2. D.W. Cannon, T. Oja and G.L. Petersen, "A method for measuring the electrokinetic properties of a solution", patent #4497207, United States (1985).

3. S.G. Malghan, V.A. Hackley and P.S. Wang, "Intelligent processing of ceramic powders and slurries", *Ceramic Eng. and Sci. Proc. (AcerS)*, **15**, 527-535 (1994).

4. V.A. Hackley, P.S. Wang and S.G. Malghan, "Effects of soxhlet extraction on the surface oxide layer of silicon nitride powders", *Mater. Chem. Phys.*, **36**, 112-118 (1993).

5. V.A. Hackley and S.G. Malghan, "The surface chemistry of silicon nitride powder in the presence of dissolved ions", *J. Mat. Sci.*, **29**, 4420-4430 (1994).

6. V.A. Hackley, R. Premachandran and S.G. Malghan, "Surface chemical interactions of Si_3N_4 with polyelectrolyte deflocculants", *Key Eng. Mater.*, **89-91**, 679-682 (1994).

7. R.W. O'Brien, "Electro-acoustic effects in a dilute suspension of spherical particles", *J. Fluid Mech.*, **190**, 71-86 (1988).

8. R.W. O'Brien, "The electroacoustic equations for a colloidal suspension", *J. Fluid Mech.*, **212**, 81-93 (1990).

9. R.W. O'Brien, W.N. Rowlands and R.J. Hunter, "Determining charge and size with the acoustosizer", pp 1-22, *NIST Special Publication 856,* edited by S.G. Malghan, Department of Commerce, Technology Administration, Gaithersburg, MD (1993).

10. R.W. O'Brien and L.R. White, "Electrophoretic mobility of a spherical colloidal particle", *J.C.S. Faraday Trans. 2*, **74**, 1607-1626 (1978).

EMBEDDED OPTICAL FIBER SENSORS IN CMC AND MMC MATERIALS

M.J. de Vries, K.A. Murphy, M.F. Gunther, R.O. Claus, T. Jessen*

Fiber & Electro-Optics Research Center
Bradley Department of Electrical Engineering
Virginia Polytechnic Institute & State University
Blacksburg, VA 24061-0111

*Naval Research Laboratory
Washington, D.C. 20375-5320

ABSTRACT

Optical fiber sensors are being embedded in materials for monitoring the structural integrity of materials during and after material fabrication. This paper discusses how small gage length optical fiber sensors can be embedded in CMC and MMC materials for intelligent process monitoring and structural evaluation.

INTRODUCTION

Optical fiber sensors, unlike conventional electrically-based sensors, are fabricated with high-strength silica which neither corrodes nor is affected by electromagnetic interference. Their small size and geometric flexibility allows them to be unobtrusively embedded or surface-attached to the host structure. Fiber optic sensors have been used for the quantitative, non-destructive evaluation of advanced materials and structures for over ten years.[1,2] Fiber sensors have been demonstrated to be feasible for the measurement of internal material changes during fabrication, the in-service lifetime measurement of strain, temperature, vibration, and other physical perturbations, and the eventual detection of damage or property degradation.[3,4] Recent advances in fiber optics communication technology and the widespread use of fiber communications have made available a large selection of reliable and inexpensive opto-electronic devices and components for use in sensors and sensor instrumentation support systems.

Fiber sensors have already been embedded in low temperature polymer matrix composites for measuring strain, temperature, and vibrations for performing material analysis and for detecting delamination.[5] Few conventional sensors, however, that can be used to monitor strain or temperature can survive the harsh environmental and temperature conditions used to fabricate the metal and ceramic matrix composites. That is why silica and sapphire sensor fibers, with various coating materials, were embedded in CMC and MMC specimens to determine whether they could survive the hostile material processing environment. Besides demonstrating that these fiber sensors survived the embedding process, examples are given that show that these sensors can also be used for further material evaluation.

SENSOR DESCRIPTION

The fiber sensors employed for both the CMC and MMC embedding are extrinsic Fabry-Perot interferometric (EFPI) strain or displacement sensor elements developed and demonstrated during the past four years.[6,7,8] For application here, a high temperature polyimide coated single mode fiber (λ= 1300 nm, 9 µm core diameter, 125 µm cladding diameter) acts as the input/output fiber and a multimode fiber (85 µm core, 125 µm cladding) is used as the reflector. In such sensors, a reflective cavity is formed between the two fiber ends placed in a hollow tube. Coherent laser light propagates down the fiber and into this cavity in such a way that the signal returning from the second fiber/air interface (the sensing signal) interferes with the signal reflected from the first fiber air interface (the reference signal) as shown in the detail in Figure 1.

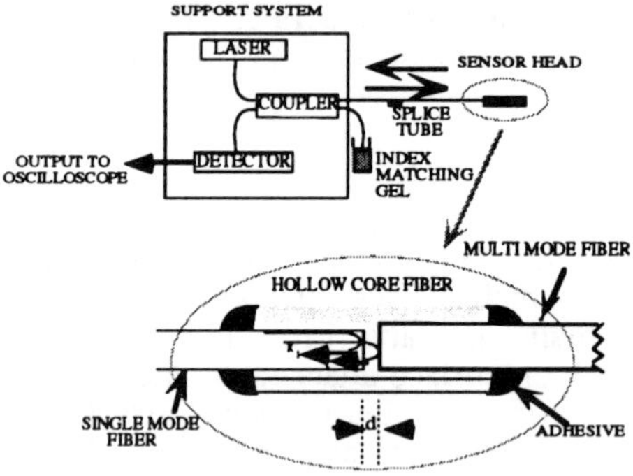

Figure 1. The extrinsic Fabry-Perot interferometer (EFPI) system.

These signals interfere constructively and destructively as their differential phase varies, leading to intensity modulation at the detecting end of the fiber. Cavity length changes caused by external mechanical stress therefore leads to a change in the phase of the reflected sensing signal, and due to the resulting interference with the constant phase reference signal, is an intensity modulation at the output. This output modulation may be interpreted quantitatively in terms of the strain at the location of the sensor element: Note here that stress applied along the lengths of the input and output fibers do not cause such differential phase modulation or result in additional modulation of the output signal.

CERAMIC MATRIX COMPOSITES

CVI represents one of the most promising, yet intrinsically difficult, reaction forming processes for making CMCs. The CVI process for densification of porous ceramic preforms consists of reactant diffusion and decomposition followed by solid product deposition in the porous body. Researchers at the Oak Ridge National Laboratory (ORNL) have made important improvements in the CVI process.[9] At ORNL optical fiber sensors were placed on different layers of interwoven Nicalon fibers. Once the fibers had been placed on a layer, the layers were stacked and packed into a porous graphite disc. The sample was first coated with approximately 0.2μm of graphitic carbon and was then infiltrated using the CVI process.[9] Once the sample was infiltrated it was cut to form rectangular sides (5 x 5 cm) which were carefully cleaned and polished. The optical fibers that were embedded were successfully identified under a microscope. Figure 2 shows a micrograph of one of the embedded optical fibers.

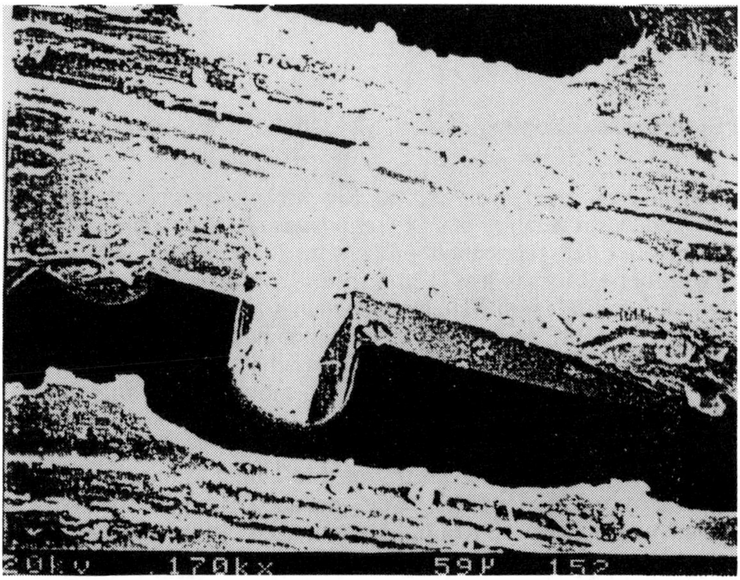

Figure 2. SEM picture of the optical fiber sensor that suvived embedding.

Using one of the hollow core sapphire fibers that was embedded in the CMC, an EFPI type optical fiber sensor was constructed. To demonstrate that this sensor can be used to monitor strain induced by temperature, a heating rod was placed in close proximity to the Nicalon sample and a thermocouple was used to monitor the temperature of the Nicalon block. Figure 3 shows the output response of the EFPI sensor as the block cooled from 93 °C to 40 °C. Ongoing research at AVX demonstrates that this sensor when embedded in ceramics can be used for the intelligent processing of multilayer ceramic actuator elements.[10]

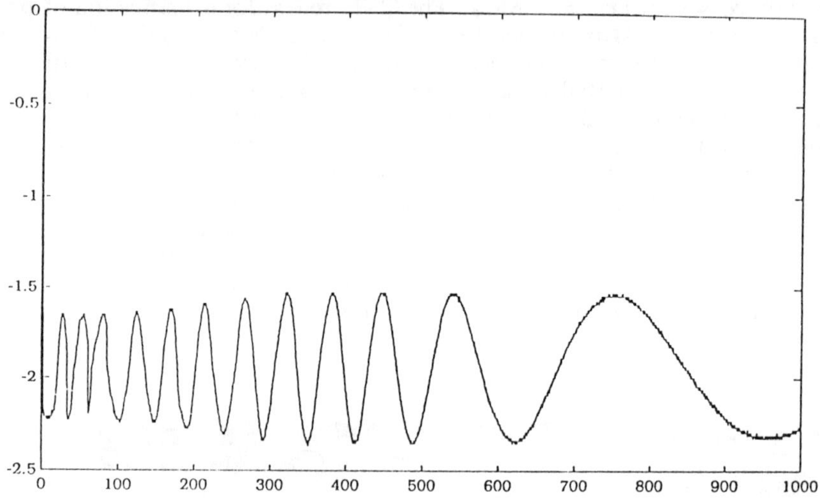

Figure 3. EFPI response to cooling of the CMC sample.

Another experiment performed at the Naval Research Labs concerning sapphire optical fibers embedded in ceramic matrix composites showed some difficulties in light transmission after the fabrication process was complete. Optical quality sapphire fiber with optical losses less than 5 dB/m length, and dimensions of 100 µm in diameter and 6 cm in length were embedded in a Nicalon fiber reinforced, Zr TiO_4 matrix CMC. The CMC coupon after fabrication had dimensions of 60 x 10 x 5 mm, with the sapphire fiber embedded approximately 2 millimeters from the top and centered in the width. The sapphire fiber moved very little during the consolidation process.

After consolidation the CMC coupon was trimmed in length to approximately 6 cm using a diamond saw and the endfaces were polished using 1 µm grit polish. Photographs of the endfaces are shown in Figure 4, showing the sapphire fiber and the Nicalon reinforcement fibers. The light guiding properties of the sapphire fiber were tested using a HeNe laser with a wavelength of 633 nanometers. The sapphire fiber guided no light for the 6 cm length. Smaller slices were taken of the sample and tested for light transmission until the slice was only 3 mm in length.

The 3 mm length of embedded sapphire fiber still transmitted no light. It is believed that the combination of high temperature and compaction pressure caused twinning in the sapphire single crystal optical quality fiber, which may be evident in the endface photograph, leading to zero transmission.

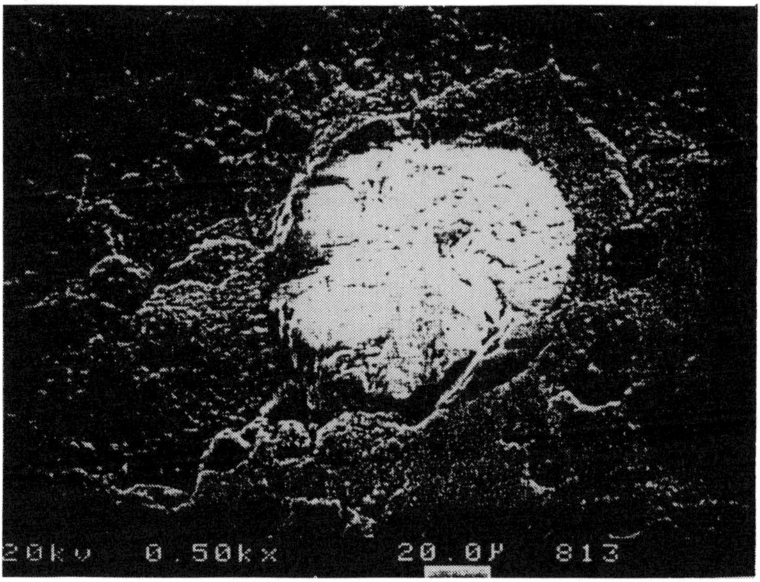

Figure 4. SEM photograph of the CMC sample with embedded sapphire fiber.

METAL MATRIX COMPOSITES

MMCs, which can have high strength capabilities and withstand temperatures greater than 1500 °C, are ideal candidates for high temperature environments, typically found in turbine engines. It is important, just as it was for the CMCs, to monitor the distribution of temperature and strain throughout the entire material. In order to accomplish this optical fibers need to be developed that can survive the temperature limitations placed on traditional gages. The first approach was to use a gold coated silica fiber that could survive in environments with temperatures less than 1000 °C. An EFPI-type optical fiber sensor was constructed using the gold coated optical fibers and was attached to the MMC using a glass slurry anti-oxidation coating, which was used to protect the MMC at high temperatures.[11]

To test the output response of the gold coated sensor, a high frequency Matec amplifier was used to couple an input pulse into the MMC material at 1.2 Mhz. Figure 5 shows the detected signal from the EFPI sensor. Optical fiber sensors made out of sapphire are currently being developed, and have been demonstrated in the laboratory for monitoring strain in high temperature environments. Strain measurements using these sapphire-based EFPI sensors have been made at temperatures as high as 1500 °C.[12] Sapphire-based sensors are now being developed for embedding in MMC.

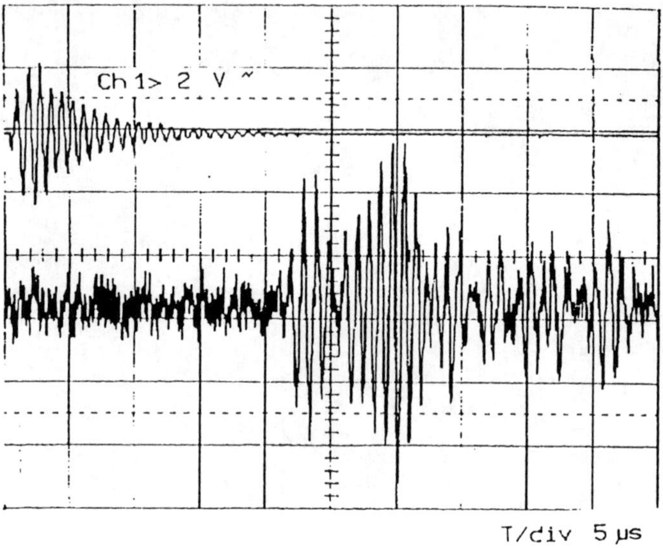

Figure 5. Output response from the EFPI sensor (bottom trace); and the high frequency Matec amplifier (top trace).

Summary

The principal goal of this research was to show that optical fiber sensors can be embedded in CMC and MMC specimens to monitor the material fabrication process and to subsequently monitor material performance. In the future, this sensor information can be used to improve the fabrication process such that the process is efficient, repeatable and more reliable.

ACKNOWLEDGEMENTS

This work was supported in part by the Virginia Center for Innovative Technology and Fiber & Sensor Technologies.

REFERENCES

[1] J. Dakin, B. Culshaw, Optical Fiber Sensors: Principles and Components, Artech House, Boston, MA 1988.

[2] T.G. Giallorenzi, J.A. Bucaro, A. Dandridge, G.H. Sigel, J.H. Cole, S.C. Rashleigh, R.G. Priest, "Optical fiber sensor technology," IEEE J. of Quantum Electronics, vol. 18, pp. 626 - 665, 1982.

[3] E. Udd, Overview of fiber optic applications to smart structures, 1988, New York: Plenum.

[4] R.O. Claus, Fiber Optic Sensor-Based Smart Materials and Structures, IOP Publishing, Bristol and Philadelphia, April 1992.

[5] C.A. Schmid, M.S. Miller, G. Carmen, K.A. Murphy, R.O. Claus, "Delamination detection scheme incorporating fiber sensors," Proc. IEEE Region 3 Conf., April 1993.

[6] K. Murphy, M. Gunther, A. Vengsarkar and R. 0. Claus, "Quadrature phase-shifted, extrinsic Fabry-Perot optical fiber sensors," Optics Letters, Vol. 16, No. 4, February 1991, pp. 273-275.

[7] A.D. Kersey, D. A. Jackson, and M. Corke. 1983. "A Simple Fibre Fabry-Perot Sensor," Opt. Comm., v. 45, 71.

[8] T. Yoshino, K. Kurosawa, K. Itoh, and T. Ose. 1982. "Fiber-Optic Fabry-Perot Interferometric Sensor," IEEE J. Quantum Electron., v. QE-18, 1624.

[9] M.J. de Vries, M. Nasta et al., "Survivability of optical fiber sensor elements embedded in silicon carbide ceramic matrix composites," SPIE Vol. 1918, *Smart Sensing, Processing, and Instrumentation,* Albuquerque, New Mexico, February 1993, pp. 450.

[10] M.F. Gunther et al., "Intelligent Process Monitoring of Multilayer Ceramic Actuators Using High Temperature Optical Fiber Displacement Sensors," SPIE Vol. 2191, Orlando, Florida, February 1994.

[11] S.E. Baldini et al., "Cooperative implementation of a high temperature e acoustic sensor," SPIE Vol. 1588, *Fiber Optic Smart Structures and Skins IV,* Boston, MA, September 1991, pp. 125 - 131.

[12] A. Wang, S. Gollapudi, R.G. May, K.A. Murphy, R.O. Claus, "Advances in sapphire-based intrinsic interferometric sensors," Optics Letters, Vol. 17, No. 21, November 1992, pp. 1544 - 1546.

AUTOMATED LASER SCATTER DETECTION OF SURFACE AND SUBSURFACE DEFECTS IN Si_3N_4 COMPONENTS

J. Scott Steckenrider
Energy Technology Division
Argonne National Laboratory
9700 South Cass Avenue
Argonne, IL 60439

ABSTRACT

Silicon nitride (Si_3N_4) ceramics are currently a primary material of choice to replace conventional materials in many structural applications because of their oxidation resistance and desirable mechanical and thermal properties at elevated temperatures. However, surface or near-subsurface defects, such as cracks, voids, or inclusions, significantly affect component lifetimes. Because these defects are currently difficult to detect, a technique is desired for the rapid automated detection and quantification of both surface and subsurface defects. To address this issue, we have developed an automated system, based on the detection of scattered laser light, that provides a two-dimensional map of surface or subsurface defects. This system has been used to analyze of flexure bars and button-head tensile rods of several Si_3N_4 materials. Mechanical properties of these bars have also been determined and compared with the laser scatter results.

INTRODUCTION

The need to achieve higher gas firing temperatures in new stationary gas turbines has led to the requirement for stronger, longer-lasting materials capable of functioning in environments more severe than traditional materials permit. The higher operating temperatures seen in these environments (>1300 °C) would substantially increase energy efficiency and reduce emissions. Because conventional materials are not suitable, alternate materials are being considered for turbine components. Included among these are the Si_3N_4 ceramics. In an effort to improve and extend the working lifetime of the components, improvement in the detection of critical defects (such as cracks, voids, inclusions, or microstructural variations) in their surface and near-subsurface regions is being investigated. Although all of the defects can occur either during manufacturing or in operation, the specific type of defect most likely to occur in a given component depends on a various factors, including the material system, processing and machining parameters, component geometry, and application environment.

Defects generated in the near-subsurface region must be detected as early as possible to minimize cost [1] or reduce or eliminate in-service failure of the ceramic components. Although several techniques can be used to characterize surface condition and defects [2-4], nondestructive characterization of the near-subsurface is relatively complicated, particularly for two-dimensional (2-D) surfaces. Traditional methods of nondestructive evaluation (NDE), such as radiography and ultrasonics, are not well suited for detection of subsurface defects. Radiography is primarily sensitive to interior (bulk) defects and is therefore relatively insensitive to surface/near-surface defects [5]. Ultrasonics, on the other hand, can be highly sensitive to the surface/near-surface region. However, to obtain sensitivity to the near-subsurface region, higher frequencies must be used, and these require a coupling agent (e.g., water, oil), which is often undesirable in ceramic applications where the ceramic may not be 100% dense. Ultrasonic methods are also generally best suited for flat surfaces, so the complex geometries associated with turbine components can be problematic [6].

A technique is therefore desired for rapid detection, characterization, and quantification of defects in the near-subsurface (and surface) regions of an array of Si_3N_4 ceramics of multiple geometric configurations. Previous researchers have focused considerable effort on the noncontact characterization of machined surfaces of metals by elastic-optical-scattering (predominantly laser-based) techniques [3,7,8]. However, these techniques do not directly provide information on the subsurface condition of these materials.

The ability to characterize the subsurface material in some ceramics (e.g., zirconia) for density variations has been shown [9]. For many Si_3N_4 (and other) ceramics, the optical penetration depth can be ≥100 μm in the visible spectrum, depending on grain size, second-phase composition, and material absorption [10]. Because Si_3N_4 ceramics can partially transmit visible (and IR) light, we have previously applied static polarization-based laser scattering to the analysis of various Si_3N_4 materials fabricated into several component shapes, and have detected defects as deep as 100 μm below the surface [11]. In an effort to apply this technique to the inspection of gas turbine components, several candidate Si_3N_4 materials systems have been investigated with an automated 2-D scanning laser scatter imaging system, which provides a 2-D map of defects and indicates their relative positions and severities. To determine applicability to multiple geometries, and to correlate laser scatter results with actual mechanical properties, both flat (flexure bars) and curved (button-head tensile rods) specimens were examined.

EXPERIMENTAL PROCEDURE

Surface Analysis

The most important features in the inspection of ceramic surfaces are the size and orientation of surface defects. Detection of such defects is somewhat simplified in materials whose "normal" surface condition is uniform and well defined. Surface irregularities can then be detected and quantified by optical Fourier filtering; in this technique, a spatial Fourier transform of a surface is recorded using precision optics and coherent (i.e., laser) illumination. For a well-defined uniform surface, the Fourier pattern of an undamaged surface can be recorded or computed and then subtracted from the actual scatter pattern obtained from the surface. In regions where no surface defect is present, the difference is essentially zero. However, when a defect enters the inspection system's field of view, additional spatial frequencies appear in the Fourier plane. Because these frequencies are not present in the undamaged surface, they produce an increase in signal in the pattern difference. Furthermore, if the distribution of spatial frequencies is analyzed, the specific shape and orientation of the surface defect can be determined.

Subsurface Analysis

Subsurface laser scatter analysis is a relatively straightforward technique in which a specimen is directly illuminated by a laser beam (either focused or unfocused, depending on the specific application). The only material requirement is that the specimen be translucent (i.e., able to diffusely transmit or scatter light), as is the case with most polycrystalline ceramics. The degree of scatter depends on the difference between the index of refraction of the grain boundary region and the bulk crystal. Si_3N_4 materials are highly scattering (so much so that they generally appear opaque) because a small percentage (usually ≈5%) of a second-phase material such as Y_2O_3, Al_2O_3, or MgO is added to the Si_3N_4 to achieve maximum densification. This second-phase material, which has a significantly different index of refraction than does Si_3N_4 and may even absorb light, concentrates at the grain boundaries. Thus, the Si_3N_4 opacity results from the absorption and/or relatively high reflection that occurs at these grain boundaries. When a relatively thick specimen of Si_3N_4 is illuminated by a beam of light, whatever light is not absorbed is back-scattered in an almost purely diffuse pattern (i.e., an equal intensity is scattered in all directions). If a defect (e.g., a pore, crack, inclusion) is placed in the subsurface material, the amount of back-scattered light will be changed. If the defect is, for instance, an iron inclusion, it will absorb significantly more light than would the displaced Si_3N_4 material, giving a decrease in back-scattered intensity. If, on the other hand, the defect were a lateral crack or pore, the difference in index of refraction between the air that would fill such a defect ($n \approx 1$) and the Si_3N_4 itself ($n \approx 1.9$) is significantly higher than that between the Si_3N_4 and the densification aiding material, so the back-scattered intensity would increase for such a defect.

The relative depth of a given defect can also be approximated by determining the ratio of the scattered intensity in the region of the defect to that of the normal material. The more this ratio differs from unity, the closer the defect is to the surface. Although it is not possible to determine an exact relationship between scattered intensity and defect depth, the general trend does permit an approximation with regard to defect depth. (In addition, any depth determination requires that the size of the beam spot on the surface be smaller than the detected defect. Otherwise, the intensity change will be moderated by undamaged material surrounding the defect.) Thus, by measuring the type, degree, size, and shape of a defect's laser scatter signature, a general description of the defect can be derived, along with its location on the specimen.

System Design and Operation

The experimental arrangement used for the analyses presented here is illustrated in Fig. 1. A vertically polarized laser beam in directed through a polarizing beam-splitter (PBS) cube onto the specimen surface by focusing optics so that light specularly reflected from the surface will be collimated. Any light reflected from the component surface will undergo no change in polarization (unless the surface is extremely rough), so all surface scattered light will again be reflected in the PBS and directed back toward the laser. However, the polarization of any light that is scattered from subsurface material will be randomized (and the scattered light will be completely diffuse). Thus, half of the subsurface scattered light will also be reflected by the PBS and directed back to the laser. However, the other half of the subsurface scattered light will be transmitted by the PBS into the detection train.

If surface-scattered light is of interest, a properly oriented quarter-wave ($\lambda/4$) plate is inserted between the PBS and the component surface. This rotates the light's polarization by 90° on each pass so that all of the surface scattered light and half of the subsurface-scattered light is transmitted to the detection train. Because in most cases, the surface-scattered light is much more intense than the subsurface-scattered light, it dominates the detected signal so that the subsurface scatter can be neglected.

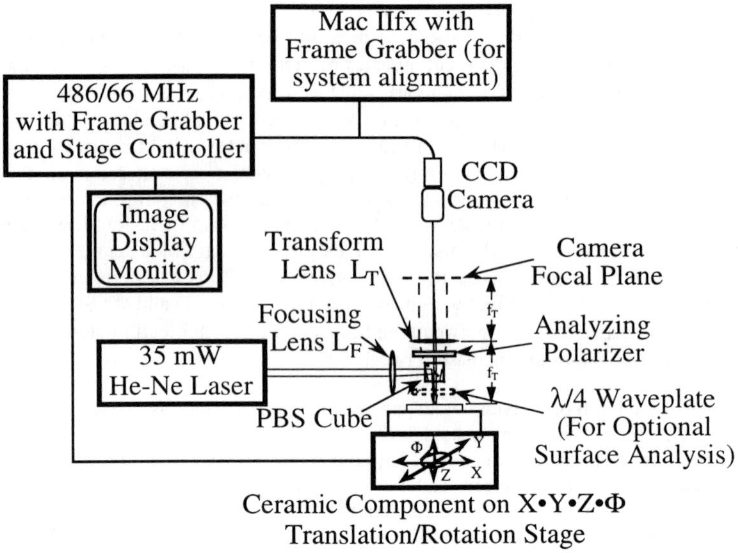

Figure 1. Illustration of the basic setup for the laser scatter inspection technique.

In both arrangements, the detection train is the same. A linear sheet analyzing polarizer (with a polarization efficiency higher than that of the PBS) is placed behind the PBS to maximize the signal-to-noise ratio. A high-quality positive lens with focal length f_T (the transforming lens) is then placed just beyond the polarizer. A CCD camera is focused at a distance of f_T behind the transforming lens. What is imaged on the CCD array is the power spectrum of the spatial Fourier transform of the surface or subsurface of the component. The specific system used in this effort records the full 2-D optical Fourier power spectrum of back-scattered laser light from a single point on the component, including spatial frequencies up to 0.2 μm^{-1}. This Fourier scatter pattern is then digitized for further analysis with a 486/66 MHz PC. Thus, by examining the intensity distribution in the camera's image, one can determine the character of roughness on the component's surface or subsurface. To improve acquisition speed, the system initially computes only the average magnitude of all such frequencies. The average magnitude is then mapped in a gray-scale value as a function of specimen position The result is scatter "images" of the surface or subsurface material. In cases where specific defects were detected with this scanning approach, the Fourier power spectrum of the actual scattered light was inspected in more detail.

Machined Step-Wedges

The first specimens examined were machined step-wedges 4 mm wide and 25 mm long, with nominal step thicknesses of 800, 400, 200, 100, and 50 μm. Wedges were made from various different Si_3N_4 materials, and the top (flat) surface of each wedge was polished. These wedges were used for two purposes. First, to determine optical transmittance at a wavelength of 632.8 nm with an integrating sphere for various different Si_3N_4 ceramics. The results of these measurements are shown in Fig. 2, where the transmission characteristics vary across the several Si_3N_4 materials by more than an order of magnitude. This means that the total depth of penetration (i.e., the depth to which defects can be detected) will also vary among materials. The step-wedges were also used to simulate subsurface lateral cracks. In this case, the wedge was again illuminated normal to its polished surface, and the intensity of back-scattered light was monitored. Thus, while the front surface appeared uniform across the entire wedge, the steps machined into the back surface provided a ceramic/air interface in the subsurface material to simulate a subsurface lateral crack. The back-scattered intensity at each point was digitized and recorded. These scatter intensities were normalized into gray scale values and assembled into a composite image of the step wedge. Examples of such images are presented in Fig. 3, with the thinnest steps on the left. If these steps had been actual cracks, the back-scattered intensity would have been even greater, owing to additional scatter from the material beneath the defect.

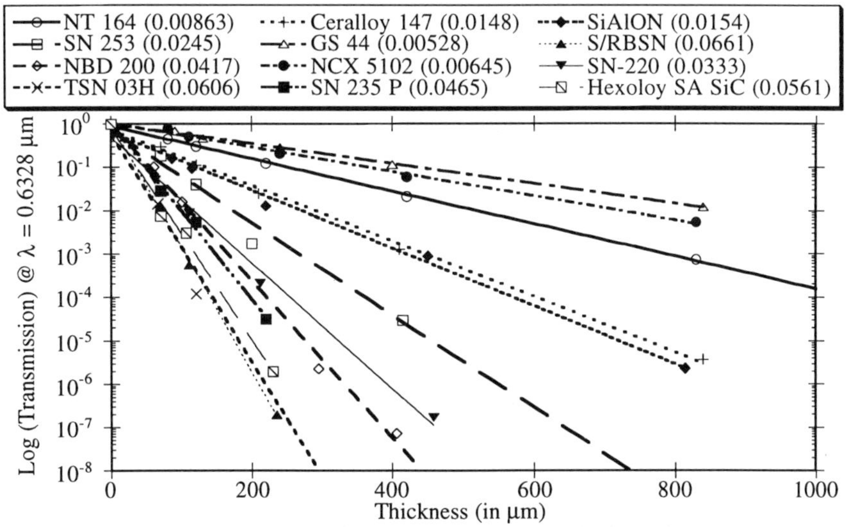

Figure 2. Plot of optical transmission as a function of specimen thickness at a wavelength of 632.8 nm. Lines are best-fit approximations to: $T = I_T/I_I = e^{-\alpha t}$, where t is thickness, I_T and I_I are transmitted and incident intensity, respectively, and α (attenuation) is given parenthetically in the legend.

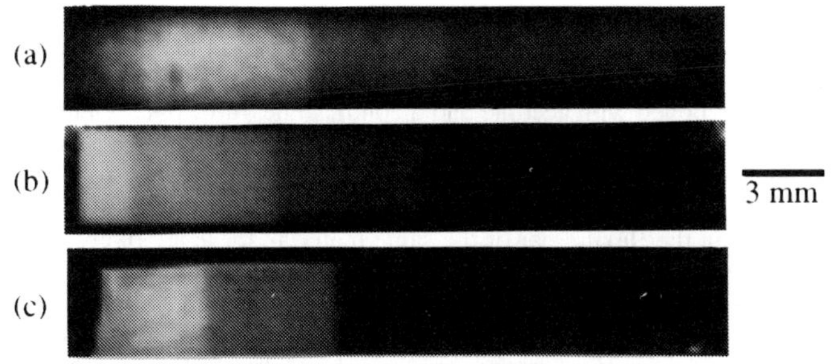

Figure 3. Subsurface laser scatter images of three Si_3N_4 step-wedges. (a) GS44, (b) NT164, and (c) SiAlON.

RESULTS AND DISCUSSION

In this study, 80 flexure bars from various manufacturers and 8 tensile rods of four different materials were examined. The surfaces of the flexure bars were both machined and as-fired. Surface-sensitive and subsurface-sensitive analyses were performed. Although presentation and discussion of all results would be excessive, several noteworthy examples are discussed below.

The response from one material type to another varied widely. Three examples are shown in Figs. 4 and 5. Figures 4 a and b show the laser scatter images for both surface and subsurface scans of 10 AlliedSignal Ceramic Components (ASCC) GN-10 flexure bars. (All flexure bars were 4 mm wide, and a 20 mm long gauge section of each bar was examined at 200 μm resolution. Defects smaller than 200 μm were detectable; however, the intensity of their signatures was reduced.) Two features are clearly evident in Fig. 4. First, labels written in ink on all 10 bars are visible in both the surface and subsurface images. Although numerous attempts were made to remove this ink, all were unsuccessful. Thus, a defect beneath the ink would not be detected. The second feature seen in the laser scatter data is an overall mottling of the subsurface image. This mottling is not seen in the surface image: therefore, it is entirely a subsurface phenomenon. In addition, because the same mottled appearance is evident in X-ray radiographic images, the mottling must be a variation in local density. (A similar phenomenon has been seen in bearing components made from another ASCC ceramic, GS44. In that case, destructive analysis indicated a locally high porosity in the regions of reduced laser scatter [11].)

Figures 4 c and d show the laser scatter images for both surface and subsurface scans of 10 Norton NT-164 flexure bars. The five bars on the left were machined, while the five on the right were in the as-fired condition. The most severe of the defects in this group is the dark spot in the lower extreme of the subsurface image of bar 11 (Fig. 4d). Based on experience (through the visual-inspection methodology currently being applied for flexure bar qualification), this appears to be an iron inclusion of some kind, and it is readily seen with high resolution radiography. However, several similar, though less severe, occurrences were seen with the laser scatter technique in other NT164 bars, as shown with the black circles in Fig. 4d, which were not detected by either visual inspection or X-ray radiography. Initial mechanical testing has revealed that these iron inclusions act as stress concentration points where failure initiates, but the specimens shown in Fig. 4 have not yet been mechanically tested.

Figure 5 shows subsurface and surface laser scatter images for an NT-164 tensile rod. These images were acquired with a 100-μm resolution and represent a portion of the rod that is 40 mm long and has a circumference of 20 mm. As with the GN-10 flexure bars presented above, this tensile rod was written on with a "permanent" ink pen in the gauge region. Fortunately, we were able to remove all of the ink from the surface (as seen in Fig. 5a). Although all of the surface ink was removed with acetone and alcohol, some of the ink remained in the subsurface after cleaning, as can be seen in the subsurface image (with boxes drawn around the ink marks to delineate them from actual defects). Despite the presence of this ink, two defects are still evident in this rod. First, a significant number of distributed defects in the NT164 rods appear as dark spots. Again, as in the flexure bar of the same material, these are likely iron inclusions, which are common in the NT-164 material. Furthermore, because they are not visible in the surface scan, they are completely enclosed within the rod. The second feature, a linear defect on the left side of the image about halfway up its length, cannot be attributed to residual ink. Because it appears in both the surface and subsurface images as a bright and dark indication, respectively, this is likely a surface scratch that is sufficiently severe to dominate the laser scatter, even in subsurface-sensitive analysis.

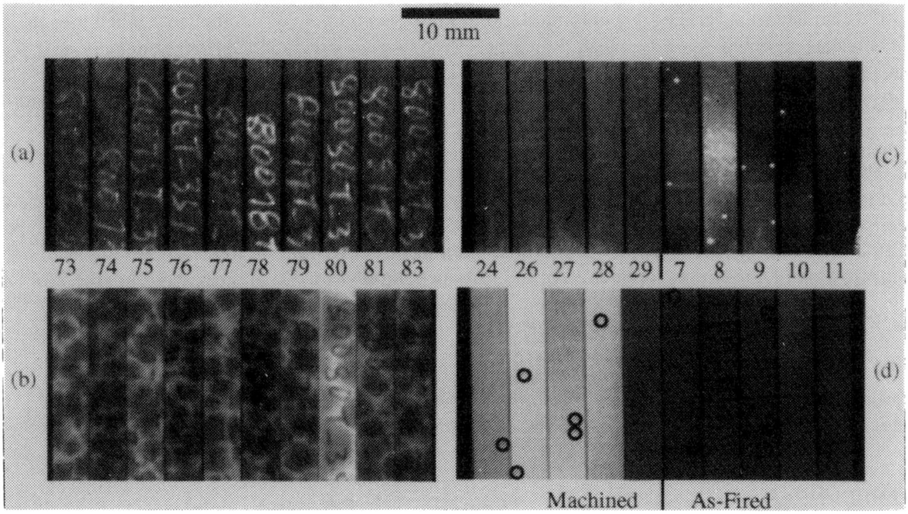

Figure 4. Images of (a and b) AlliedSignal Ceramic Components GN-10 flexure bars and (c and d) Norton NT-164 flexure bars showing (a and c) surface-sensitive laser scatter images and (b and d) subsurface-sensitive laser scatter images.

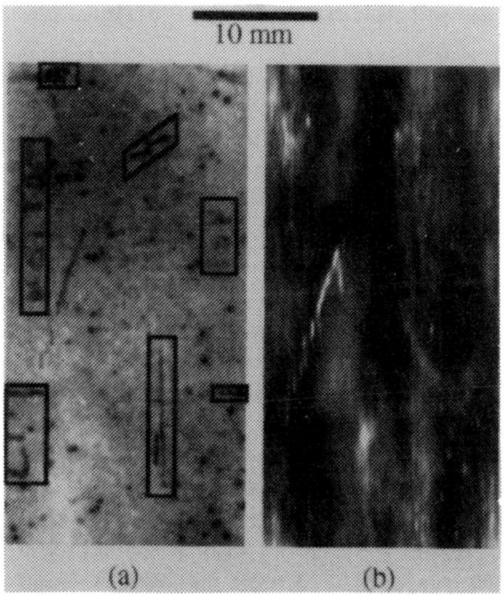

Figure 5. Images of Norton NT-164 tensile rod showing (a) surface- and (b) subsurface-sensitive laser scatter images.

CONCLUSIONS

By taking advantage of the ability of Si_3N_4 to partially transmit visible (and IR) light, we have developed a laser scatter inspection technique and applied it to the detection of defects and/or damage in several Si_3N_4 ceramics and in multiple component geometries. The types of defects detected include iron inclusions, surface scratches, variations in subsurface density, and even subsurface residual ink. Using polarization techniques, we could separate the effects of surface and subsurface defects to depths of several hundred micrometers. Furthermore, through the application of a two-dimensional scanning system, conformally mapped images of the scattering surface or subsurface were generated for flat and cylindrical specimens with a resolution of 10 μm to 1 mm. Mechanical testing of all inspected components is currently underway, but results are not yet available. However, limited results from previous analyses indicate that the type of defects that were detected (especially iron inclusions and density variations) are those that would cause premature failure. Thus, the laser scatter method holds promise for automated inspection and qualification of ceramic components for gas turbine components.

ACKNOWLEDGMENTS

The author thanks Steven Hester for his diligence in data acquisition, and Vijay Parthasarathy of Solar Turbines for preliminary mechanical testing. The work was performed under a subcontract to Solar Turbines, Inc., as part of the Advanced Turbine Systems program, under DOE Contract DE-AC02-92CE40960.

REFERENCES

1. Konig, W. and M. Popp, "High Precision Machining of Advanced Ceramics," in Third International Symposium on Ceramic Materials and Components for Engines, Las Vegas, NV, Nov. 27-30, 1988, V. J. Tennery, ed., published by Am. Ceram. Soc., pp. 1159-1224.
2. Françon, M., *Laser Speckle and Applications in Optics*. New York, Academic Press (1979).
3. Stover, J. C., *Optical Scattering: Measurement and Analysis*, New York, McGraw-Hill (1990).
4. Church, E. L., J. C. Dainty, D. M. Gale, and P. Z. Takacs, "Comparison of Optical and Mechanical Measurements of Surface Finish," *Proc. SPIE*, 954-27 (1988).
5. Sivers, E. A., D. L. Holloway, W. A. Ellingson, and J. Ling, "Development and Application of Local 3-D CT Reconstruction Software for Imaging Critical Regions in Large Ceramic Turbine Rotors," in *Review of Progress in Quantitative Nondestructive Evaluation*, Vol. 12, eds. D. O. Thompson and D. E. Chimenti (Plenum, New York, 1993), pp. 357-364.
6. Chou, C.-H., P. Parent, and B.T. Khuri-Yakub, "Acoustic Microscopy with Mixed Mode Transducers," in *Review of Progress in Quantitative Nondestructive Evaluation*, Vol. 8, eds. D.O. Thompson and D.E. Chimenti (Plenum, New York, 1989), pp. 865-871.
7. Vorburger, T. V. and E. C. Teaque, Optical Techniques for On-Line Measurement of Surface Topography, Precis. Eng. **3** (1981), pp. 61-83.
8. Bennett, J. M. and L. Mattson, *Introduction to Surface Roughness and Scattering* (Optical Society of America, Washington, DC, 1989).
9. Cielo, P., X. Maldague, S. Johar, and B. Lauzon, "Some Laser-Based Techniques for the Characterization of Sintered Ceramics," *Mater. Eval.*, **44** (May 1986), pp. 770-774.
10. Ellingson, W. A., D. M. Ayaz, M. P. Brada, and W. O'Connell, "Detection of Subsurface Defects in Machined Silicon Nitride Ceramics by Optical Scattering Methods," in *Machining of Advanced Materials*, NIST Special Publication 847, Washington, DC (1993), pp 147-157.
11. Steckenrider, J. S. and W. A. Ellingson, "Application of Laser Scattering to Detection of Surface and Subsurface Defects in Si_3N_4 Components," *Ceramic Engineering and Science Proc. Vol. **15**: 18th Annual Conference on Composites and Advanced Ceramic Materials*, Cocoa Beach, FL, January 9-14, 1994, published by Am. Ceram. Soc., pp. 382-389 (1994).

CHARACTERIZATION OF REFRACTORY MATERIALS USING SAPPHIRE STRAIN GAGES

T. A. Tran, J. A. Greene, and M. A. Alcock
Fiber & Sensor Technologies, Inc.
P.O. Box 11704
Blacksburg, VA 24062-1704
Phone: (703)382-7556

K. A. Murphy, R. G. May, G. Wang, and R. O. Claus
Fiber & Electro-Optics Research Center
Virginia Polytechnic Institute and State University
648 Whittemore Hall
Blacksburg, VA 24061
(703)231-7203

1. ABSTRACT

A high-temperature sapphire strain gage based on the optical fiber extrinsic Fizeau interferometric sensor (EFI) was used to measure strain on a compressive loaded silicon carbide rod at a temperature of 1100°C. Experimental strain sensitivities on the order of $1\mu\varepsilon$ were obtained.

2. INTRODUCTION

Fiber optic sensors are becoming more widely used for the non-destructive evaluation (NDE) of materials. They have been shown to be equally effective as, and in some cases outperform, comparable existing technologies for characterizing and evaluating materials. Due to their diminutive sizes and high sensitivity, fiber optic strain gages such as the extrinsic fiber-optic Fizeau interferometer (EFI) are ideal for embedment in composites for structural evaluation and cure monitoring. Subsequently, they can be used to detect surface acoustic waves generated by acoustic emissions during material failure [1,2]. For embedded applications, fiber optic sensors are the only options for obtaining data since resistive strain gages typically cannot withstand the temperatures used to cure epoxy. Silica fiber sensors can be used in medium temperature applications up to 750 °C. With a gold coating, such silica optical fiber sensors can operate up to 800 °C.

In recent years, new classes of materials have emerged and are challenging the upper temperature limits of silica fiber sensors. Materials such as metal matrix composites (MMCs), ceramic matrix composites (CMCs) including silicon carbide are processed at temperatures that push the limits of conventional fiber optic sensors. The need to characterize these materials at high temperatures is now becoming apparent, for instance, during the development of the National Aerospace Plane (NASP). Last year, the Fiber and

Electro-Optics Research Center (FEORC) in a joint cooperation with Fiber and Sensors Technologies (F&S) developed the first high temperature strain sensors composed of sapphire waveguides [3]. Sapphire melts at a temperature of 2048 °C, well beyond the temperatures at which the candidate NASP materials are required to operate. The sapphire EFI sensors are capable of matching the sensitivities of silica fiber sensors. FEORC and F&S have demonstrated sapphire Michelson interferometers used to detect strain, temperature, and surface acoustic waves[4].

With the direct assistance and cooperation from the Systems Support Materials Laboratory at Wright Patterson AFB, a high temperature surface attached sapphire strain gage was recently demonstrated. A sapphire EFI strain gage was mounted on a silicon carbide rod and was used to measure strain imparted by a 4000 lb compressive load while operating at a temperature of 1100°C. Strain sensitivity on the order of 1 $\mu\varepsilon$ was observed for the sapphire EFI. This paper describes the design challenges of a high temperature sapphire strain sensor and experimental results after its implementation.

3. SENSOR DESCRIPTION

The sapphire EFI, first described by Murphy, et al., consists of two polished sapphire rods inserted into both ends of a sapphire tube [4]. The endfaces of the two rods form a low-finesse Fabry-Perot cavity. The EFI sensing head is fabricated entirely from sapphire materials and is therefore well designed for high temperature applications. The gage length can be set by either permanently affixing the sapphire rods with high temperature adhesives and measuring the distance between the two bonding sites or, alternatively, by measuring the distance between the bonding sites where the sensor is attached to the test material.

A silica-lead-in fiber to the sensor was also employed because the sapphire rod was relatively rigid due to its relative large diameter (1 mm) and thus can only be inserted straight into the test area. The flexibility of the silica fiber allowed the sensor to be easily negotiated around bends and curves and to be routed into the testing area. Due to the present high-cost of sapphire rods and since the sensing head is the only part exposed to the hot zone, it was more cost effective to situate the remote end of the sapphire sensor in the cool zone where a silica fiber splice could be employed.

The experimental arrangement for testing EFI strain sensors is illustrated in Figure (1a). The setup uses a Fiber & Sensor Technologies EFI support system model EFPI-SS-1300 which houses an isolated 1300 nm laser diode (LD) pigtailed to a single-mode fiber and spliced onto the input leg of a bidirectional 2x2 coupler. One output leg of the coupler is spliced onto the sapphire EFI sensor head through a GRIN lens. The other output leg is submersed in index matching gel to prevent unwanted spurious back reflections. Intensity modulations are monitored at the return leg of the coupler using a germanium photodiode, and are proportional to the phase difference between the reflected and reference signals. The detected wavefront can be modeled using a plane wave representation of the complex electric field **E**(r,t), given by

$$\mathbf{E}(r,t) = A_1\, e^{j\omega t + \phi_1} + A_2\, e^{j\omega t + \phi_2}, \tag{1}$$

where A_1, A_2 represent the reference and sensing signal amplitudes. The sensing amplitude has been approximated by the following relation

$$A_2 = A_1 \left[\frac{Ta}{a + 2d\tan[\sin^{-1}(NA)]}\right], \tag{2}$$

where T is the transmission coefficient of the air-sapphire interface, d is the gap separation, a is the radius of the sapphire rod, and NA is the numerical aperture of the sapphire waveguide. The detected intensity is then

$$I = |\mathbf{E}(r,t)|^2 = A_1^2 + A_2^2 + 2\,A_1A_2 \cos(\phi_1 + \phi_2)\,, \tag{3}$$

or

$$I = A_1^2\left[1 + \frac{2Ta}{a + 2d\tan[\sin^{-1}(NA)]}\cos\left(\frac{4\pi d}{\lambda}\right) + \left(\frac{Ta}{a + 2d\tan[\sin^{-1}(NA)]}\right)^2\right], \tag{4}$$

where λ is the wavelength, $\phi_1 = 0$ is assumed and $\phi_2 = 4\pi d/\lambda$. Changes in the gap separation distance between the sapphire sensing head surfaces produce a sinusoidal modulation of the intensity output signal. Experimental results for a silica-based EFI show the signal degradation due to the air gap loss to be insignificant for a small gap displacement which is consistent with the modeling. This then allows for a very accurate and direct measurement of strain by resolving the sinusoidal output with a high precision data acquisition instrument.

4. DESCRIPTION OF THE EXPERIMENT

The high temperature experimental setup shown in Figure (1b) consisted of an RF induction heater mounted on an MTS load frame. To perform a high temperature strain measurement, the sapphire EFI sensor head was mounted parallel to the axial length of a 20 cm x 1.5 cm diameter silicon carbide rod using a high-temperature ceramic adhesive designed to withstand temperatures up to 2000 °C. As a basis for comparison, a gold-coated silica EFI sensor was attached alongside the sapphire EFI. The gold-coated sensor was not expected to operate much above the softening temperature of the silica (1000° C). The melting temperature of the gold coating occurs at 1063 °C. However, we can compare results obtained from both sensors up to the temperature at which the silica fiber began to fail.

A water cooled RF induction heater was used to produce a uniform heating zone of 1100°C localized to a 10 cm area around the silicon carbide rod. Since the silica-sapphire splice was attached with a low temperature epoxy, a specially designed water jacketed cooling cap was placed on the ends of the silicon carbide rod to protect both the splice and the load frame push rods. Coolant water passed through the cap and successfully maintained room temperature at the interconnection point.

Strain data was recorded on both a two channel XY plotter and on a digitizing LeCroy oscilloscope by slowly loading the silicon carbide rod with a 4000 lb hydraulic press. Figure 2 shows the raw intensity output signal as a function of the load for an experiment performed at room temperature with the sapphire sensors. The same data plotted as strain versus load in Figure 3 shows a very linear response as expected. At 1100 °C the sapphire sensor assembly was again used to measure strain. At this temperature, however, the silica sensor began to melt and subsequent data collection for this sensor was not possible from the remaining duration of the experiments. The response of the sensor is also shown in Figure 3. The experimental relationship between stress and measured strain shown in Figure 4 agrees very well with theoretical data.

In another experiment the survivability and repeatability of the sapphire EFI during high-cycle fatigue loading was determined. The present experiment cyclically compresses and releases the silicon carbide rod at a constant rate of 25 Hz with the load ranging from 30 lbs. to 1500 lbs. In Figure 4 we plot the stress-time pattern of the SiC rod and the measured strain-time responses of sapphire EFI sensor after it was fatigue loaded for 60,000 cycles. Using a modulus of 37 Mpsi, we calculated 105 microstrain for an applied load of 3880 psi. Experimentally, the strain was determined to be 100 microstrain - an error margin of 5%. One source of the error may be attributed to the Young's modulus of the SiC having an approximate range of 29 - 58 Mpsi. Because the first order strain is related to stress by the simple equation

$$\sigma = \varepsilon E \, , \tag{5}$$

where σ is the stress, ε is the strain and E is the Young's modulus, different values of the modulus will result in maximum variation in the calculated strain.

After having fatigue loaded to 200,000 cycles at 1727 °F at the same load, the waveform from the sensor shown in Figure 6 was obtained. The averaged strain was again measured to be 100 microstrain. The collected data suggests that the sapphire EFI sensor is repeatable regardless of the number of times it has been fatigue loaded. Variation in the amplitude of the strain-time pattern is attributed to the coefficient of thermal expansion of the SiC rod.

The survivability of the sensor has also been demonstrated. Although the load frame was shut-off after 200,000 cycles, one would expect to be able to fatigue load the sensor for millions of cycles since the components of the EFI sensing head are free to move inside the alignment tube and fiber components are essentially not loaded. This is consistent with recent data obtained for similar silica optical EFI sensors that were fatigue loaded up to million cycles .

5. CONCLUSION

A sapphire extrinsic Fabry-Perot interferometric sensor has been demonstrated for high temperature applications. The sapphire EFI was used successfully to measure the strain on a silicon carbide rodwhich was compressively loaded to 4000 lbs at temperatures up to 1100°C (limited by the induction heater and not the sapphire sensor). The signal from the sapphire sensor shows an exact correlation with the strain response of a conventional silica EFI sensor for both static and fatigue loading. Additional tests have successfully demonstrated the sapphire EFI up to 1760°C for both temperature and strain detection. The sensor measures strain, both thermal and mechanical, which can be separated by using two EFIs. One of the EFIs would be attached to measure strain and the other co-located but not attached could be used to measure temperature only.

6. REFERENCES

1. M. S. Miller, S. W. Case, R. G. May, G. P. Carman, and R. O. Claus, SPIE OE/Fibers '92, Boston, Mass., Sept. 8-11.
2. S. E. Baldini, E. Nowakowski, and H. Smith, SPIE OE/Fibers '91, Boston, Mass., Sept 5-6.
3. A. Wang, S. Gollapudi, K. A. Murphy, R. G. May, R. O. Claus, Opt. Lett. 17, 1023 (1992).
4. K. A. Murphy, G. Z. Wang, B. R. Fogg, A. M. Vengsarkar, and R. O. Claus, Proc. Soc. Photo-Opt. Instrum. Eng. 1588, 117 (1991).
5. K. A. Murphy, M. F. Gunther, A. M. Vengsarkar, and R. Claus, Opt. Lett. 16, 273 (1991).
6. G. P. Carman, K. Murphy, C. A. Schmid, and J. Elmore, SEM, Spring Conf. 1993, pp. 1079-1087.

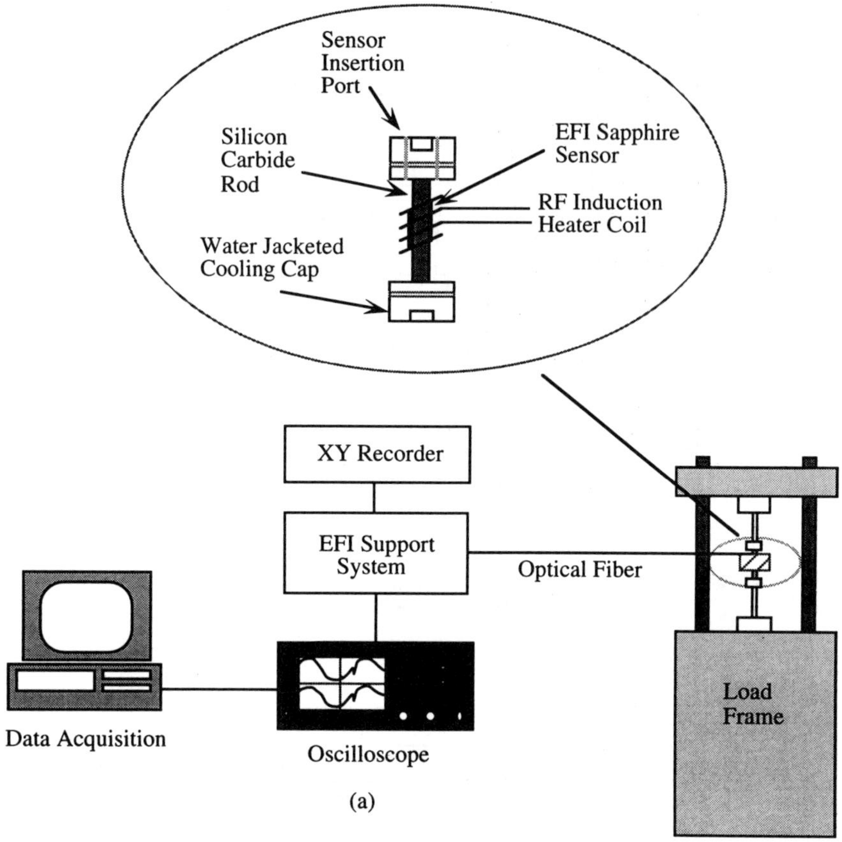

(a)

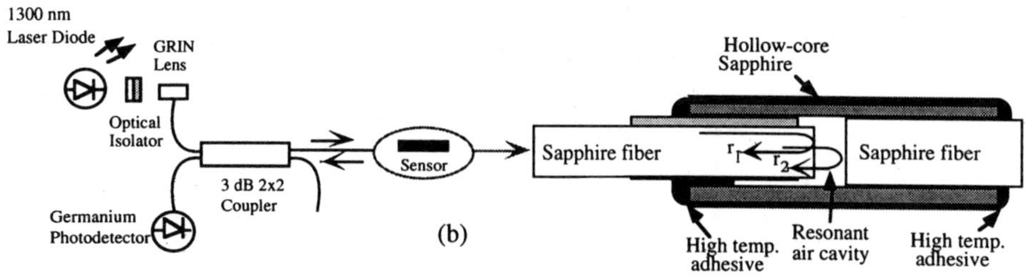

(b)

Figure 1. (a) Sapphire Extrinsic Fizeau Interferometricr Strain Gage
(b) High Temperature Strain Sensor Setup for EFI.

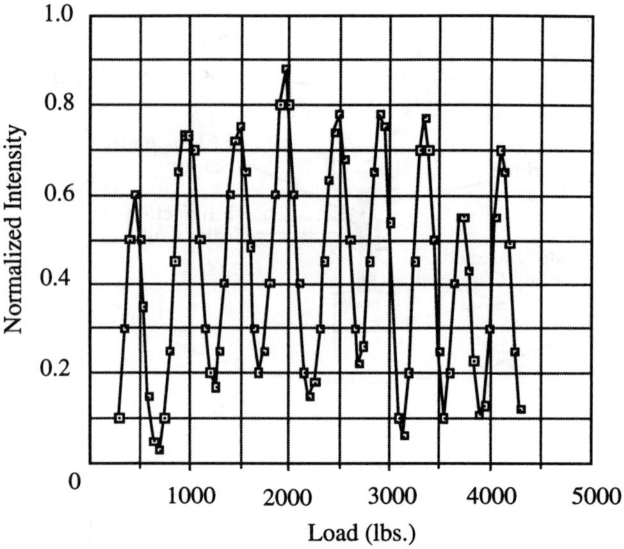

Figure 2. Variation of the EFI output intensity as a function of applied load on a silicon carbide rod.

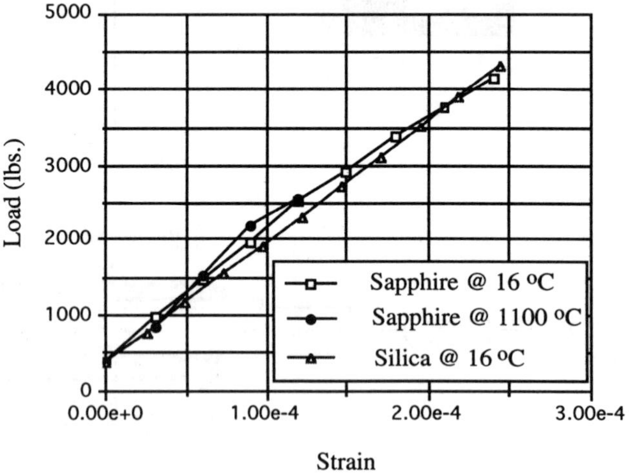

Figure 3. Strain versus load curve for silicon carbide rod measured using sapphire and silica fiber sensors.

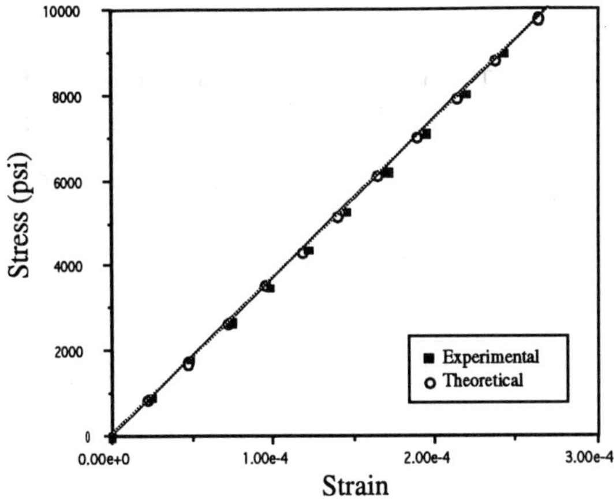

Figure 4. Stress versus strain for SiC loaded at 2000° F.

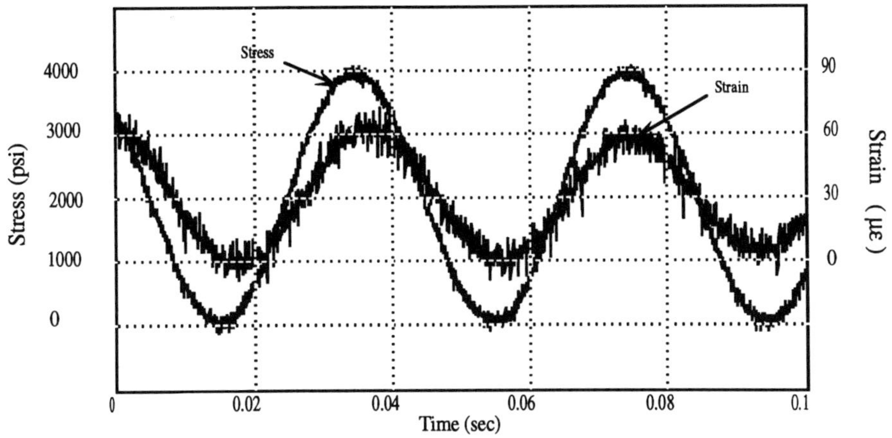

Figure 5. Data from the sapphire EFI strain sensor attached to SiC rod. Sample of fatigue loading history and measured strain-time output at 60,000 cycles, 1940°F.

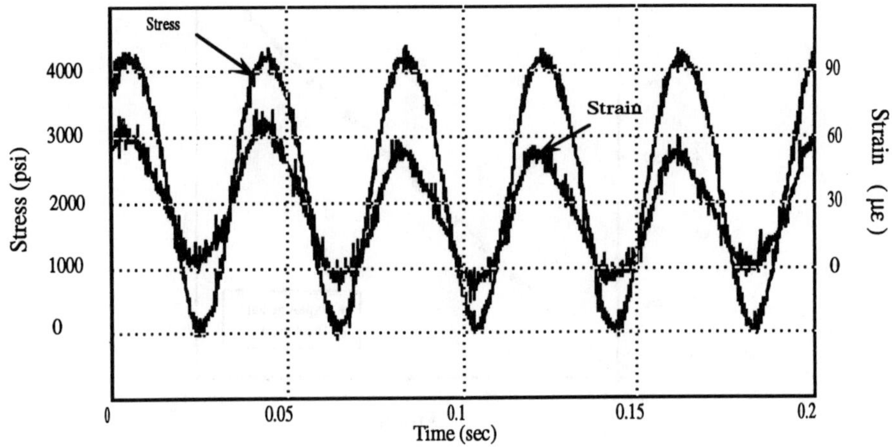

Figure 6. Data from the sapphire EFI strain sensor attached to SiC rod. Sample of fatigue loading history and measured strain-time output at 200,000 cycles, 1727° F.

OXIDE FILMS AND NANO-SIZED POWDERS FABRICATED BY MODIFIED PLASMA CVD

C. Q. Shen, K. D. Vuong, J. A. A. Williams, and X. W. Wang
Alfred University, Alfred, New York 14802

ABSTRACT

We report new results on nano-sized oxide films and powders deposited by a modified plasma CVD process. An RF aerosol plasma technique was used to deposit various oxides including Y_2BaCuO_5 (211) for $YBa_2Cu_3O_{7-x}$ (123) additions, BiSrCaCuO superconductors, indium-tin oxide (ITO) transparent conductive films, yttria stabilized zirconia(YSZ) as buffer layer and fuel cell electrolyte material, gadolinium iron oxide for magnetic heat pumps, as well as others. The deposited films or powders consisted of nano-sized crystallites as revealed by XRD peak broadening (SHADOW Program). The films and powders were also characterized by SEM, EDS, Mössbauer, optical, and resistivity measurements.

INTRODUCTION

Nanosized ceramic oxide powders and films were prepared using an atmospheric, RF aerosol plasma deposition technique. Nanoscaled materials are believed to exhibit better thermal and mechanical properties compared to micrometer sized materials.[1] These advantages include possible lowered sintering temperatures and an increased fraction of atoms in grain boundary areas.

Currently, many deposition processes for nanoscale materials occur in a low pressure atmosphere. Some of these include sputtering, vacuum evaporation, laser, and low pressure CVD methods.[2] The atmospheric, RF aerosol plasma deposition technique has advantages over vacuum deposition methods which include; substrate size is not limited by vacuum chamber size, deposition time is not affected by vacuum chamber pump down time, decreased initial capital investment, and lower operation and maintenance costs. This RF plasma process simplicity and lack of a vacuum system gives it an excellent potential for use as a large scale industrial production process.[3]

Films deposited using this process were characterized using XRD, crystallite size, SEM, EDS, resistivity, optical, and Mössbauer measurements. The Shadow program was used to calculate film and powder crystallite sizes via XRD peak broadening.[4]

EXPERIMENTS

Ceramic oxide thin films and powders were prepared using an RF aerosol plasma unit, illustrated in figure 1. The unit consists of an RF (4 MHz) power supply (not shown), an induction plasma torch[5], a starting material assembly, and a substrate assembly. The induction plasma torch (35 KW maximum generator power) consists of an RF coil, plasma gas inlets, a mist inlet, and a quartz tube. The starting material assembly consists of an ultrasonic nebulizer, a solution container, a carrier gas inlet (Ar), and a mist outlet. The substrate assembly consists of a metal block with heating elements and a K-type thermocouple.

Argon is used as the plasma forming gas and the carrier gas. Oxygen is added as a secondary plasma gas to enhance the oxidation of deposited films or powders. The mist, generated from the starting solution via an ultrasonic nebulizer (1.63 MHz), is fed into the plasma region under the pressure of the Ar carrier gas. The cations in the starting solution are vaporized in the plasma region, and then deposited on the substrate (thin films) or collector (powders).

To prepare the starting solutions, nitrate and/or chloride based powders were mixed with the appropriate ratios and dissolved in distilled water. Typical deposition parameters are similar to those of ITO which are as follow: Axial flow 5 l/min. (Ar), Swirl flow 7 l/min. (Ar), Swirl/Tangential flow 10 l/min. (O_2), mist feed rate 1-2 l/min., nebulizer power 40%, and generator power 30%. The substrate temperature varies from 350°C to 900°C, dependent on the material deposited.

As deposited films were characterized by X-ray diffraction (XRD), scanning electron microscopy (SEM), and energy dispersive spectrometry (EDS) elemental analysis. For some films, optical, Mössbauer, and resistivity (standard 4-point method with silver paint contacts) measurements were performed. Table I lists the crystallite sizes of films and powders produced using this atmospheric plasma process.

RESULTS

$BiSrCaCuO_x$ (Bi-2223) Films:

Bi-2223 thin films were formed on single crystal MgO substrates. Figure 2 shows the XRD pattern for a Bi-2223 film. Only

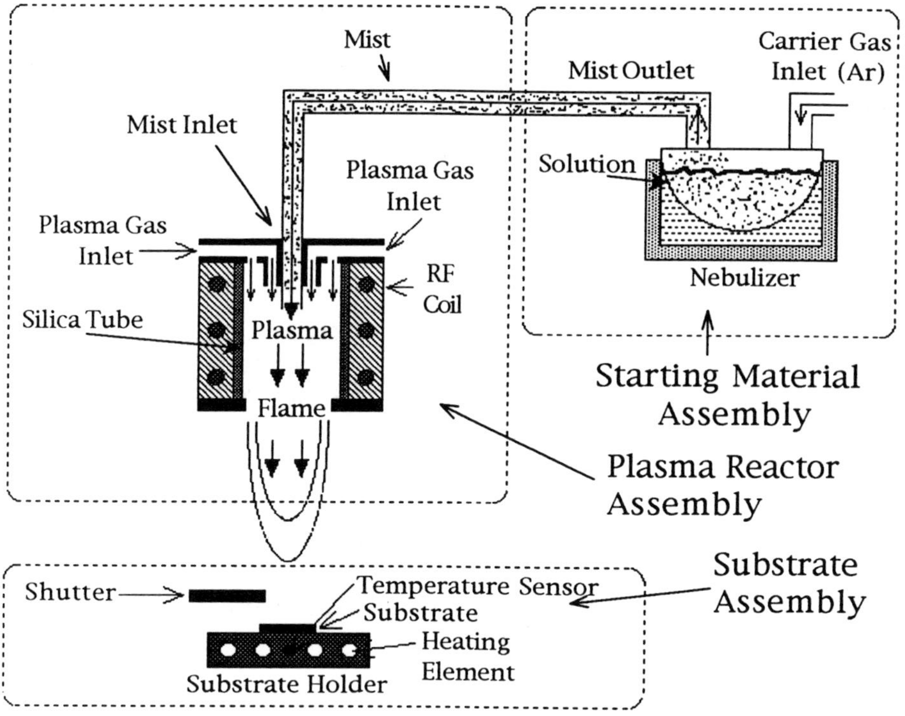

Figure 1. Atmospheric RF Aerosol Plasma Deposition Apparatus

Table I. Crystallite size of plasma deposited films and powders

Material	Crystallite Size (nm)
Bi-2223	6 - 14
ITO	6 - 22
YSZ - Powder - Film	22 - 31 28 - 50
211	
- Collector Temp. 500°C	53
- Collector Temp. 570°C	70
- Collector Temp. 650°C	105
GdFeO	
As-Deposited	37-50
Heat Treated	122-160

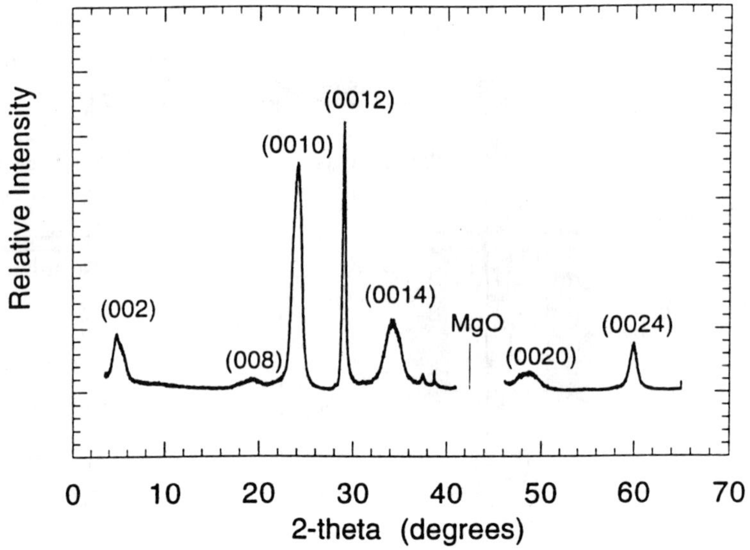

Figure 2. XRD pattern for a Bi-2223 as-deposited film. [Only (00l) peaks are visible]

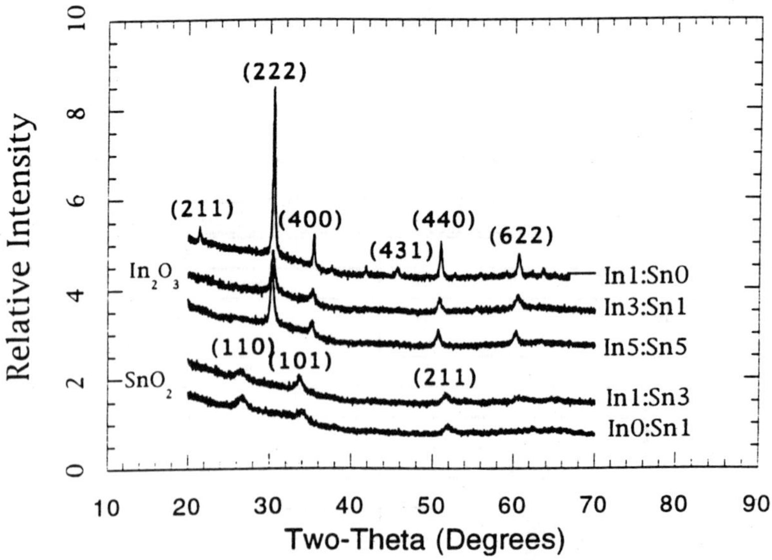

Figure 3. XRD patterns of indium-tin-oxide thin films on SLS float glass

the 001 peaks for Bi-2223 were visible in the XRD pattern, suggesting that the deposited Bi-2223 film is single phase and strongly oriented with the c-axis perpendicular to the substrate. X-ray peak broadening revealed the crystallite sizes range from 6 to 14 nm.

Variations to Bi-2223 deposition parameters allow the formation of films with different morphologies. Current density of the films were measured at 40 K. For a film with poorly oriented grains, a J_c of 2.0×10^2 A/cm^2 was measured. A slightly increase in grain orientation resulted in a Jc of 8.4×10^3 A/cm^2. A highly oriented film has a $J_c = 1.2 \times 10^5$ A/cm^2. The thickness of the films were observed, using the SEM, to be 0.2 - 0.7 μm. The J_c values for the deposited films were strongly dependent on the film microstructure, increasing with grain orientation and crystallization.

Indium-Tin Oxide Films:

Transparent, conductive indium-tin oxide thin films were deposited on soda-lime silicate float glass at a rate of 0.5 to 1.0 μm/hour using this plasma technique. Films were deposited with varying In/Sn starting solution ratios (1/0, 9/1, 3/1, 2/1, 5/5, 1/3. 1/9, and 0/1). The starting solutions were formed from $SnCl_4 \cdot 5H_2O$ and $In(NO_3)_3 \cdot 5H_2O$ powders dissolved in distilled water.

XRD analysis of the films showed that at ratios of 5/5 or greater indium concentration, In_2O_3 (JCPDS Card # 6-416) is the only visible phase. At indium concentrations less than this, SnO_2 (JCPDS Card # 41-1445) is the only visible phase. Figure 3 shows the XRD patterns for the as-deposited films. Through the use of XRD peak broadening and the SHADOW program, the crystallite size of the ITO films were found to range from 6 to 22 nm. In general, the crystallite size of the high tin content films was smaller than that of the high indium content films. This trend may be due to the difference in the lattice parameters of tin oxide and indium oxide, a = 4.7382 Å and c = 3.1871 Å for tetragonal tin oxide, and a = 10.118 Å for cubic indium oxide. Tin oxide has a smaller unit cell than indium oxide which may permit the formation of smaller crystallites.

Films deposited with a starting solution concentration of ≥ 50% indium were conductive, with room temperature resistivities ranging from 0.12 to 5.0 Ω-cm, for In1:Sn0 and In5:Sn5 films respectively. The optical properties of the ITO films were not dramatically different than the glass substrate throughout the complete testing wavelength ranges of 200 nm - 25 μm, as determined using IR-UV-Vis transmission, IR-UV-Vis reflective, and FTIR reflective spectrometry. A SEM micrograph of a typical ITO film magnified 50,000x is shown in figure 4. The particle size ranges from 50 to 150 nm. There are also no sodium chloride crystals visible on the surface of the films. Sodium chloride contamination is a negative

Figure 4. SEM micrograph of an ITO film (50,000X)

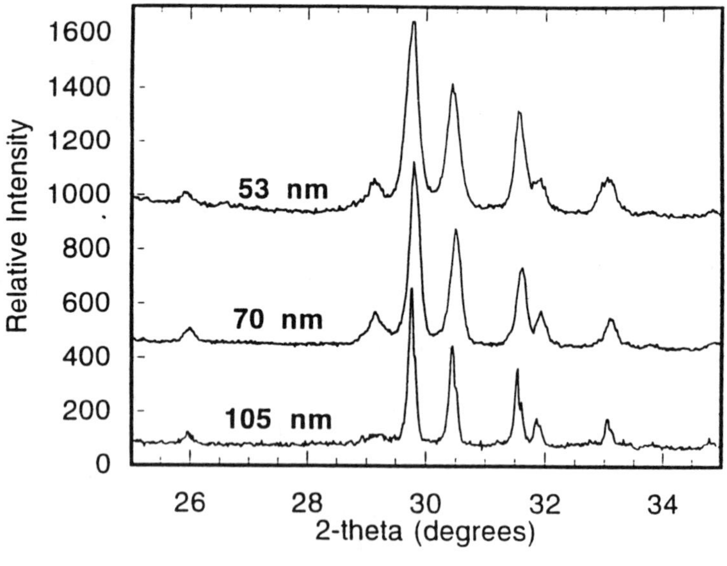

Figure 5. XRD peak broadening effect, showing crystallite sizes for three different samples

characteristic of ITO films deposited using other techniques, where sodium migrated from the glass substrate and reacted with chlorine contaminants from the chloride based starting solutions.[6]

Yttria Stabilized Zirconia (YSZ) Films:

YSZ thin films were deposited on Ni containing substrates. All films were deposited using a 10 mole % yttrium solution. As-deposited films were yttria stabilized zirconia (cubic) as determined by XRD. The crystallite size of the YSZ powder ranged from 22 - 31 nm. The density and morphology of the film were dependent on the substrate temperature and deposition rate. Correspondingly, the deposition rate is controlled by the mist feed flow rate, reactor-substrate distance, nebulizer setting, and solution concentration. As-deposited YSZ films are uniform, consisting of 28 - 50 nm crystallite size. Lower heat treatment temperature can be used as a result of the very high reactivity of the nanometer crystallite size film.

Y_2BaCuO_5 (211) Powders:

Y_2BaCuO_5 (211) powders were scrapped off a bulk, single crystal MgO collector (5 cm X 5 cm). When the starting material ratios were close to 0.194:0.158:0.158 for Y:Ba:Cu, and the substrate temperature remained between 500°C and 600°C, phase pure 211 powders were obtained. The window for solution ratios was about 5%. If the ratios fell outside this region, impurities (usually Y_2O_3, $YBa_2Cu_3O_{7-x}$, and $Y_2Cu_2O_5$) were present. The crystallite size of the as-deposited powder was affected by the MgO collector temperature. Higher collector temperatures yielded larger crystallite sizes in the 211 powder. Figure 5 shows the XRD patterns for three 211 powders deposited at different collector temperatures. The peak broadening decreased for powders with increasing collector temperatures of 500, 570, and 650°C, which correspond to crystallite sizes of 53, 70, and 105 nm.

Gadolinium Garnet Powders:

Nanocrystalline GdFeO powder and thin films were prepared. As-deposited powder consists of Gd_2O_3 and Fe_2O_3. While as-deposited films consist of $GdFeO_3$, plus impurities. These phases were determined via XRD analysis. Post-deposition heat treatment of the powders yielded a $Gd_3Fe_5O_{12}$ garnet phase, along with $GdFeO_3$ Orthoferrite phase. The crystallite size of the films was affected by post-deposition heat treatments. Prior to heat treatment, the crystallite size ranged from 37 - 50 nm, while the heat treated crystallite size increased to 122 - 160 nm.

CONCLUSIONS

Films and powders with nanosized crystallites were deposited via an atmospheric, RF aerosol deposition method. Superconductive $BiSrCaCuO_x$ (2223) films were deposited with a Jc dependent on grain orientation. Transparent, conductive indium-tin oxide films were deposited on glass substrates. The crystalline phase present in the ITO films was either In_2O_3 or SnO_2, dependent on the In:Sn ratio in the starting solution. Yttria stabilized zirconia (Cubic) films were deposited for use as buffer layers. Film adhesion and morphology was dependent on substrate temperature. Phase pure Y_2BaCuO_5 (211) powders were formed for use as additions to $YBa_2Cu_3O_{7-x}$ (123) superconductors. Higher collector temperatures yielded larger 211 crystallite sizes. Gadolinium garnet powders and films were also deposited. Post-deposition heat treatments of the GdFeO materials affected the crystallite phase and crystallite size.

This atmospheric, RF aerosol deposition process has excellent potential for use as a large scale industrial production process due to its simple operation, material versatility, and lack of a vacuum system.

ACKNOWLEDGMENTS

This work was partially supported by the Center for Advanced Ceramic Technology, the Center for Glass Research, the U.S. Air Force, and Argonne National Laboratory. We would also like to acknowledge the assistance of Dr. Robert Snyder.

REFERENCES

1. R.P. Andres, R.S. Averback, W.L. Brown, L.E. Brus, W.A. Goddard, III, A. Kaldor, S.G. Louie, M. Moscovits, P.S. Peercy, S.J. Riley, R.W. Siegel, F. Spaepen, and Y. Wang, "Research Opportunities on Clusters and Cluster-Assembled Materials- A Department of Energy, Council on Materials Science Panel Report," J. Mater. Res., 4, [3], 704-736, (1989).
2. H.K. Pulker, *Coatings on Glass,* pp. 92-275, Elsevier Science Publishing, NY, NY, 1984.
3. T. Okada, H. Hamatani, and T. Yoshida, "Radio-Frequency and Hybrid Plasma Spraying of Ceramics," *J. Am. Ceram. Soc.*, 72 [11] 211-16 (1989).
4. S.A. Howard, *SHADOW: A System for X-Ray Powder Diffraction pattern Analysis; Annotated Program Listings and Tutorial,* University of Missouri-Rolla, 1990.
5. Tafa Model 56 Plasma Torch. TAFA Metallisation Inc., Concord, NH.
6. W. LaCourse and H. Black, "Coatings Research: Search for Enhancement," *The Glass Researcher,* 1 [2] pp. 1,3&11 (1991).

Sol-Gel Technology

SOL–GEL PROCESSING

Sumio Sakka
Department of Applied Physics and Chemistry
Fukui University of Technology
3-6-1 Gakuen, Fukui, 910 Japan

ABSTRACT

Sol-gel processing has been reviewed from the standpoint of preparing materials in the shape of a bulk cylinder or plate, fiber or coating. Also, microstructures achieved by sol-gel processing have been reviewed. It is indicated that sol-gel processing is versatile and produces a wide variety of high technology materials having optical, electronic, chemical and biological functions. Processing of some optical coating films is discussed in more detail.

INTRODUCTION

Usually, materials are useful when they are prepared in certain shapes: bulk, fiber and coating film. Especially for inorganic materials like glasses, ceramics and composites, preparation of shaped, high performance materials is not easy because of their brittle nature. Such difficulty in forming useful shapes can be overcome through sol-gel processing of materials [1,2].

As shown in Fig.1, the sol-gel method starts with solution containing metal compounds. The solution becomes a sol which contains fine, solid particles produced as a result of hydrolysis and polycondensation of the starting compound. It is easy to form coating films and draw fibers from the sol. It is known that formation of ceramic coating films and fibers of desired compositions is not easy in the conventional processing which starts with powder forming. Bulk gels can be made by molding.

Gels may be final products which can include organic materials because of low temperature processing. Heating gels to high temperatures produces glasses, ceramics and composites with various microstructures[3]. It should be noted that the final heated materials can be obtained at much lower tem-

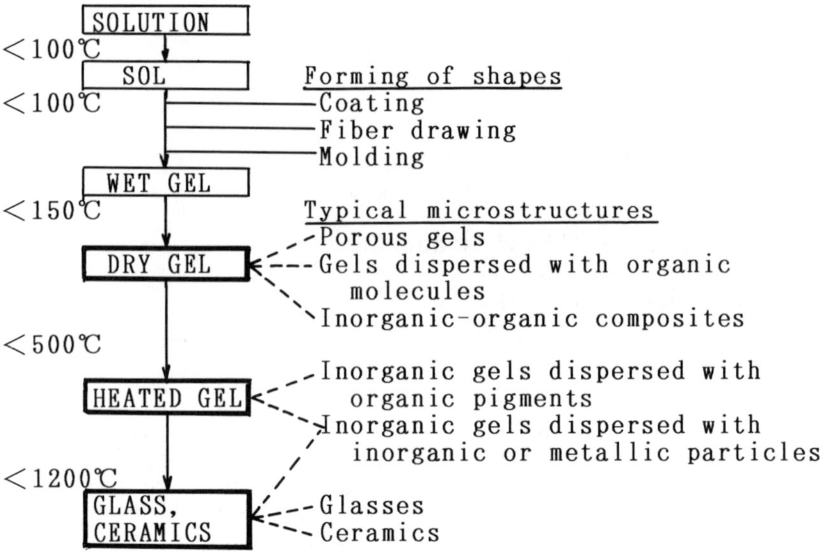

Fig.1 Sol-gel processes. Forming processes and microstructures are also shown. Bold-lined frames show final products.

peratures than in conventional processing of glasses and ceramics.

FORMATION OF SHAPES

There are three kinds of useful shapes: bulk,fiber and coating film. All these shapes are formed in the course of sol-to-gel transition. The bulk bodies are made by molding the sol. Much effort has been made in preparing bulk silica gels and glasses. The most serious problem encountered in preparing bulk bodies is the formation of cracks and fracture which occurs during drying of wet gel bodies containing solvents. The capillary force arising in the pore solution in the gel may be the cause for the occurrence of cracks. The common principle underlying the techniques so far adopted[4] is to suppress the capillary force acting on the silica backbones at the final stage of drying. Three series of silica gel bodies, which have an average pore diameter 2.5 nm [5], 16 nm [6] and 5 μm [7], respectively, have been made without cracks.. Bulk gel bodies may be useful as micr ofiltersand catalyst supports. Bulk silica glasses may be used as an optical glass preform and photomask substrate for IC preparation.

In the direct sol-gel fiber formation, gel fibers are drawn from the viscous sols [8]. For silica fibers, we start from a solution containing $Si(OC_2H_5)_4$, water, ethanol and hydrochloric acid. Drawing from the solution of viscosities higher than ca 10 poises gives silica gel fibers, which are converted to silica glass fibers on heating to 800℃. It should be noted that not all solutions become drawable. Fibers can be drawn when a starting solution consisting of low content water and acid catalyst is used so that long-shaped sol particles might be formed in the viscous solution. These conditions for the occurrence of spinnability also apply to solutions for other oxide fibers than silica [9]. Thus, various kinds of insulating, electroconducting and superconducting oxide fibers have been made by the sol- gel method [10].

Coating of glasses, ceramics, metals and plastics with oxide films can be easily made by the sol-gel method [12,13]. Preparation of coating films of glasses and ceramics consists of (1) application of coating film by dip coat, spin coat or dripping and (2) heating to 300-500℃ for adherence of the film to the substrate. Additional heating to higher temperatures may be necessary for sintering or precipitation of desired crystals. In order to make uniform crack-free inorganic film adhering firmly to the substrate , the coating conditions have to be adjusted, so that the final thickness might be smaller than the critical thickness of 0.1-0.3 μm. Such conditions are achieved with low concentrations of the coating solution, low rates of withdrawal in dip coating and use of low viscosity solutions. The thickness of the coating film can be increased by repeating the whole coating procedure. The critical thickness is larger than 0.7 μm when silica films are prepared from a solution containing $CH_3Si(OC_2H_5)_3$ [14]. The critical thickness may be much larger ($>$10 μm) for organic-inorganic hybrid films[15]. This is attributed to the small shrinkage and low stiffness of the film, which reduces the possible stress by plastic deformation.

Coating films can protect the substrate chemically or mechanically and provide the substrate with characteristic functions, such as optical, electronic and magnetic, chemical, catalytic and biological functions.

MICROSTRUCTURES OF SOL-GEL PREPARED MATERIALS

Microstructures listed in Table I can be achieved in dried gels, heated gels and glasses and ceramics [3]. In early

Table I. Microstructures of materials achieved by sol-gel preparation

Microstructure	Shape	Example
Gels		
Inorganic porous	Bulk, Coating	Insulator Filter Catalyst Catalyst support
Gel with dispersed organic molecules	Bulk, Coating	Optical lens Luminescent material Laser PHB material Photochromics Nonlinear optical material
Inorganic-organic hybrid	Bulk, Coating, Powder	Optical material Water repellent film Medical powder
Heated gels		
Amorphous oxide	Bulk, Coating, Fiber	SiO_2 and many other oxides
Metal colloid-dispersed oxide	Bulk, Coating	Metals Au, Pt, Pd, Ag in matrix SiO_2, Al_2O_3, TiO_2
Glasses and ceramics		
Dense glass	Bulk, Coating, Fibers	SiO_2 (Preform for optical fibers) Other glasses
Randomly crystalline	Bulk, Coating, Fiber	Optical material Nonlinear optical material Ferroelectrics Superconducting oxides
Preferentially crystal-oriented	Coating	Ferroelectrics Optical material Superconducting oxides

stage of sol-gel technology (1970-1985), glasses and ceramics were the prime targets and final products. These materials are still very important, but at present, many kinds of dried gels of various microstructures are regarded as one of the final products. This has markedly enlarged the range of materials and made the sol-gel method quite versatile.

Inorganic, as-dried gels are usually characterized by porous microstructure. With the same oxide composition, e.g. SiO_2, the size and shape of particles and pores are markedly dependent on the composition of the starting solution, the temperature of gelation, the atmosphere during sol-gel conversion, drying condition and so on[7,16,17]. Thus, the pore diameter can be changed from ~1.0 nm to 10 μm. The porous gels can be used as filter and support for catalysts and enzymes. Transparent silica gels may be used as optical elements [18].

Silica gels can be prepared at temperatures lower than 100℃ to 150℃. Based on this advantage, Avnir et al. [19] and Makishima and Tani [20] introduced functional organic molecules, such as rhodamine B, resorcin, quinizarin, and spiropyrane, which show optical activities like laser action, luminescence, photochemical hole burning, photochromism and nonlinear optical properties. In this case, organic molecules are added to the starting solution. When a disc thicker than several mm is used as host, the disc should be transparent with fine pores and crackless [5].

Since the gel is produced at low temperatures, inorganic-organic hybrid materials can be prepared by using a mixture of inorganic and organic species as starting solution. The resultant materials have a microstructure in which inorganic species like -Si-O-Ti-O- are combined with organic polymers at the molecular level [21]. These materials have an excellent combination of functional properties along with good mechanical properties.

Heated gels found in Table 1 have microstructures achieved by heating gels to intermediate temperatures which are lower than those required for formation of glasses and ceramics. An example of heated gel of amorphous oxide is given by SiO_2 gel [22]. When the silica gel prepared from a solution of $Si(OCH_3)_4$-$(CH_3)_2NCHO$-CH_3OH-H_2O-HCl system is heated, the gel network becomes strong up to 700℃ and above 840℃ the porosity and the average pore size of the gel rapidly decrease. The gel becomes a full dense silica glass on heating at 1050 ℃ for 2h. This means that we can produce silica gels of which the porosity and pore size and, accordingly, mechanical and chemical properties vary in wide ranges [16].

Heating of dried gels containing reacting chemical species to higher temperatures than 150℃ provides solids with microstructures in which fine particles of metals and semiconductors, as listed in Table I, are dispersed in an oxide matrix [23]. The resultant materials are important nonlinear optical materials with high values of $\chi^{(3)}$ [24]. The sol-gel processing of SiO_2 and TiO_2 matrix doped with metal colloids will be described in some detail as examples.

As indicated in Table I, glasses and ceramics are formed when oxide gels are heated to sufficiently high temperatures, which are still lower than the temperatures for conventional melting of glass and solid state sintering preparation of ceramics. It is remarkable that the sol-gel process makes it possible to prepare oxide coating films in which crystallites are highly oriented. Hirano and Kato [25] prepared homogeneous and stoichiometric $LiNbO_3$ films with preferred orientation by crystallizing the sol-gel coating film at 400 ℃ on α-Al_2O_3 single crystal substrate. Preferred crystal orientation may be quite important for ferroelectric and superconducting crystalline coating films. It has been shown that for the coating film of $Li_2O \cdot 2B_2O_3$ composition, glassy, randomly crystallized and preferentially crystal-oriented films can be prepared by changing the composition of the starting solution [26].

Besides those listed in Table I, the following microstructures can be successfully prepared by the sol-gel method: (1)microsturcture providing functionally gradient materials[27], (2)microstructure giving gradual porosity change for antireflection coating film [28], (3)microstructure for asymmetric ultrafine filter [29] and (4)microstrucdture of composites produced by impregnation of porous gel. For the details, consult with references mentioned above.

EXAMPLES OF OPTICAL COATING FILMS

Among materials prepared by the sol-gel methods, coating films of optical functions are especially important due to the present trend of high technology. Preparation and properties of two kinds of coating films with different microsructures will be mentioned in the followings.

Ceramic Coating Films of α-Fe_2O_3 and TiO_2

It is easy to prepare oxide coating films by the sol-gel method. As seen in Fig.2, we start with iron(III) nitrate enneahydrate $Fe(NO_3) \cdot 9H_2O$ for fabricating α-Fe_2O_3 coating film [31]. The iron nitrate was dissolved in a mixture of 2-methoxyethanol $CH_3OCH_2CH_2OH$ and acetylacetone $CH_3COCH_2COCH_3$ and was kept standing for two hours at 30℃. Dip-coating was carried out by immersing a silica glass substrate into the coating solution, pulling up at a rate of 3 cm/min and heating the resultant gel film at 800℃ for 10 min for α-Fe_2O_3 formation.This procedure was repeated four times to

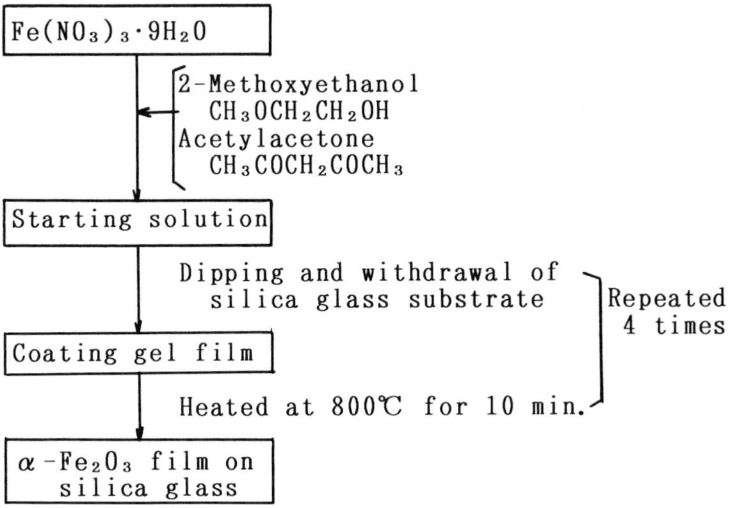

Fig.2 Sol-gel preparation of α-Fe_2O_3 film

obtain the film thickness of about 0.07 μm appropriate for the measurements.

X-ray diffraction pattern of the resultant film shows that the film consists of single phase α-Fe_2O_3. No noticeable orientation is observed and the film is transparent. With this coating film, the third-order nonlinear optical susceptibility $\chi^{(3)}$ has been determined by the third-harmonic generation (THG) method. The $\chi^{(3)}$ of the α-Fe_2O_3 film was 5.8×10^{-11} esu. As shown in Fig.3, this value is the highest among inorganic compounds and is comparable with those of organic materials with high nonlinearity.

In Fig.3, the $\chi^{(3)}$ data on sol-gel derived TiO_2 coating films [32] are also shown. It is seen that the rutile phase shows larger $\chi^{(3)}$ than the anatase phase and that the denser film has higher $\chi^{(3)}$ than the porous film.

SiO_2 and TiO_2 Coating Films Containing Dispersed Noble Metal Fine Particles

Glasses and transparent ceramics containing fine noble metal colloids are attracting much attention because of their high third-order optical nonlinearity due to surface plasmon excitation. We have made several such systems by combining Au, Ag, Pt and Pd colloids with SiO_2 and TiO_2 matrix [33-36]. The

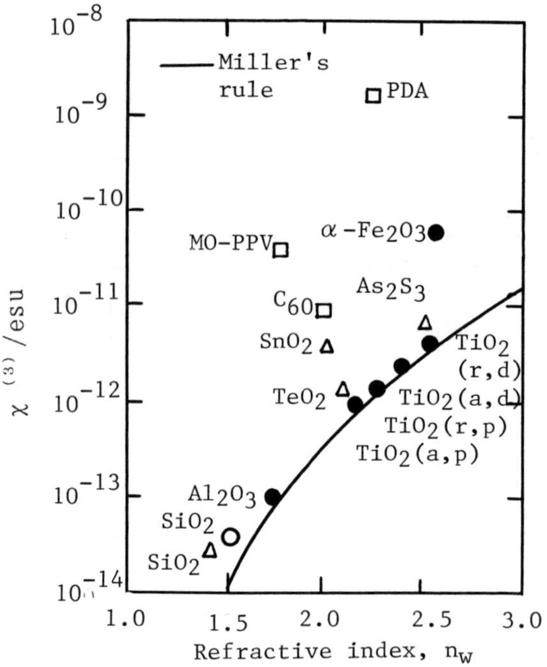

Fig.3 Relationship between $\chi^{(3)}$ and refractive index, n_w, at 1900 nm for α-Fe_2O_3 thin films and other nonlinear optical materials[32]. Circle, triangle and square denote inorganic crystal, inorganic amorphous material, and organic material, respectively. Closed circles are the data by the present authors. The letters of r, a, d, and p in parenthese denote rutile, anatase, dense, and porous, respectively. PDA: polydiacetylene, MO-PPV: poly(2,5-dimethoxy p-phenylene vinylene).

sol-gel method has been adopted by aiming at incorporating a high concentration of metal colloids and controlling the size and shape of colloidal particles. For this purpose we started from the homogeneous solution; for example, homogeneous $Si(OC_2H_5)_4$-H_2O-HCl-C_2H_5OH-$HAuCl_4 \cdot 4H_2O$ solutions were employed as starting solution, gelled and heated to 500℃-800℃ for the precipitation of gold particles. By this method 2.8 mol% Au can be incorporated and Au particles precipitated in SiO_2 gels can be changed from several to several tens nanometers [33].

We have also made Au colloids in TiO_2 coating films. It was expected by calculation [34,35] that the absorption peak of Au colloids due to surface plasmon excitation would shift to longer wavelengths when the matrix with higher dielectric constant is employed. Fig.4 shows that this is the case. It is seen that the absorption peak lies at 550 nm for SiO_2 matrix and at 650 nm for TiO_2 matrix.

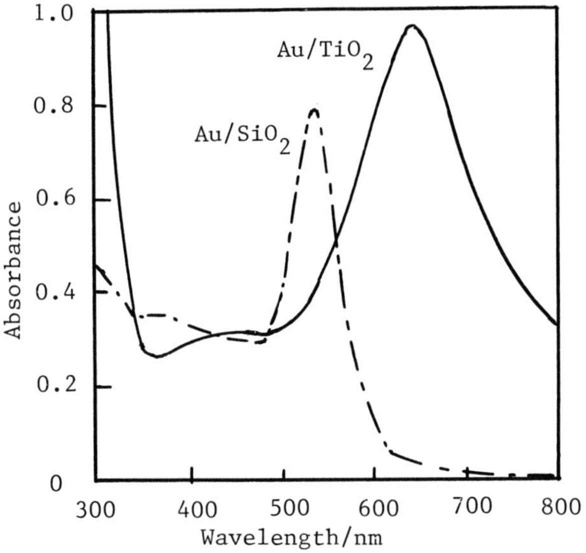

Fig.4 Optical absorption spectra of the Au/TiO_2 and the Au/SiO_2 composite films containing films containing round-shaped Au particles of about 20 nm in size.

This relation was also confirmed in Pd colloids in the matrix prepared by the sol-gel method. It was found [36] that Pd colloids in the SiO_2 film do not absorb in the visible region, while Pd colloids in TiO_2 film show large absorption in the visible region ranging up to about 600nm or longer wavelengths.

SUMMARY

It was shown that the sol-gel technique produces the shaped materials without passing through the powder process and that various microstructures can be easily achieved due to its low temperature nature. Coating films having nonlinear optical properties were mentioned as examples which show the advantage of the sol-gel method in providing advanced materials.

REFERENCES

1. S.Sakka, Science of Sol-Gel Methods, Uchida-Rokakuho Publishing Company, Tokyo (1988) pp.1-221. [in Japanese]
2. C.J.Brinker and G.W.Scherer, Sol-Gel Science, Academic Press, San Diego (1990) pp.1-908.
3. S.Sakka, "The Current State of Sol-Gel Technology,"J.Sol-Gel Sci.Tech. 3, 69-81 (1994).
4. S.Sakka, "Organometallic-Derived Ceramics,"Ceramics Today -Tomorrow's Ceramics, ed. P.Vincenzini, Elsevier Science Publishers (1991) p.813-828.
5. S.Sakka, K.Aoki, H.Kozuka and J.Yamaguchi, "Sol-Gel Synthesis of Silica Gel Disc as Applied to Supports of Organic Molecules with Optical Functions," J.Mater.Sci., 28, 4607-4614 (1993).
6. T.Adachi and S.Sakka, "The Role of N,N-Dimethylformamide, a DCCA,in the Formation of Silica Gel Monoliths by Sol-Gel Method," J.Non-Cryst.Solids, 99,118-128 (1988),
7. H.Kozuka and S.Sakka, "Formation of Silica Gels Composed of Micrometer-Sized Particles by the Sol-Gel Method," ACS Advances in Chemistry Series No.234, The Colloid Chemistry of Silica, ed.H.E.Bergna, American Chemical Society (1994) 129-141.
8. S.Sakka and H.Kozuka, "Rheology of Sols and Fiber Drawing," J.Non-Cryst.Solids, 100, 142-153 (1988).
9. S.Sakka and H.Kozuka, "Rheology of Sols in the Sol-Gel Processing," Chimica Chronica,New Series,23,137-146(1994)
10. S.Sakka, "Sol-Gel Processing of Insulating, Electroconducting and Superconducting Fibers", J.Non-Cryst.Solids, 121, 417-423 (1990).
11. S.Sakka, "Preparation and Properties of Sol-Gel Coating Films," J.Sol-Gel Sci.Tech., 2, 451-455 (1994).
12. S.Sakka and T.Yoko, "Sol-Gel-Derived Coating Films and Applications," Structure and Bonding 77, Springer-Verlag Berlin (1992) p.89-118.
13. S.Sakka, "Preparation and Properties of Sol-Gel Coating Films," J.Sol-Gel Sci.Tech., 2, 451-455 (1994).
14. P.Innocenzi,H.Kozuka and S.Sakka,"Preparation of Coating Films Doped with Gold Metal Particles from Methyltriethoxysilane-Tetraethoxysilane Solutions," J.Sol-Gel Sci.Tech. 305-318 (1994).
15. H.Krug, N.Merl and H.Schmidt, "Fine Patterning of Thin Sol-Gel Films,"J.Non-Cryst.Solids,147&148, 447-450 (1992).
16. S.Sakka, H.Kozuka and T.Adachi, "Preparation of Porous Materials By the Sol-Gel Method," Ceramic Transactions, Vol 31, Porous Materials, The American Ceramic Society (1993) p.27-40.

17. K.Nakanishi,R.Takahashi and N.Soga,"Dual-Porosity Silica Gels by Polymer-Incorporated Sol-Gel Process,"J.Non-Cryst. Solids, 147&148, 291-295(1992).
18. L.L.Hench, F.G.Araujo, J.K.West and G.P.LaTorre, "Gel-Silica Optics: Theory and Applicaation," J.Sol-Gel Sci. Tech., 2, 647-655 (1994).
19. D.Avnir,D.Levy and R.Reisfeld, "The Nature of the Silica Cage as Reflected by Spectral Changes and Enhanced Photo-stability of Trapped Rhodamine 6G," J.Phys.Chem.,88, 5956-5959 (1984).
20. A.Makishima and T.Tani,"Preparation of Amorphous Silicas Doped with Organic Molecules by the Sol-Gel Process,"J.Am. Ceram.Soc., 69, C-72-C-74 (1986).
21. H. Schmidt, "Inorganic-Organic Composites by Sol-Gel Technique," J.Sol-Gel Sci.Tech., 1, 217-231 (1994).
22. T.Adachi and S.Sakka, "Dependence of the Elastic Moduli of Porous Silica Gel Prepared by the Sol-Gel Method on Heat-treatment," J.Mater.Sci. 25, 4732-4737 (1990).
23. S.Sakka, "Current States of the Preparation of Optical Solids by the Sol-Gel Method,"SPIE Vol.1758 Sol-Gel Optics II (1992) p.3-12.
24, J.D.Mackenzie, "Nonlinear Optical Materials by the Sol-Gel Method," J.Sol-Gel Sci.Tech., 1, 7-19 (1993).
25. S.Hirano and K.Kato, "Preparation of Crystalline $LiNbO_3$ Films with Preferred Orientation by Hydrolysis of Metal Alkoxides," Adv.Ceram.Mater., 3, 503-506 (1988).
26. H.Yamashita, T.Yoko and S.Sakka, "Preparation of LiB_4O_7 Thin Films with Preferential Orientation by Sol-Gel Method," J.Am.Ceram.Soc., 74, 1668-1674 (1991).
27. M.Yamane,H.Kawazoe,A.Yasumori and T.Takahashi,"Gradient-Index Glass Rods of $PbO-K_2O-B_2O_3-SiO_2$ System Prepared by the Sol-Gel Process," J.Non-Cryst.Solids, 100, 506-51 (1988).
28. B.E.Yoldas and D.P.Partlow, "Formation of Broad Band Antireflective Coatings on Fused Silica for High Power Laser Applications," Thin Solid Films, 129, 1-14 (1985).
29. L.Cot, A.Larbot and C.Guizard, "Ultrafilters by the Sol-Gel Process," Ultrastructure Processing of Advanced Ceramics, eds. J.D.Mackenzie and D.R.Ulrich, John Wiley & Sons, New York (1988) p.211-221.
30. E.J.A.Pope,M.Asami and J.D.Mackenzie,"Transparent Silica Gel-PMMA Composites," J.Mat.Res., 4, 1018-1026 (1989).
31. T.Hashimoto, T.Yoko and S.Sakka, "Third-Order Nonlinear Optical Susceptibility of α-Fe_2O_3 Thin Film Prepared by the Sol-Gel Method," J.Ceram.Soc.Japan, 101, 64-68 (1993)
32. T.H.Hashimoto, T.Yoko and S.Sakka, "Sol-Gel Preparation and Third-Order Nonlinear Optical Properties of TiO_2 Thin Films," Bull.Chem.Soc.Jpn., 67, 653-660 (1994).

33. H.Kozuka and S.Sakka, "Preparation of Gold Colloid-Dispersed Silica Coating Films by the Sol-Gel Method," Chem Mater.,5, 222-228 (1993).
34. H.Kozuka, G.Zhao and S.Sakka,"Control of Optical Properties of Gel-Derived Oxide Coating Films Containing Fine Metal Particles," J.Sol-Gel Sci.Tech., 2, 741-744 (1994).
35. S.Sakka, H.Kozuka and G.Zhao, "Sol-Gel Preparation of Metal Particle/Oxide Nanocomposites,"SPIE Vol.2288 Sol-Ge Optics III, 108-119 (1994).
36. G.Zhao, H.Kozuka and S.Sakka, "Preparation of TiO_2 Coating Films Containing Pd Fine Particles by Sol-Gel Method," J.Sol-Gel Sci.Tech., 4, 37-47 (1995).

OPTIMIZATION OF THE GELCASTING OF A SILICON NITRIDE FORMULATION.

O. O. Omatete, Oak Ridge National Laboratory, Oak Ridge TN, J. P. Pollinger and Kevin O'Young, AlliedSignal Ceramic Components, Torrance, CA.

ABSTRACT

An optimum gelcasting condition for a silicon nitride formulation was determined using the Taguchi statistical method. An $\mathbf{L}_8(4^1 \times 2^4)$ design, in which the effects of one factor at four levels and four factors at two levels were evaluated in only eight experiments, was used. The factors at two levels were: the total monomer concentration, the monomer/crosslinker ratio, the initiator concentration, and the initiator/catalyst ratio; the factor at four levels was the initiator concentration per mass of the slip. The primary criterion used to determine optimum design was the green strength of the dried part. Three other parameters were also considered: initial slip viscosity, time for the slip viscosity to reach 300 mPa.s at 25°C, and time for the slip to gel at 50°C. The optimum gelcasting conditions from the designed experiments predicted 80% increase in green strength (4.3 MPa versus 2.4 MPa, the initial value). The confirmation runs showed only a 60% increase in strength (3.9 MPa).

INTRODUCTION

Gelcasting [1] is a generic ceramic forming process developed at the Oak Ridge National Laboratory. In gelcasting, a suspension of commercial ceramic powder in a solution of organic monomers is cast into a mold and polymerized to form a green body which takes the shape of the mold. If the solvent for the monomers is organic, it is nonaqueous gelcasting [2]; if it is water, it is aqueous gelcasting [3,4]. In aqueous gelcasting, the solution consists of one or more monofunctional (one double bond) monomers and at least one multifunctional (at least two double bonds) cross-linking monomer and an initiator. Typically, the monomers are from the vinyl and allyl groups, and their homologues or derivatives.

In gelcasting a ceramic powder, the optimal selection of monomer, crosslinker and initiator can be onerous. For example, Table I [5] lists some monomers, crosslinkers and initiators that may be used in the gelcasting of silicon nitride. In this case, to determine an optimum combination by examining one factor at a time would require 224 (8x4x7) experiments. Consequently, a fractional factorial design was used to reduce the number of experiments. Even after the combination of a monomer, a crosslinker, and an initiator has been identified, optimal concentrations and processing conditions must be determined. Thus, the goal of this study was to determine the optimal gelcasting condition for the AlliedSignal Ceramic Components' GN-10 silicon nitride formulation in a near-production environment.

Table I. Typical Monomers[1], Crosslinkers[1] and Initiators[1] for Gelcasting Si_3N_4.

MONOMERS*	**CROSSLINKERS***	**INITIATORS**
MAM	MBAM	APS
NVP	PEGDMA	APS/TEMED
MPEGMA	TAA	AIBN
DiMAEMA	DADTA	AZIP
DMAPMAM		AZAP
AA		H_2O_2
MA		TBHP
AM		

*Including their homologues and derivatives

[1]List of Acronyms

MONOMERS:

MAM = methacrylamide NVP = N-vinyl pyrrolidone

MPEGMA = methoxy poly(ethylene glycol) methacrylate

DiMAEMA = dimethyl aminoethyl methacrylate

DMAPMAM = 3-N,N-dimethylaminopropyl methacrylate

AA = acrylic acid MA = methacrylic acid AM = acrylamide

CROSSLINKERS (CROSSLINKING MONOMERS):

MBAM = N,N'-methylene bisacrylamide

PEGDMA = poly(ethylene glycol)1000 dimethacrylate

TAA = triallylamine DATDA = diallyl tartardiamide

INITIATORS:

APS = ammonium persulfate AIBN = azobis (isobutyronitrile)

TEMED = N,N,N',N'-tetramethyl ethylene diamine

AZIP = azobis [2-(2-imidazolin-2yl)propane] 2HCl

AZAP = azobis (2-amidinopropane) HCl H_2O_2 = hydrogen peroxide

TBHP = t-butyl hydroperoxide (90%, 5% H_2O, 5% BuOH)

Gelcasting is a well defined process that has been applied to several ceramic powders [1]; for example, the detailed gelcasting of alumina has been reported [6]. Therefore, Taguchi method (quality engineering) [7] is appropriate for the optimization of the gelcasting of silicon nitride under industrial conditions. In increasing the amount of gelcasting slurry from a typical laboratory quantity of 100 g to industrial values of 1 kg or more, simple scale-up resulted in insufficient processing time before the slurry gelled. The objective here, therefore, was to increase the processing time while optimizing the green strength of the dried parts.

PROCEDURE

The system chosen for gelcasting silicon nitride in this study consisted of: monomer = MAM, crosslinker = MBAM, and initiator system = APS/TEMED. (See Table I for an explanation of the acronyms). After brainstorming, five controlling factors (independent variables) were selected: the total monomer (MAM + MBAM) concentration, the ratio of the monomer to the crosslinker (MAM/MBAM), the initiator (APS) concentration, the ratio of the initiator to the catalyst (APS/TEMED), and the amount of APS added per mass of slip. Four of the factors were tested at two levels, high (H) and low (L). Because the amount of APS per unit mass of slip strongly determines the gelation rate, it was tested at four levels, lowest (LL), low, high, and highest (HH). The experimental design selected was $\mathbf{L}_8(4^1 \times 2^4)$ in which the effects of one factor at four levels and four factors at two levels were evaluated in only eight experiments [7,8]. Table II shows the factors, their levels and combinations in the experimental design. Four response parameters (dependent variables) were measured: initial slip viscosity, η(mPa.s), at ambient conditions; time, TG(min), for the slip to gel at 50°C; time, TV(min), for the slip viscosity to reach 300 mPa.s at 25°C; and the strength, σ(MPa), of dried gelcast Si_3N_4 discs.

Table II. Factors, levels and their combination for the $\mathbf{L}_8(4^1X2^4)$ design.

FACTORS	LEVELS			
	1	2	3	4
A (MAM+MBAM) conc	L	H		
B μL APS/100 g slip	H	L	HH	LL
C MAM/MBAM	L	H		
D APS conc.	H	L		
E APS/TEMED	L	H		

Experiment No.	Factors and the level combinations: A	B	C	D	E	Measured Variable
1	1	1	1	1	1	Y1
2	1	2	1	2	2	Y2
3	1	3	2	1	2	Y3
4	1	4	2	2	1	Y4
5	2	1	2	2	2	Y5
6	2	2	2	1	1	Y6
7	2	3	1	2	1	Y7
8	2	4	1	1	2	Y8

For each experiment, approximately 1 kg of a 50 vol % solids suspension was prepared by measuring the calculated amount of the components (Si_3N_4 powder, sintering aids, dispersants, monomer, crosslinker and water) which were then ball milled for a given number of hours. The amount of each component was determined by the experiment number in Table II; No. 1 was the standard process set-up. After milling, the slurry was de-aired and the calculated amount of initiator and catalyst as indicated in Table II were mixed into the slip. After another de-airing step, the viscosity, specific gravity, and pH of the slip were measured at ambient conditions. To determine the time they took to gel, samples were placed in an oven set at 50°C. A sample was placed in a Brookfield Digital Rheometer Model DV-II to measure the time for its viscosity to rise to 300 mPa.s at 25°C. Several disc molds (25 mm diameter x 5 mm thick) were filled with the slip and gelled in a second oven set at 50°C. The gelled Si_3N_4 discs were dried and their tensile strength was measured by a diametral compression test [9]. The remaining slip was used to cast other parts such as plates or turbine wheels. To compare with the predicted value, two confirmation runs were made at the optimum levels determined from the experimental design. These levels are shown in Table IV, the result of the runs in Table III, and both are discussed in the next section.

RESULTS AND DISCUSSIONS

Table III shows a set of typical measured variables for the eight experiments. The slip properties measured at ambient conditions exhibited minimal variations indicating the slip preparation was consistent. However, the gelation times and the strength varied with the experimental conditions. The average strengths in the table were calculated from the measurement of a minimum of five to a maximum of twenty dried gelcast Si_3N_4 discs, with a pooled standard deviation of 0.25 MPa.

Table III. Typical Measured Variables for the Eight Experiments.

Expt.	Slip Properties			Gelation Time		Av Strength
No.	pH	SG	η(mPa.s)	TV(min)	TG (min)	σ(MPa)
1	8.96	2.068	76	6.5	45	2.41
2	9.21	2.069	62	12.5	145	1.05
3	8.83	2.055	88	6.5	45	2.24
4	9.35	2.060	55	80.0	220	0.50
5	9.22	2.041	57	27.0	130	1.61
6	8.80	2.050	74	9.5	90	3.90
7	9.02	2.043	55	23.5	135	4.26
8	9.30	2.041	55	33.0	135	3.47
Confirmation Runs						
# 1	8.93	2.047	62	14.5	110	3.75
# 2	9.30	2.049	72	11.0	95	3.89

The effect of a factor at a given level was calculated by averaging the data over that factor level. For example, in Table II, the effect of factor **A** at level 1 is $(Y_1+Y_2+Y_3+Y_4)/4$, at level 2, it is $(Y_5+Y_6+Y_7+Y_8)/4$. Similarly, the effect of factor **E** at level 1 is $(Y_1+Y_4+Y_6+Y_7)/4$, at level 2, $(Y_2+Y_3+Y_5+Y_8)/4$. Factor **B** with four levels has only two data points at each level. Thus, the effect of factor **B** at level 1 is $(Y_1+Y_5)/2$, at level 2, it is $(Y_2+Y_6)/2$ and so on. Table IV is the response table showing the calculated effect of each factor at the different levels and the absolute value of their differences. The effects at all the four levels of factor **B** (the amount of the initiator per mass of slip) are shown and the difference value here is between level 1 and the level with highest value.

Table IV. The Response Table for Measured Parameters (σ, TV, & TG).

Measured Variable, Strength (σ, MPa)					
FACTOR	**A**	**B**	**C**	**D**	**E**
Level 1	1.55	2.01	2.80	3.00	2.77
Level 2	3.31	2.48	2.06	1.86	2.09
Level 3		3.25			
Level 4		1.99			
$\|(\Delta)\|$	1.76	1.24	0.74	1.14	0.68

Measured Variable, Time (TG, min) to Gel at 50°C.					
FACTOR	**A**	**B**	**C**	**D**	**E**
Level 1	113.8	87.5	115.0	78.8	122.5
Level 2	122.5	117.5	121.3	157.5	113.8
Level 3		90.0			
Level 4		177.5			
$\|(\Delta)\|$	8.8	90	6.3	78.7	8.8

Measured Variable, Time (TV, min) to reach 300 mPa.s at 25°C.					
FACTOR	**A**	**B**	**C**	**D**	**E**
Level 1	26.4	16.8	18.9	13.9	29.9
Level 2	23.3	11.0	30.8	35.8	19.8
Level 3		15.0			
Level 4		56.5			
$\|(\Delta)\|$	3.1	39.7	11.9	21.9	10.1

Confirmation Levels					
FACTOR	**A**	**B**	**C**	**D**	**E**
Level	2	2	1	1	1

The confirmation levels were based primarily on optimizing the strength (with its larger-the-better characteristic); that is, the largest possible value of strength is desired. Thus, for each factor, the level with the highest strength (**A**-2, **B**-3, **C**-1, **D**-1, and **E**-1) should have been selected. However, another requirement of the design is that there should be sufficient processing time before the slip gelled. This implies that the times TV and TG should be larger than a minimum value. As the actual data in Table III show, the worst strength was in experiment No. 4 which had the largest times while the highest strengths were at intermediate times. The confirmation levels shown in Table IV, indicate that **B**-2 was used instead of **B**-3 in order to obtain intermediate gelation times; **B**-1 and **B**-3 were fast, and **B**-4 was very slow. The times for the actual confirmation runs in Table III show this; they are intermediate between the times for the eight experiments.

In the Taguchi method, once the confirmation levels have been determined, the predicted values may be calculated based either on simple averaging or on the signal-to-noise ratio [7,8]. In this study, the prediction was by simple averaging and, to minimize over-estimation, only the three strongest effects (**A**-2, **C**-1 and **D**-1) were used. The predicted strength was 4.3 MPa which is an 80 % increase over the value for experiment No.1, the standard conditions. The average result of the two confirmation runs shown in Table III was 3.9 MPa, a 60% increase in strength over the standard conditions. The confirmation values were, however, less than the predicted value which was based on only three effects. If all the effects of the five factors were additive, the confirmation values would have been at least equal to the predicted value. This discrepancy implies that there are interactions among the factors. In the confirmation runs, the gelation times were intermediate between the highest and the lowest values obtained in the experiments, as was predicted. Also in these runs, the slip properties were consistent. A surprising finding of this study was that increasing the crosslinker ratio did not correspond to an increase in the green strength of the dried gelcast parts.

CONCLUSIONS

A strategy for the optimization of the gelcasting of the AlliedSignal Ceramic Components' GN-10 silicon nitride formulation in large, production-size batches using the Taguchi method has been demonstrated. The slip properties of all batches were consistent, indicating that the processing was reproducible. The two desired goals of the optimization were achieved. At the optimum conditions, the gelation time was increased thus providing adequate processing time before gelation commenced. The green strength of the dried gelcast parts was increased by nearly 60% to 3.9 MPa. However, because this was less than the 80% increment predicted, there are interactions between the factors that were selected. Nevertheless, this initial study has identified improved processing conditions which provide the basis for further optimization. In addition, this investigation has shown that increasing the crosslinker-to-monomer ratio in the gelcasting process does not produce a corresponding increase in the green strength, a finding which was unexpected and needs to be studied further.

ACKNOWLEDGEMENT

Research is sponsored by the US Department of Commerce, National Institute of Science and Technology Advanced Technology Program and the US Department of Energy, Assistant Secretary of Energy Efficiency, office of Transportation Technologies, as part of the Ceramic Technology Project of the Materials Development Program, under contact DE-AC05-84OR21400 with Martin Marietta Energy Systems, Inc. The authors thank M. A. Janney, S. D. Nunn and R. J. Lauf for reviewing the manuscript and Jeffry Wang for experimental assistance.

REFERENCES

1. O. O. Omatete, M. A. Janney & R. A. Strehlow, "Gelcasting- A New Ceramic Forming Process," Ceramic Bulletin, 70 [10], Oct 1991.
2. M. A. Janney, "Method for Forming Ceramic Powders into Complex Shapes," U.S. Patent No. 4,894,194, January 16, 1990.
3. M. A. Janney & O. O. Omatete, "Method for Molding Ceramic Powders using Water-Based Gel Casting," US Patent No. 5,028,362, Jul. 2, 1991
4. M. A. Janney & O. O. Omatete, "Method for Molding Ceramic Powder Using a Water-Based Gelcasting Process," U.S. Patent No. 5,145,908, Sept. 8,1992.
5. M. A. Janney & O. O. Omatete, "Development of Low-Toxicity Gelcasting Formulations - I," Paper SVIII-67-94, ACerS's 96th Annual Meeting, Indianapolis, IN, April, 1994.
6. A. C. Young, O. O. Omatete, M. A. Janney, & P. A. Menchhofer, "Gelcasting of Alumina," J. Am. Ceram. Soc., 74 [3] 612-18 (1991).
7. G. Taguchi, "Introduction to Quality Engineering," Asian Productivity Organization, Tokyo, Japan (1986).
8. "Taguchi Methods - Implementation Manual," American Supplier Institute, Dearborn, MI, 1992.
9. S. D. Nunn, O. O. Omatete, C. A. Walls and D. L. Barker, "Tensile Strength of Dried Gelcast Green Bodies," Ceram. Sci. & Engr. Proceed., 15, [4], 493-8, 1994.

SINTERING OF PZT CERAMICS FROM HYDROTHERMALLY SYNTHESIZED POWDERS

C.H. Lin*, S.C. Pei*, T.S. Chin*, J.Y. Huang**, and C.H. Li**
*Dept. of Materials Sci., Tsing Hua Univ., Hsinchu, Taiwan, R.O.C
**Inst. of Electro-Optical Eng., Chiao-Tung Univ. Hsinchu, Taiwan, R.O.C.

ABSTRACT

PZT ($PbTi_{1-x}Zr_xO_3$) powders were hydrothermally produced at 200°C for 24 hrs, from a solution containing $Ti(OH)_4 \cdot xH_2O$, $ZrOCl_2$, $Pb(CH_3COO)_2$, and NaOH. PZT powder was also produced by the solid-state sintering method. The powders from both the hydrothermal method and the solid-state reaction method were pressed and subsequently sintered ranging between 1100°C and 1250°C for 2 hrs. The microstructures, density, grain size, K value, D value, Kp value, and Qm value of the sintered PZT ceramics were measured and compared with each other. Large density and grain size differences were found in ceramics produced between 1100°C and 1250°C from the hydrothermal powders and by the solid-state reaction powders. The hydrothermal powder was much more sinterable than powders produced from the solid-state reaction method and grain growth occurred more easily. The hydrothermal PZT ceramics were observed to have a higher K value, higher Kp value, lower D value, and lower Qm value, and were electrically 'softer' than the ceramics produced from the solid-state reaction powders at the same chemical composition.

INTRODUCTION

PZT powders used for producing PZT ceramics are conventionally produced by solid-state reaction method. Hydrothermal method was reported to be capable of producing PZT powders(1-16). The advantages of hydrothermal method are that (1) powders produced by this method are homogeneous in chemical composition and particle size, (2) presintering can be avoided, and (3) low sintering temperature is possible because of the finer particle size. Few investigations have been reported on the sintering behavior and subsequent electrical properties of PZT ceramics produced from the hydrothermal powders. In this study, both the hydrothermal method and solid-state reaction method were used to produce PZT powders, thereby producing PZT ceramics from these powders. A comparison was also made of the difference in sintering behaviors and subsequent electrical properties between PZT ceramics produced from the hydrothermal powders and those from the solid-state reaction method. A composition of x=0.54 in $PbTi_{1-x}Zr_xO_3$ was selected in this study because a rhombohedral-tetragonal phase boudary change occurs and PZT ceramics demonstrate high dielectric constant at this compositions(16).

EXPERIMENTAL

PZT($PbTi_{1-x}Zr_xO_3$, x=0.54) powders were hydrothermally produced at 200°C for 24 hrs from a slurry solution containing $Pb(CH_3COO)_2$, $Ti(OH)_4 \cdot xH_2O$, $ZrOCl_2$, and NaOH. PZT(x=0.52) powders were also produced by solid-state reaction method by reacting PbO, ZrO_2, and TiO_2 together at 800°C for 4 hrs. The powders from both the hydrothermal method and the solid state reaction method were pressed at 1.75 ton/cm^2 and subsequently sintered at 1100°C, 1150°C, 1200°C, and 1250°C for 2 hours, respectively.

Laser particle size analyzer, STEM, X-ray diffractometer, and impedance analyser were used to examine the characteristics of PZT powders and ceramics. Detailed preparation processes and examination procedures of PZT powders and ceramics have been reported elsewhere(14,15).

RESULTS AND DISCUSSION

Electron micrograph in Figure 1 reveals that the PZT powders produced from the hydrothermal method are cubic in shape and around 0.1-0.4 u_m in size. X-ray diffraction patterns in Figure 2a and 2b revealed that the PZT powders produced from the hydrothermal method are cubic in structure, and the PZT powders produced from the solid-state reaction method are tetragonal. After the powders were pressed and sintered, the structure of the powders have changed. X-ray diffraction patterns in Figure 2c and 2d indicated that the structure of sintered PZT ceramics produced from the hydrothermal powders changed to a mixture of tetragonal and rhombohedral, and the sintered PZT ceramics produced from solid-state reaction powders is still tetragonal. Both tetragonal and rhombohedral phases are stable because x=0.54 is at tetragonal and rhombohedral phase change boundary.

The microstructure of PZT ceramics produced from the hydrothermal powders and from the solid-state reaction powders sintered at 1100°C, 1150°C, 1200°C, and 1250°C for 2 hrs are shown in Figures 3a to 3h, respectively. Large density and grain size differences were found in Figures 3 and 4 for ceramics produced between 1100°C and 1250°C from the hydrothermal powders and by the solid-state reaction powders. At a lower sintering temperature, e.g., 1100°C, the microstructure of the PZT ceramics in Figure 3 revealed that the PZT ceramics from the hydrothermal powders were dense, well crystallized and had a clear grain boundary; however, the PZT ceramics from the solid-state reaction powders were porous, not well crystallized. At a higher sintering temperature, e.g., 1250°C, the PZT ceramics both from the hydrothermal powders and from the solid-state reaction powders were dense and well crystalline; however, the grain size of the PZT ceramics from the hydrothermal powders was much larger than that from the solid-state reaction powders. The hydrothermal powder was much more sinterable than the solid-state powders and grain growth occurred more easily in hydrothermal powders. The PZT ceramics from the hydrothermal powders sintered at 1100°C have developed good density, which could not be obtained in the PZT ceramics from the solid-state reaction method, even when they were sintered as high as 1250°C. A higher dielectric constant and a lower dissipation factor (D value), which are desirable for a potential application in capacitor, were obtained from the PZT ceramics produced from the hydrothermal powders than those from the solid-state reaction powders. As shown in Figure 4, the superiority of the hydrothermal powders over the solid-state reaction powders in dielectric constant and dissipation factor is more obvious when they have to be used for low temperature sintering. A low sintering temperature is desirable for PZT powders because less expensive palladium could be used in Pd-Ag alloy for internal electrodes when the powders are used in producing actuators and multilayer capacitors. The evaporation of lead oxide at a higher sintering temperature not only caused the deterioration of the density and the dielectric constant of the PZT ceramics but also produced poisonous lead oxide dust everywhere. A higher sintering temperature could be avoided if hydrothermal powders were used. Judging from high density, well crystillized grain size, and high dielectric constant of the PZT ceramics sintered at 1100°C, the possibility that the hydrothermal powders could be sintered around 1000°C is relatively high although no experimental data were taken below 1100°C. The hydrothermal PZT ceramics were observed to have a higher Kp value and lower Qm value than those values of the ceramics produced from the solid-state reaction powders. A higher Kp value indicates that the PZT ceramics has higher electromechanical coupling factor, i.e., less energy is lost during electrical and mechanical energy interchange. A higher Qm value implies a higher mechanical quality factor, i.e., sharper mechanical resonance frequency can be obtained from these PZT ceramics. The PZT ceramics produced from the hydrothermal powders possessed a higher Kp value and a lower Qm value because the grain size in these PZT are larger and have better crystallinity. The electrical dipole moments in the PZT produced from the hydrothermal powders are more easily moved by an external field, i.e., these PZT ceramics are electrically 'softer' because they possessed more 'softer' rhombohedral phase, larger grain size, and well crystallinity. The PZT ceramics from the hydrothermal powders are suitable for (1) Buzzers, in which high Kp (>60%) and low Qm(<200) are required, (2) delay lines, in which high Kp is required, and (3) impact type gas igniter, in which a not specifically high Qm is required. For wave filters, piezoelectric transformer, and ultrasonic transducer, in which high Qm is required, x value in $PbTi_{1-x}Zr_xO_3$ must increase and sintering temperature must be decrease to make the PZT ceramics electrically 'harder'. The PZT ceramics from hydrothermal powders, having a larger grain size, are electrically 'softer' than that from the solid-state reaction powders at the same chemical composition.

CONCLUSIONS

A hydrothermal method has been successfully developed in this work to produce PZT powders for high quality PZT ceramics. The operation of this process is rather simple and only inexpensive chemical reagents were used. The hydrothermal powder was much more sinterable than powders produced from the solid-state reaction method and grain growth occurred more easily. The hydrothermal PZT ceramics were observed to have a higher K value, higher Kp value, lower D value, and lower Qm value, and were electrically 'softer' than the ceramics produced from the solid-state reaction powders because the hydro-PZT ceramics possessed more rhombohedral phase, larger grain size, and well crystallinity.

REFERENCES

(1) S. Kaneko and F. Imoto, Bull Chem. Soc. Japan, 51, pp.1739(1978).
(2) K. Takai, S. Shogi, H. Naito, and A. Sawaoka, in: Proc. 1st International Symposium on Hydrothermal Reactions pp.877(1983).
(3) T.R.N. Kutty and R. Balachandran, "Direct Predipitationb of Lead Zirconate Titanate by Hydrothermal Method," Materials Research Bulletin, 19, pp.1479 (1984).
(4) K.C. Beal, "Precipitation of Lead Zirconate Titanate Solid Solutions under Hydrothermal Conditions," in Advances in Ceramics, Vol. 21, P.33, ed. by G.L. Messing, K.S. Mazdiyasni, J. W. McMauley and R.A. Haber, The American Ceramic Society, Inc., ISBN 0-916094-83-9, 1987.
(5) K.H. Lee, B.H. Lee, M. Daimon, K. Asaka, W.H. Kang, and H.S. Park, J. Korean Ceram. Soc., 24, pp.397(1987).
(6) K. Abe et al, Japanese Patents 61-31345, 63-144115, and 63-144116.
(7) W.J. Dawson, "Hydrothermal Synthesis of Advanced Ceramic Powders," Ceramic Bulletin, 10, pp.1673-1678(1988).
(8) D.J. Watson, C.A. Randall, R.E. Newnham, and J.H. Adair, "Hydrothermal Formation Diagram in the Lead Titanate, in: Ceramic Trans., Vol. 1, pp. 154-164(1988).
(9) M. Suzuki, S. Uedaira, H. Masukyka, and H. Tamura, "Hydrothermal Synthesis of Lead Titanate Fine Powders, in: Ceramic Trans., Vol. 1, pp.163-170, (1988).
(10) T. Yamamoto, R. Tanaka, K. Okazaki, and T. Ueyama, Japanese J. Appl. Physics, 28, pp.81(1989).
(11) J.H. Adair, T.S. Shrout, and K. Osseo-Asare, Hydrothermal Synthesis of Perovskite Ceramic Powder, in: Proc. MRS International Meeting on Advanced Materials, pp. 141-155(1989).
(12) T. Ichihara, T. Tsurumi, K. Asaga, K.H. Lee, and Daimon, J. Ceram. Soc. Japan, 98, pp.155(1990).
(13) C.E. Millar, L. Pedersen and W.W. Wolny, "Hydrothermally Processed Piezoelectric and Electrostrictive Ceramics," Ferroelectrics, Vol. 133, pp. 271-276(1992).
(14) C.H. Lin, S.C. Pei, T.S. Chin, "Hydrothermal Synthesis of $PbTi_{1-x}Zr_xO_3$ Ceramic Powders," in Ceramics Transactions Vol. 32, Dielectric Ceramics: Processing, Properties, and Applications, pp.261-274(1993), ISBN 0-944904-60-2.
(15) C.H. Lin and S.C. Pei, "PLZT Ceramics from Hydrothermally produced Powders," Proc. IUMRS Inter. Conference in Asia, China, Sept. 6-10, 1993, Yantze Gorges, China.
(16) C.H. Lin, T.S. Chin, S.C. Pei, J.Y. Huang, and C.H. Li, "PZT Ceramics from Hydrothermally Synthesized Powders," in Proc. Ninth Inter. Sym. Appl. Ferroelectrics, ISAF-94, Univ. Park, Penn. USA,August 7-10, 1994.

Figure 1 Electron micrograph of hydrothermal PZT powders

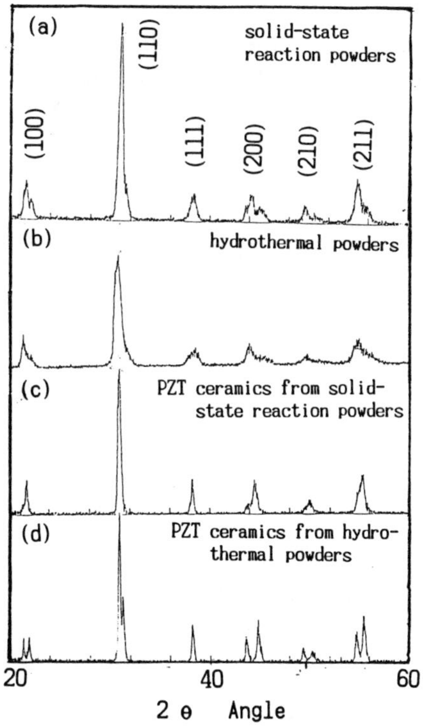

Figure 2 X-ray diffraction patterns of PZT powders and ceramics

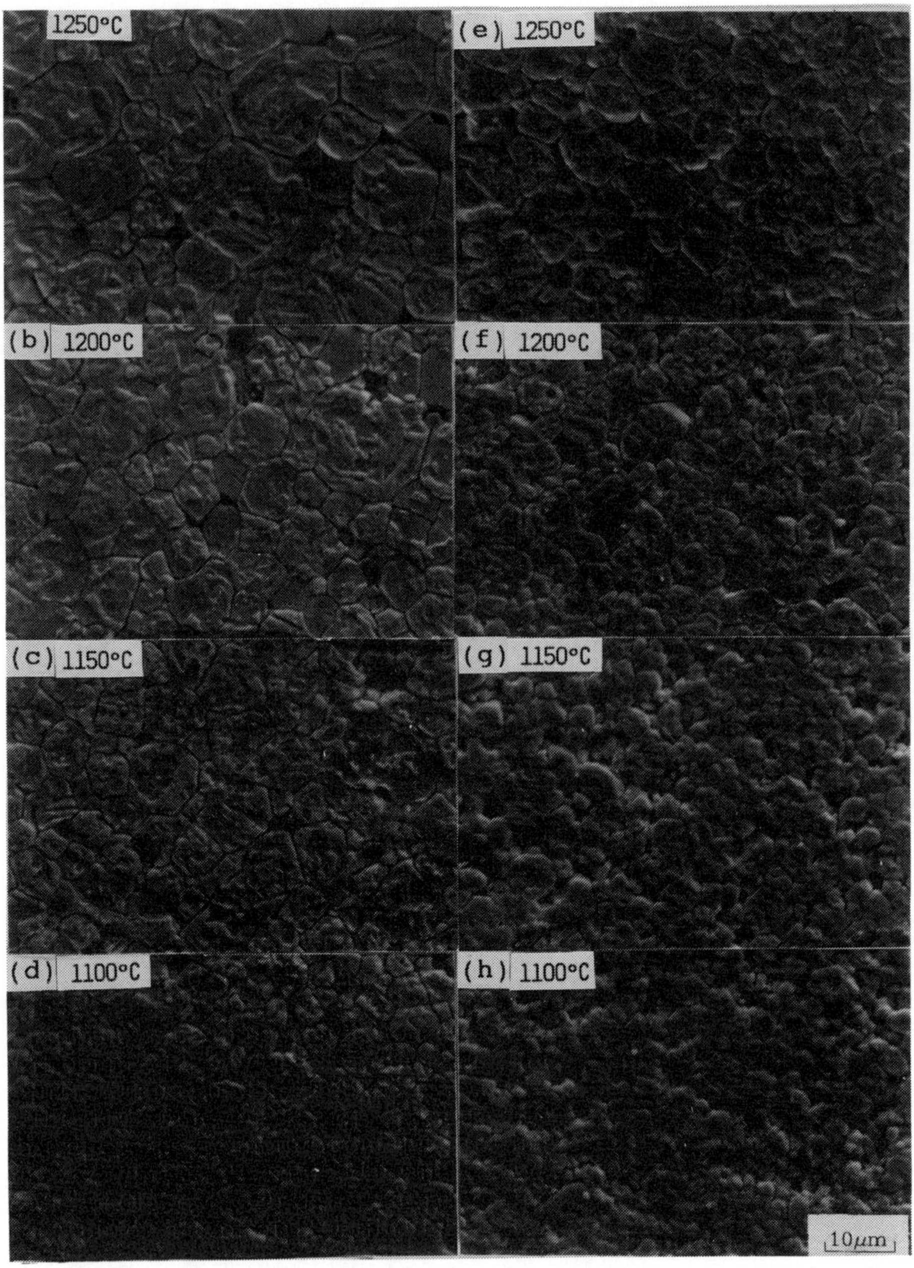

Figure 3 Microstructure of PZT ceramics, (a) to (d) from hydrothermal powders, (e) to (h) from solid-state reaction powders.

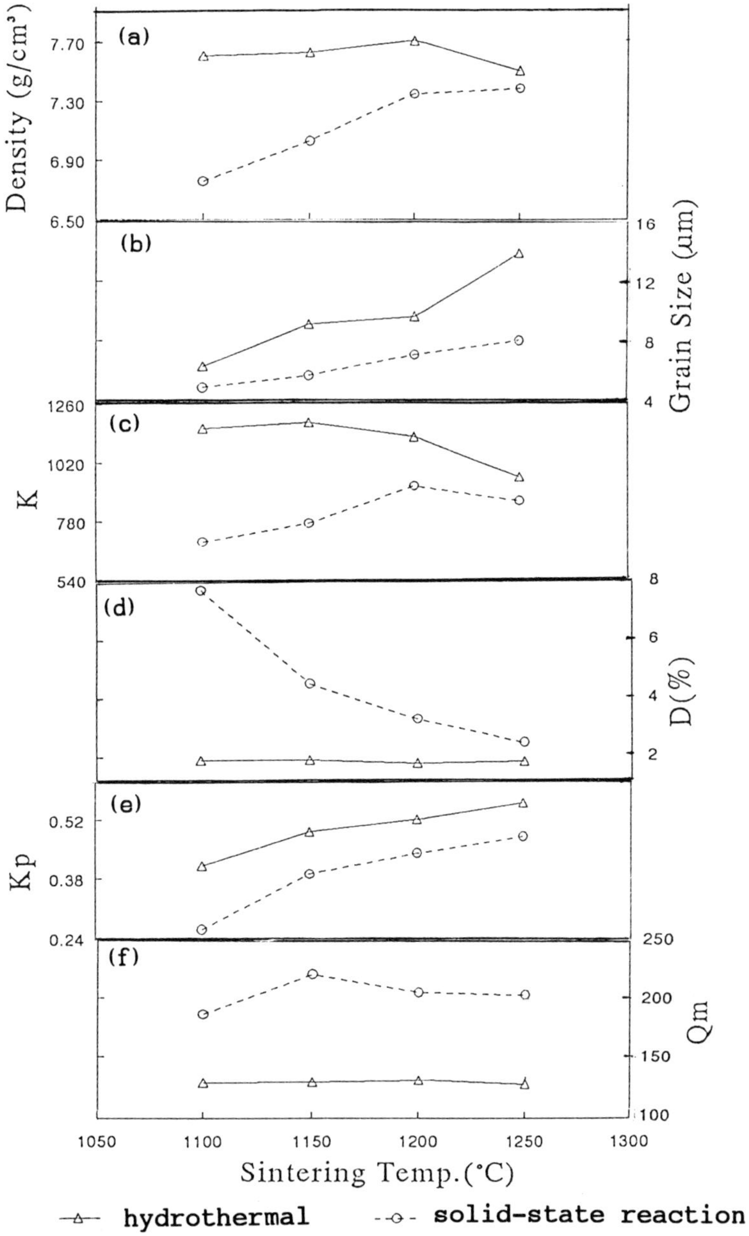

Figure 4 Density, grain size, K value, D value, Kp value, and Qm value of the PZT ceramics produced from the hydrothermal and from solid-state reaction powders

FABRICATION AND MECHANICAL PROPERTIES OF ZrO_2 SOLID SOLUTION AND COMPOSITE CERAMICS IN THE SYSTEM $ZrO_2(Y_2O_3)$-Al_2O_3

O. Yamaguchi and K. Hirota, Doshisha University, Tanabe Kyoto 610-03, Japan
S. Inamura and H. Miyamoto, Osaka Prefectural Institute of Industrial Technology, Osaka 550, Japan

ABSTRACT

In the system ZrO_2-Al_2O_3, dense ZrO_2 solid solution ceramics containing 25 mol% Al_2O_3 are fabricated by the HIP technique using metastable ZrO_2 solid solution powders prepared from alkoxides. The ceramics with nanometer grain size show excellent high fracture toughness of ≈23 $MPa{\cdot}m^{1/2}$, although they exhibit weak bending strength of ≈600 MPa. To improve the strength, metastable ZrO_2 solid solutions containing 25 mol% Al_2O_3 and (0.75-3.75) mol% Y_2O_3 are hot isostatically pressed at 1130°C and 1230°C. Two kinds of materials are fabricated: (i) ZrO_2 solid solution ceramics and (ii) composite ceramics of ZrO_2 solid solution and α-Al_2O_3. Their mechanical properties are examined in connection with the microstructures and *t/m* ZrO_2 ratios.

1 INTRODUCTION

Many studies have been conducted to improve the fracture toughness of ceramic materials (1). Much attention has been given to partial stabilized ZrO_2 ceramics. It has previously reported that in the system ZrO_2-Al_2O_3, ZrO_2 solid solutions ZrO_2(ss) containing ≈up to 40 mol% Al_2O_3 were metastably formed from amorphous materials prepared by the simultaneous hydrolysis of zirconium and aluminum alkoxides (2). In this paper, we report very high fracture toughness in metastable ZrO_2(ss) ceramics containg 25 mol% Al_2O_3 fabricated by hot isostatic pressing; although they exhibit weak bending strength. Then, in order to improve the strength we have tried to fabricate the ZrO_2 ceramics containing 25 mol% Al_2O_3 and a small amount of Y_2O_3.

Zirconia ceramics with various dopants have traditionally fabricated by sintering mixtures of ZrO_2 powders with the desired stabilizing oxide powders. This process generally results in the formation of the inhomogeneous materials. It is therefore advantageous to produce the ceramics using specially prepared powders. The best and most widely used preparation method has been the chemical reaction route, using for instance co-precipitation and hydrolysis of alkoxides; development in this direction led to the production of highly reactive powders with controlled particle size. The hydrolysis of acetylacetonates, because of the simpler process and lower cost than alkoxide, is an effective method for the powder production. Therefore, zirconium, aluminum and yttrium acetylacetonates have been used as the starting materials to prepare ZrO_2(ss) powders containing Al_2O_3 and Y_2O_3.

2 EXPERIMENTAL PROCEDURE

Zirconium isopropoxide ($Zr(OC_3H_7)_4$, >99.9% pure) and aluminum isopropoxide ($Al(OC_3H_7)_3$, >99.99% pure) were used to prepare a powder with the composition of ZrO_2/Al_2O_3=75/25 mol%. The mixed alkoxides were refluxed for 24 h at 82°C and then hydrolyzed by adding aqueoues ammonia (28 wt%) at room temperature. The temperature was slowly increased for ≈1h to 75°C while the resulting suspension was stirred. The hydrolysis product was separated from the suspension by centrifugation, washed 10 times in hot water, and dried at 120°C under reduced pressure.

Three acetylacetonates, $Zr(C_5H_7O_2)_4$, $Al(C_5H_7O_2)_3$ and $Y(C_5H_7O_2)_3$ were used to prepare powders with the compositions of $ZrO_2/Al_2O_3/Y_2O_3$ =(74.25-71.25)/25/(0.75-3.75) mol%. They were dissolved in an excess of analytical-grade 2-propanol. Under almost the same synthetic process with the ZrO_2 powder containing 25 mol% Al_2O_3 as mentioned above, the starting powders in the $ZrO_2(Y_2O_3)$-Al_2O_3 system were prepared.

Differential thermal analysis (DTA) was conducted in air at a heating rate of 10°C/min; α-Al_2O_3 was used as the reference. Crystalline phases of the samples prepared under the various conditions were examined by X-ray diffraction using Ni-filtered Cu*K*α radiation. Interplanar spacings were measured with the aid of an internal standard of high-purity silicon, and unit-cell values were determined by a least-squares refinement.

Before hot isostatic pressing, the calcined powders were pressed into rectangular blocks at 196 MPa and then isostatically cold-pressed at 392 MPa. The green compacts covered with BN powders were sealed in a Pyrex-glass tube under vacuum. Hot isostatic pressing experiments were performed for 1 h at 1000°-1230°C under 196 MPa. Test samples were then cut from the sintered bodies and lapped with an SiC disc and then a diamond disc. The bulk densities were verified by the Archimedes method. The fraction of monoclinic phase for test samples was determined from the intensity ratio of the monoclinic (111) and ($\bar{1}$11) lines to the tetragonal (111) line by XRD analysis (3). The fracture toughness K_{Ic} was evaluated by the micro-indentation method with a load of 490 N using the equation reported by Niihara et al. (4). Three-point bending strength σ_b was measured using 16 mm span and a crosshead speed of 0.5 mm/min.

3 RESULTS AND DISCUSSION

3.1 Characterization of the Powder and Evaluation of the Material with the Compositon of ZrO_2/Al_2O_3=75/25 mol%

The starting powder was amorphous. DTA curves revealed an exothermic peak resulting from the crystallization of cubic (*c*)-ZrO_2 (ss) at 735°-805°C. This phase was transformed into tetragonal (*t*)-ZrO_2 (ss) at ≈1000°C, although no thermal activity was detected in the DTA curve. No transitional phase of Al_2O_3 was detected throughout the heating process. The formation of α-Al_2O_3 from *t*-

ZrO_2(ss) and transformation from *t*- to *m*-ZrO_2 phase were observed above 1225°C. Calcining at 800°-900°C for 1h yielded *c*-ZrO_2(ss) with the lattice parameter of *a*=0.5115 nm. *t*-ZrO_2(ss) were formed at 1000°-1100°C. The lattice parameters were estimated at *a*=0.5119 nm and *c*=0.5262 nm; the values were larger than those of pure *t*-ZrO_2 (5). *m*-ZrO_2(ss) having *a*=0.5159 nm, *b*=0.5242 nm, *c*=0.5322 nm, and β=98.9° was produced when calcined at 1200°C. These values, as well as in *t*-ZrO_2(ss), were large in comparison with those of pure *m*-ZrO_2 data (*a*=0.51463 nm, *b*=0.52135 nm, *c*=0.53110 nm, and β=98.2°)(6). The phases in the specimens prepared at each calcined temperature are given in Table I.

Table I Experimental conditions and bulk density, *t/m* ratio, and mechanical properties of isostatically hot-pressed ZrO_2 solid solution containing 25 mol% Al_2O_3

Specimen	Calcining temp.(°C)	Phase of calcined powder	HIPing temp.(°C)	*t/m** ratio(%)	Bulk density (g/cm^3) Relative density(%)	K_{Ic}** ($MPa{\cdot}m^{1/2}$) Bending strength (MPa)**
1	800	Cubic(ss)+	1000	70/30	4.78(88)	9.5/380
2			1100	40/60	5.10(97)	18/470
3			1125	25/75	-	-
4	900	Cubic(ss)	1000	80/20	4.75(86)	11.5/330
5			1100	55/45	5.14(97)	23/570
6			1125	35/65	-	-
7	1000	Tetra.(ss)	1000	95/5	4.98(92)	4.5/700
8			1100	70/30	5.26(98)	17/700
9			1125	50/50	5.25(99)	16.5/470
10			1150	30/70	-	-
11	1100	Tetra.(ss)	1000	85/15	4.90(90)	4.5/130

* *t*=*t*-ZrO_2, *m*=*m*-ZrO_2.
** K_{Ic} and bending strength are average values of three test samples.
\+ ss=solid solution.

Table I shows the characteristics of the samples fabricated under the various experimental conditions. The fraction of monoclinic phase, and bulk densities, except for specimen 9, increased with increasing HIPing temperatures. Relative densities were estimated using the theoretical densities calculated from the lattice parameters of *t*-ZrO_2(ss) and *m*-ZrO_2(ss). The specimens sintered at 1000°C had low densities less than 92% irrespective of the calcining conditions. However, higher HIPing temperatures above 1100°C gave the specimens high relative densities more than 97%, especially high densities of 98-99% were obtained by using the calcined powders with tetragonal phase heated at 1000°C. Visible cracks were formed in the sintered bodies containing the monoclinic phase above 65%. No presence of Al_2O_3 in all sintered bodies, as well as in calcined powders, was confirmed. These results suggest that metastable ZrO_2(ss) ceramics could be prepared. As shown in Table I, specimen 5 gave the maximum K_{Ic} of ≈23 MPa•m$^{1/2}$. It is considered that the K_{Ic} obtained from the present method is related to crack propagation. Therefore, the results indicate that a large amount of stress-induced transformation led to a higher value of K_{Ic}. The σ_b are also shown in Table I. The highest value was 700 MPa for specimen 8. The strength was weak in comparison with that of Y_2O_3 partially stabilized ZrO_2 containing Al_2O_3 (7). Microstructural observation by SEM revealed that the texture of the specimen consisted of a homogeneous microstructure with an average grain size of ≈50 nm (Fig.1). No ZrO_2 ceramics consisting of nanometer grain size have ever been reported in the literature. From above results, metastable ZrO_2(ss) ceramics in the system ZrO_2-Al_2O_3 were found to give very high K_{Ic}.

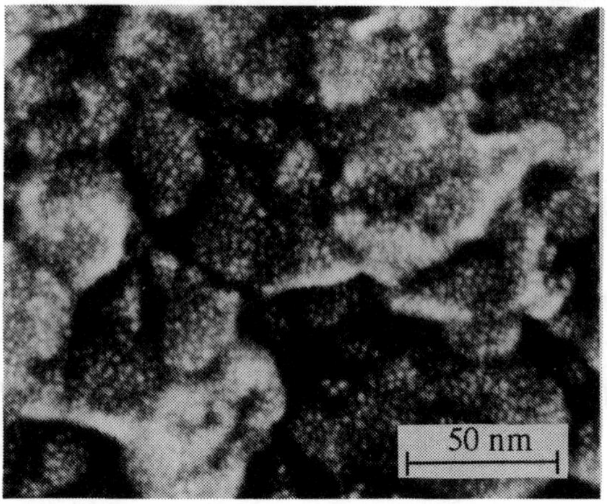

Fig. 1 Scanning electron micrograph of a solid solution ceramics, specimen 5 with the composition of ZrO_2/Al_2O_3=75/25 mol%.

3.2 Characterization of the Powders and Evaluation of the Materials with the Compositions of $ZrO_2/Al_2O_3/Y_2O_3$=(74.25-71.25)/25/(0.75-3.75) mol%

All starting powders, as shown in Table II, were amorphous. Microscopic observation by TEM revealed that the powders were highly agglomerated and had a primary particle size of ≈10 nm. Exothermic peaks resulting form the crystallization of *t*-ZrO_2(ss) were observed at ≈780°-860°C. In the binary system *c*-ZrO_2(ss) crystallized, however, in the system ZrO_2(Y_2O_3)-Al_2O_3, *t*-ZrO_2(ss) were formed from amorphous materials. The specimens heated at 860°-1200°C showed the *t*-ZrO_2 phase. Intermediate alumina, such as γ- and θ-Al_2O_3, and α-Al_2O_3 were not confirmed through the heating process up to 1200°C. Although no peaks were detected in the DTA, *t*- to *m*-ZrO_2 transformation occurred at

≈1250° to ≈1300°C for powder A. Precipitation of α-Al_2O_3 from the solid solution was observed above 1250°C. The appearance of α-Al_2O_3 was shifted at higher temperatures with increasing Y_2O_3 addition.

Calcined powders consisting of the *t*-ZrO_2 phase were prepared by heating for 1h at 1000°C . Table II shows the HIPing temperatures and *t/m* ratios in the materials fabricated by HIP. No presence of intermediate alumina and α-Al_2O_3 was observed in the materials sintered at 1130°C; they consisted of either mixtures of *t*- and *m*-ZrO_2 phases or only *t*-ZrO_2. The results indicate that perfect ZrO_2(ss) ceramics were obtained. On the other hand, materials sintered at 1230°C contained α-Al_2O_3. The material of well-precipitated α-Al_2O_3 was prepared by heating powder A in air for 3 h at 1700°C. The fractional precipitation was determined from the ratio of the intensity using the (113) line. The fraction of α-Al_2O_3 of the materials sintered at 1230°C decreased from ≈90 to ≈53 % with increasing Y_2O_3 additon; in particular, it dropped sharply between 1.88 and 3 mol% Y_2O_3.

Composites of ZrO_2(ss) and α-Al_2O_3, in which their contents could be controlled by the amount of Y_2O_3 addition and the sintering temperature, could be easily fabriacated. The microstructure was characterized by SEM. Figure 2 (a) shows the SEM (secondary electron image) photograph for the fracture surface of the dense solid solution ceramics from powder D. The average grain size, measured directly by SEM, was of ≈0.1 μm in all solid solution ceramics. The α-Al_2O_3 grains in composite ceramics could not be clearly resolved by the observation; however, the presence of homogeneous dispersed α-Al_2O_3 grains was proved by backscattered elelctron imaging of the polished and thermally etched surface (Fig.2(b)).

Table II Chemical composition of starting powders, HIPing temperature and *t/m* ratio in materials

Starting powder	composition		HIPing temperature (°C)	*t /m* ratio (%)
	$ZrO_2/Al_2O_3/Y_2O_3$ (mol%)	$Y_2O_3/(ZrO_2+Y_2O_3)$ (mol ratio)		
A	74.25/25/0.75	1.0	1130	47/53
			1230	24/76
B	73.50/25/1.50	2.0	1130	80/20
			1230	100/0
C	73.12/25/1.88	2.5	1130	100/0
			1230	100/0
D	72.75/25/2.25	3.0	1130	100/0
			1230	100/0
E	72.00/25/3.00	4.0	1130	100/0
			1230	100/0
F	71.25/25/3.75	5.0	1130	100/0
			1230	100/0

The proportion of the *t*- and *m*- ZrO_2 phases is shown in Table II. The *m*-ZrO_2 phase was present in the solid solution ceramics from powders A and B, and in the composite ceramics from powder A. Above 1.88 mol% Y_2O_3 addition both materials consisted of the *t*-ZrO_2 phase. The absence of the *m*-ZrO_2 in the composite ceramics from powder B might be explained in terms of the increase of the Y_2O_3 concentration in ZrO_2, as a result of the precipitaion of α-Al_2O_3. Bulk densities of both materials were measured. The solid solution ceramics increased linearly in density from $\approx$5.12 to $\approx$5.55 g/cm^3 with increasing Y_2O_3 addition. The bulk densities of the compsotie ceramics increased rapidly from $\approx$5.22 to $\approx$5.55 g/cm^3 between 0.75 and 1.5 mol% Y_2O_3 and attained a constant value ($\approx$5.55 g/cm^3) above 1.5 mol% Y_2O_3 addition. The theoretical densities of α-Al_2O_3, *t*-ZrO_2 and *m*-ZrO_2 are 3.987, 6.097 and 5.840 g/cm^3, respectively Accordingly, the lower density of the composite ceramics with 0.75 mol% Y_2O_3 can be attributed to the presence of large amounts of α-Al_2O_3 and *m*-ZrO_2.

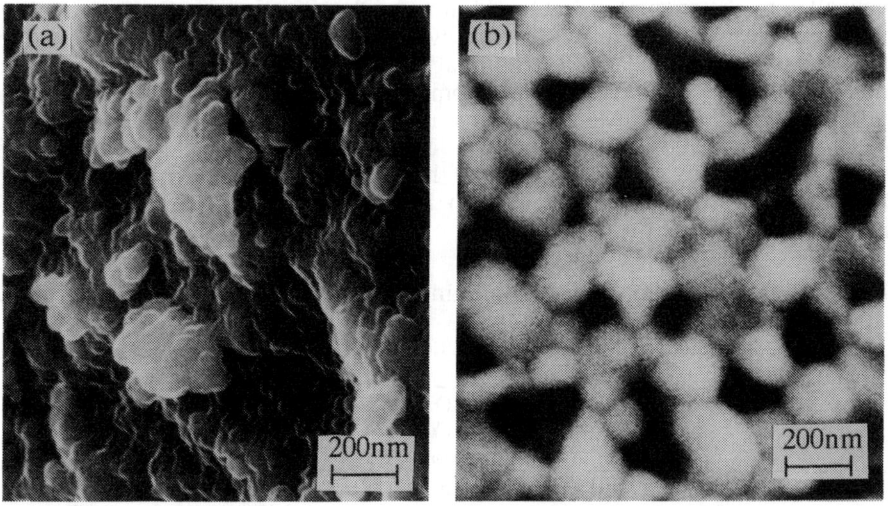

Fig. 2 SEM photographs for (a) secondary electron image of solid solution ceramics and (b) backscattered electron image of composite ceramics fabricated from powder D.

Figure 3 (a) shows the fracture toughness K_{Ic} and bending strength σ_b of the solid solution ceramics as a function of Y_2O_3 addition. Curves with similar features were obtained for both of them. The maximum values of 13.2 MPa•m$^{1/2}$ and 1045 MPa for the K_{Ic} and σ_b, respectively, were achieved in the solid solution ceramics with 2.25 mol% Y_2O_3. The σ_b was much higher than that of the material in the binary system. However, K_{Ic} gave a lower value. The *t*- and *m*-ZrO_2 particles enhance the toughness by creating a "stress-induced" transformation and by extending "pre-existing" microcracks, respectively. The solid solution ceramic in the binary system contained m-ZrO_2 phase of 30-50% as mentioned before. The most highly toughened ceramics in the $ZrO_2(Y_2O_3)$-Al_2O_3 system consisted of the t-ZrO_2 phases. The t- to m- ZrO_2 transformation depends on the critical size of particle [e.g.(8)]. The decrease of the K_{Ic} value must be due to stable t-ZrO_2 particles below the critical size being hard to transformation by Y_2O_3 addition. Excellent σ_b (1445 MPa) and high K_{Ic} of

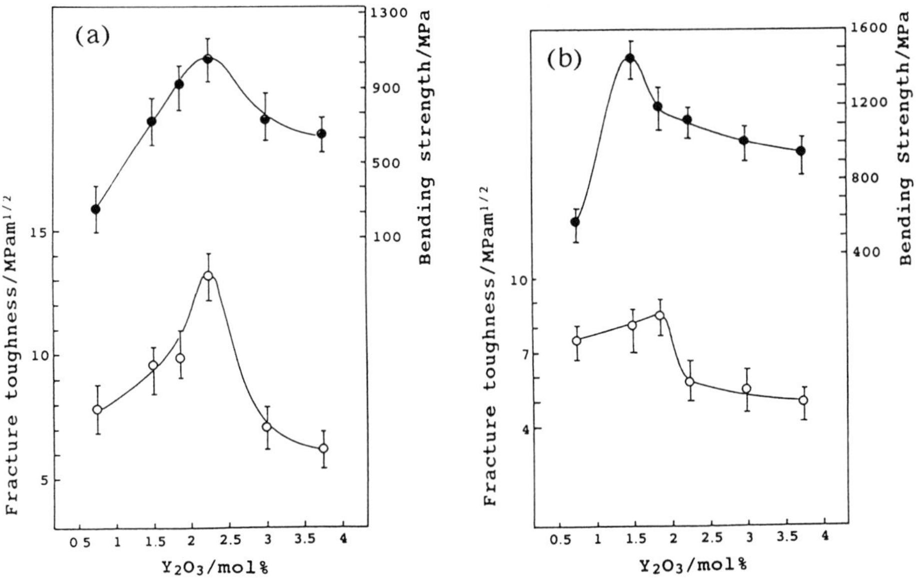

Fig. 3 (○) Fracture toughness and (●) bending strenth as a function of Y_2O_3 addtion: (a) ZrO_2(ss) ceramics and (b) composite ceramics of ZrO_2(ss) and α-Al_2O_3.

(8.5 MPa•$m^{1/2}$) were obtained in the composite ceramics with 1.5 and 2 mol% Y_2O_3, respectively (Fig.3(b)). The composite ceramics with a homogeneous dispersed α-Al_2O_3 derived from the solid solutions resulted in a remarkable improvement of σ_b. The composition dependence of the mechanical properties of toughened ZrO_2-Y_2O_3 ceramics has been widely studied [e.g.(9)]. Both K_{Ic} and σ_b are known to give a maximum value at a particular composition:2 mol% Y_2O_3 for the former and either 2.5 or 3 mol% Y_2O_3 for the latter. The data for the composite ceramics, as can be seen from the composition ratios in Table II, agreed with these results.

References

1. For example, A. G. Evans, "Perspective on the Development of High-Toughness Ceramics," *J. Am. Ceram. Soc.*, **73** [2]187-206 (1990).

2. O. Yamaguchi, M. Shirai and M. Yoshinaka, "Formation and

Transformation of Cubic ZrO_2 Solid Solution in the System ZrO_2-Al_2O_3," ibid., **71** [12] C-510 (1988) .

3. R. C. Garvie and P. S. Nicholson, "Phase Analysis in Zirconia System," ibid., **55** [6] 303-05 (1972) .

4. K. Niihara, R. Morena, D. P. H. Hasselman, "Evaluation of K_{Ic} of Brittle Solids by the Indentaion Method with Low Crack-to-Indent Ratios," *J. Mater. Sci. Lett.*, **1** 13-16 (1982) .

5. H. Nishizawa, N. Yamasaki, K. Matsuoka, and H. Mitsushio, "Crystallization and Transformation of Zirconia Under Hydrothermal Conditions," *J. Am. Ceram. Soc.*, **65** [7] 343-46 (1982).

6. Powder Diffraction File, Card No. 13-307. Joint Committee on Powder Diffraction Standards, Swarthmore, PA (1972).

7. T. Tsukuma, K. Ueda and M. Shimada, "Strength and Fracture Toughness of Isostatically Hot-Pressed Composites of Al_2O_3 and Y_2O_3-Partially-Stabilized ZrO_2," *J. Am. Ceram. Soc.*, **68** [1] C-4 -C-5 (1985).

8. R. Stevens and P. A. Evans, "Transformation Toughening by Dispersed Polycrystalline Zirconia," *Br. Ceram. Trans. J.*, **83** [1] 28-31(1984).

9. T. Masaki, "Mechanical Properties of Toughened ZrO_2-Y_2O_3 Ceramics," *J. Am. Cram. Soc.*, **69** [8] 638-40 (1986).

SYNTHESIS OF NOVEL PRECURSORS FOR PMN POWDERS AND THE THIN FILMS OBTAINED FROM THEM

Timothy J. Boyle*, Duane B. Dimos, and Gregorio J. Moore
Sandia National Laboratories, 1001 Univ. Blvd. SE, Albuquerque., NM 87106

ABSTRACT

Sol-gel processing has been widely used in the preparation of lead zirconate titanate (PZT) thin films. We have applied this methodology to the formation of lead magnesium niobate (PMN) spin-cast deposited thin films. Since there is a limited number of soluble, commercially available compounds, we have recently synthesized a series of novel metal alkoxides for use as precursors for generation of PMN thin films and powders. The process for generation of the perovskite phase of these PMN powders and films are reported.

INTRODUCTION

The perovskite phase of lead magnesium niobate, $Pb(Mg_{1/3}Nb_{2/3})O_3$ (PMN) is of interest for use in thin film capacitors and micro-electromechanical systems due to the very high dielectric constant and the electromechanical properties of PMN.[1] The perovskite phase of PMN is very disordered and considered a classical relaxor ferroelectric. A relaxor ferroelectric can be distinguished from normal ferroelectric material by several properties.[2] First, it has a broad Curie maximum point. The Curie point is the temperature at which ferroelectric material (material in which electric domains tend to be aligned parallel to one another) becomes paraelectric. For PMN this broad Curie point is reportedly due to lattice and cation disorders which result in short range order, coupled with chemical inhomogenous micro-regions.[3] Other characteristics include a diffuse phase transition and a low frequency dispersion of the dielectric constant. Single crystals of PMN have been shown to have a maximum dielectric constant of about 20,000 at 1 kHz;[2,4] however, thin film dielectric permitivities are estimated to be around 1,250 at room temperature.[3] The electrical differences have been defined by the presence of a pyrochlore phase formed during the initial stages of conversion form metal alkoxides to the ceramic form.

Sol-gel derived thin films are favored for production of PMN films due to the flexibility in the characteristics of solution precursors, the variety of deposition methodologies, and the reduction of the sintering temperatures.[5] The standard sol-gel approach to generating PMN thin films typically involves either using commercially available precursors and dissolving them in 2-methoxyethanol, which acts as both a solvent and a chemical modifier[6] or to synthesizing large metallo-organic "soap derivative" (*neo*-deconoate) compounds using organic solvents such as xylenes.[7] For the former method, water is generated as a by-product and further uncontrolled modifications occur. For the latter, identification of intermediates and higher organic pyrolysis temperatures are required due to the increased hydrocarbon content present. Therefore, we have developed an alternative system for fabricating PMN materials which will eliminate these two problems.

Specifically, we have synthesized lead, magnesium, and niobium alkoxide (*tert*-butoxide and *neo*-pentoxide) derivatives and utilized a magnesium alkyl reagent as starting materials. These reagents are soluble in a large number of organic solvents which do not generate the customary by-products observed in the 2-methoxyethanol based system. Due to this feature, the aging of the solution is minimal. The alkoxides we have utilized are lower in carbon content and therefore produce cleaner burnout than the "soap" derivatives. Furthermore, based on previous experience with *neo*-pentoxide complexes, characterization of the intermediate species should be possible. This paper reports the process for synthesizing the precursors and some preliminary optical and electronic characteristics of the films generated.

EXPERIMENTAL

All reactions were performed under dry nitrogen or argon, using standard Schlenk, vacuum line, and glove box techniques. Solvents were dried and films produced as previously described.[8] Basic film processing variables were: in air deposition of 3-layers [1-layer: 3000 rpm for 10 secs., hot-plate treatment (300°C for 5 min.) repeat] followed by thermal processing [in air to 650°C, ramp rate 25°C/min., hold time 30 min.]. FT-IR data was obtained on a Nicolet, Magna System Spectrometer-550. $PbCl_2$ (Aldrich), $Mg((CH_2)_3CH_3)_2$ 1.0M in heptanes (Aldrich), and $Nb(OCH_2CH_3)_5$ (Aldrich), $NaOCMe_3$ (Aldrich) were used as received. $NaOCH_2CMe_3$ was isolated from the reaction of Na° and $HOCH_2CMe_3$ in THF and purified by sublimation.

$Pb(OCH_2CMe_3)_2$, 1. In a Schlenk flask $PbCl_2$ (7.30 g, 26.2 mmol) is slurried in toluene (120mL). $NaOCH_2CMe_3$ (5.77 g, 52.49 mmol) is slowly added and the reaction mixture is stirred under argon overnight while heated to 60°C. The reaction is cooled to room temperature and the volatile fraction removed *in vacuo*. **1** is extracted using THF from the resulting powder. Yield 71%. FT-IR (KBr pellet) 2956(s), 2905(w), 2871(w), 2710(m), 1600(s), 1488(m), 1421(s), 1361(s), 1316(s), 1097(m), 1020(s), 955(s), 800(m), 500(w) cm^{-1}.

$Mg(OCMe_3)_2$, 2. Synthesis similar to generation of **1**. $MgBr_2$ (5.0 g, 27.1 mmol), $NaOCMe_3$ (5.22 g, 54.3 mmol) in toluene (100 mL). Yield 55%. FT-IR (KBr pellet) 2961(s), 2922(m), 2899(m), 2859(m),1472(s), 1376(s), 1351(m), 1260(m), 1206(s), 1092(w), 999(s), 961(s), 921(w), 802(w), 761(w), 570(s), 494(s), 418(w), 409(w) cm^{-1}.

$Mg(OCH_2CMe_3)_2$, 3. $Mg((CH_2)_3CH_3)_2$ (50.4 mL, 50.4 mmol) was syringed onto 200 mL of toluene and cooled to 0°C. $HOCH_2CMe_3$ (8.88 g, 10.1 mmol) was added by addition tube. The reaction was allowed to warm to room temperature and stirred overnight. The volatile products were removed *in vacuo*. 86.5 % Yield. FT-IR (KBr pellet) 2954(s), 2903(m), 2866(s), 2834 (m), 2791(w), 2753(w), 2720(m), 2665(w), 2615(w), 1479(s), 1464(s), 1399(s),

1362(s), 1310(w), 1258(m), 1217(w), 1094(s), 1057(s), 1027(s), 940(m), 933(m), 901(m), 801(w), 756(m), 591(s), 520(s), 447(s) cm^{-1}.

$Nb(OCH_2CMe_3)_5$, 4. In a Schlenk flask $Nb(OCH_2CH_3)_5$ (5.70 g, 17.9 mmol) is added to toluene (100 mL), $HOCH_2CMe_3$ (11.1 g, 89.6 mmol) is added and the reaction mixture is stirred at 60°C for 12 hours. The volatile fraction is removed *in vacuo* and the resulting off white powder can be used without further purification. Yield 44%. FT-IR (KBr pellet) 2954(s), 2902(m), 2868(m), 2693 (w), 1480(m), 1465(m), 1442(w), 1390(m), 1371(w), 1360(m), 1292(w), 1258(w), 1076(bs), 1026(s), 935(w), 918(w), 905(w), 883(m), 848(w), 803(w), 751(w), 658(s), 560(bs), 569(m), 509 (m), 482(m) cm^{-1}.

$\{[H_3C(CH_2O_3)]Nb(OCH_2CH_3)_2\}_2$, 5. In a Schlenk flask $\{[H_3C(CH_2OH)_3$ (1.89 g, 15.7 mmol) is added to $Nb(OCH_2CH_3)_5$ (5.00 g, 15.7 mmol) in 70 mL of toluene. After stirring the reaction mixture for 12 hours, the volatile fraction is removed by rotary evaporation, washed with hexanes three times, and extracted with toluene. The resulting white powder can be used without further purification. X-ray quality crystals were grown by dissolving the extracted powder in a minimum of toluene and cooling the solution to -35°C. Yield 72% . FT-IR (KBr pellet) 2973(s), 2930(s), 2882(s), 2850(s), 2845(s), 2790(w), 2765(w), 2670(w), 1555(m), 1458(m), 1399(m), 1375(m), 1147(s), 1103(s), 1065(s), 1021(s), 1000(m) 989(s), 921(s), 620(s), 535(s) cm^{-1}.

DISCUSSION

Our main goal in this project was to generate a solution that could be used to produce the ceramic perovskite phase of lead magnesium niobate (PMN). It was decided to optimize the chemical aspects of the reaction mixture (i.e., generate a soluble and stable PMN solution) before altering the processing parameters. Standard processing parameters that have been developed for lead zirconate titanate (PZT) thin films were used.[8] Since the reagents $Pb(OAc)_4$ (OAc = O_2CCH_3), $Mg(OEt)_2$ (OEt = OCH_2CH_3), and $Nb(OEt)_5$ are commercially available, we initiated our study using these reagents. A wide variety of non-reactive organic solvents were investigated, however, a homogeneous solution could not be obtained with these compounds. Acetic acid and 2-methoxyethanol are known to undergo esterification, followed by hydrolysis, and oxolation.[6,9] We chose not to investigate 2-methoxyethanol due to the toxic and teratogenic characteristics of this solvent system. Experiments utilizing acetic acid were undertaken but the commercially available components were found to be insoluble even with heating in this solvent.[6,9]

Alternative precursors were required and the number of commercially available materials is quite limited.[10] Therefore, we synthesized (Eq.1)[11] a

$$MCl_2 + 2\ NaOR \xrightarrow{Tol} M(OR)_2 + 2\ NaCl \qquad \text{Eq.1}$$

M = Pb, Mg; OR = $OCH(CH_3)_2$, $OC(CH_3)_3$, $OCH_2C(CH_3)_3$,

collection of lead, magnesium, and niobium alkoxide complexes for use in a precursor solution of PMN. The approach we used was the standard metathesis reaction between metal halides and sodium alkoxides, simple alcoholysis exchange, or alkyl exchange. Typically these reactions were warmed slightly to insure maximum exchange had occurred. Isolation of the soluble portion generated precursors which could be used without further purification. These compounds were identified as completely as possible by FT-IR, TGA/DTA, ^{1}H and ^{13}C NMR and single crystal X-ray diffraction.

Initially we decided to utilize the bulky *neo*-pentoxide (OCH_2CMe_3) ligand since it had fewer carbons than the "soap derivatives" previously used,[7] but still maintained a similar geometric arrangement. The *neo*-pentoxide (ONep) derivatives, $Pb(ONep)_2$, **1**, and $Mg(ONep)_2$, **3**, were synthesized according to Eq. 1. $Nb(ONep)_5$, **4**, was generated by the alcoholysis exchange of $NepO^-$ for EtO^- ligands. The exact number of NepO ligands substituted is still under investigation. A solution mixture which consisted of **1**, **3**, and **4** in a stoichiometric ratio of 1.1:1/3:2/3 (Eq. 2, excess lead was utilized to compensate for lead volatility which

$$Pb(OAc)_2 + 1/3\ Mg(ONep)_2 + 2/3\ Nb(ONep)_5 \xrightarrow{THF} \qquad \text{Eq.2}$$

is not indicated in the equations) was attempted. **3** proved to be the limiting reagent since it is only sparingly soluble in non-reactive solvents, such as tetrahydrofuran (THF) and toluene; however, this ternary mixture was very soluble in methanol (MeOH). This solvent system would, in solution, form the methoxy derivatives and eliminate the need to synthesize a number of the above compounds. FT-IR and NMR data indicated that a complex mixture of aceto-, alkoxy complexes were generated from solutions of Eq. 2 in either solvent. Spin-cast deposited films generated from a 0.40 M THF or MeOH solvated solutions proved to be cracked and opaque. The X-ray diffraction (XRD) spectrum of the THF system indicated that the predominate phase formed was pyrochlore, similar to $Pb_{2.3}Nb_2O_{7.3}$ (see Figure 1); however, there was a substantial amount of the perovskite phase present as well. Pyrochlore is not a desirable phase of these systems since it is non-ferroelectric.[5] Exchanging $Pb(OAc)_2$ for $Pb(ONep)_2$, **1**, in Eq.2, dramatically increased the solubility of this ternary mixture. Analytical data, once again, showed a complex mixture existed in solution as well as in the solid state. XRD confirmed the formation of a perovskite phase (see Figure 1) but several unidentified peaks were present. Films produced from this mixture showed substantially less cracking and optical micrographs (1000X) looked very uniform. However, scanning electron microscopy (SEM) studies revealed a variety of defects which appeared as "blisters" and "warts" on the surface. Cross sections of this film indicated that the platinum electrode layer was "pulling away" from the TiO_2 adhesion layer and this is probably the cause of these defects.

To increase the solubility of these compounds, we investigated the characteristics of the commercially available species, $Mg(\text{-}n\text{-}Bu)_2$, [Bu = $(CH_2)_3CH_3$)]. Addition of this species, Eq. 3, to the reaction mixture immediately

$$Pb(ONep)_2 + 1/3\ Mg(n\text{-}Bu)_2 + 2/3\ Nb(ONep)_5 \xrightarrow{Tol} \qquad \text{Eq.3}$$

$Bu = (CH_2)_3CH_3,$

turned the solution to a dark brown color; however, no precipitate was formed initially. Based on pervious experience, this color change is a common indicator that an early transition metal alkyl has been formed. NMR data revealed a simple spectrum which was consistent with a intimately mixed species. Films generated from a 0.25 M solution of Eq.3 were transparent after initial hot-plate-bake (300°C) performed for drying and pyrolysis purposes. After firing at higher temperatures (650°C), the films showed substantial cracking and peeling. SEM studies indicated that the Pt bottom layer electrode was again being "pulled away" from the TiO_2 adhesion layer. XRD results on both the powder (see Figure 1) and films showed this species to be a perovskite phase of PMN similar to $Pb(Mg_{0.33},Nb_{0.67})O_3$ (peaks are slightly offset). In an attempt to reduce cracking several modifiers were added to the solution, such as: water (H_2O) and alcohols (ROH). Little effect was observed for these modifiers at the current level of addition; however, addition of both HOAc and acetyl acetone (Acac-H) formed gels/precipitates immediately.

It was observed that the solution generated from Eq. 3 would form a precipitate after approximately 2 hours. This was hypothesized to be due to the continued reactivity of the $Mg(R)_2$ species. Since the $Mg(ONep)_2$ complex had demonstrated low solubility, Mg species with increase solubility and decreased reactivity were desired. *Tert*-butoxide (O-*t*-Bu) ligated species possess these characteristics. Therefore, the synthesis of Mg(O-*t*-Bu)$_2$, **2**, as per Eq.2, was undertaken. Addition of **2** to the PMN precursor solution mixture, as in Eq. 4,

$$Pb(ONep)_2 + 1/3\, Mg(O\text{-}t\text{-}Bu)_2 + 2/3\, Nb(ONep)_5 \xrightarrow{\text{Toluene}} \qquad \text{Eq.4}$$

$Bu = (CH_2)_3CH_3$,

produced a yellow transparent solution, which upon hot-plate-bake produced a very clear film. Conversion of the organometallic species to the ceramic phase, once again, produced a film with defects; however, these faults were substantially reduced from previous attempts. Furthermore, optical microscopy (1000X) showed that the microstructure surface was very uniform; however, electrical measurements could not be undertaken due to the occurrence of electrical shorting.

A number of unique Nb compounds have also been isolated in our laboratory, one of which is $[(THME)Nb(OEt)_2]_2$, **5** [THME = $(OCH_2)_3CCH_3$]. **5** is a dimeric species both in the solid state and solution, wherein the THME ligand act as a chelating ligand to one Nb and one of the THME oxygens bridging to the other metal center. A ball-stick diagram of **5** is shown in Figure 2. **5** was introduced into the PMN precursor solution Eq. 5a and 5b to add a compound of

$$Pb(ONep)_2 + 1/3\, Mg(n\text{-}Bu)_2 + 2/3\, [(THME)Nb(OEt)_2]_2 \xrightarrow{\text{Tol}} \qquad \text{Eq.5a}$$

$$Pb(ONep)_2 + 1/3\, Mg(O\text{-}t\text{-}Bu)_2 + 2/3\, [(THME)Nb(OEt)_2]_2 \xrightarrow{\text{Tol}} \qquad \text{Eq.5b}$$

known nuclearity and increased steric bulk to reduce uncontrolled hydrolysis by ambient humidity. As before, the resulting mixtures from 5a turned black immediately upon addition of the Mg(*n*-Bu)$_2$. For 5b, the reaction mixture remained a pale yellow color. Films generated from 5a and b were shown by XRD to be the perovskite phase of PMN. Films from 5b demonstrated reduced cracking; however, "shorts" were reported when the electrical properties were investigated. Spin cast deposited films of 5a proved to be uniform and XRD analysis demonstrated that the perovskite phase had been formed. SEM results show very little surface defects and side views of this film showed a very densely packed material (see Figure 3). Tests of the electrical properties revealed substantial electrical shorting occurred for a 3-layer film. Films generated by firing (650°C) between each layer proved to be extremely uniform and appeared defect-free. Electrical testing of these thin films is underway.

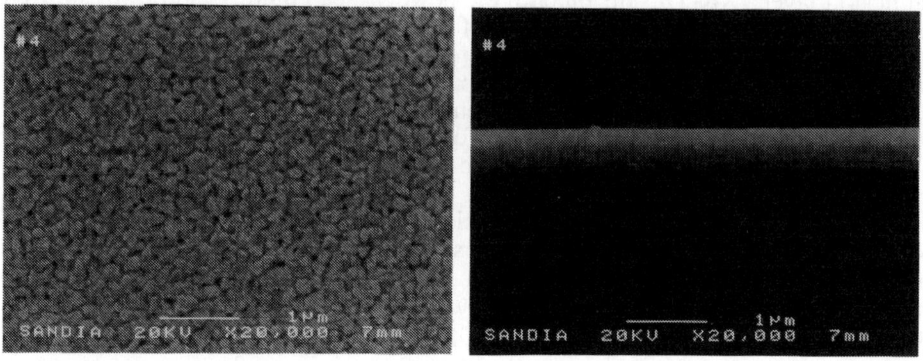

(A) (B)

Figure 3. SEM micrographs of $Pb(ONep)_2$ + 1/3 $Mg(n\text{-}Bu)_2$ + 2/3 $[(THME)Nb(OEt)_2]_2$ (A) Flat view, (B) Side view.

CONCLUSION

We have successfully demonstrated that from these unique precursor solutions the perovskite phase of PMN was consistently produced both in powder and thin film form. From the variety of precursors we have synthesized and utilized, it is apparent that the more soluble and potentially "cross-linking" "$Pb(OR)_2$" and $Mg(R)_2$ type of complexes are desired for formation of the perovskite phase of PMN; furthermore, the sterically hindered $[(THME)Nb(OEt)_2]_2$ precursor is preferred versus that of $Nb(OEt)_5$. The increase solubility insures intimate mixing whereas the steric bulk of the THME ligand insures controlled hydrolysis. Even though thin films have been shown to be the correct phase, substantial work must be undertaken to reduce cracking of the metallo-organic phase after firing. This will be accomplished by adding a number of potential "cross-linking" modifiers to our unique precursor solutions. A variety of substrates and processing parameters are being investigate to determine the ideal conditions for generating defect free thin films.

ACKNOWLEDGMENTS This work supported by the U. S. D. o E. under contract number DE-AC04-94AL85000. The authors would like to thank Robert W. Schwartz (Sandia) for helpful discussions, Mary Gonzales (Sandia) for XRD studies, Dr. Joseph W. Ziller (U.C.-Irvine) for structural determinations, and Dr. Todd M. Alam (U.N.M.) for NMR studies.

FOOTNOTES

1. This work reported in part at the *6th Ann. Joint Meeting NM Sec.- Am. Cer. Soc. and Mat. Res. Soc.*, Holiday Inn, Albuq., NM, Oct. 31, 1994. (Poster).
2. Cross, L. E. *Ferroelectrics* **1987**, *76*, 241-267.
3. Kim, Y. S. and Jang, H. M *Ferroic Materials-Ceramic Transactions Vol. 43* (The Am. Cer. Soc., Westerville Ohio, 1994) Eds. Bhalla, A. S.; Nair, K. M.; Lloyd, I. K.; Yanagida, H.; Payne, D. A., p. 123-133.
4. Yan, M. F.; Ling, H. C.; Rhodes, W. W. **1989**, *4*, 930-944.
5. Ravindranathan, P.; Komarneni, S.; Bhalla, A. S.; Cross, L. E., Roy, R. *Ferroelectric Letters* **1990**, *12*, 29-34.
6. Ramamurthi, S. D.; Payne, D. A. *J. Am. Ceram. Soc.* **1990**, *73*, 2547-2551.
7. Vest, R. W.; Zhu, W. *Ferroelectrics* **1991**, *119*, 61-75.
8. Boyle, T. J.; Schwartz, R. W.; Doedens, R. J.; Ziller, J. W. *Inorg. Chem.* **1995**, *33* (in press).
9. Alam, T. M.; Boyle, T. J.; Buchheit, C. D.; Schwartz, R. W.; Ziller, J. W. *Mat. Res. Soc. Symp. Proc.* **1994**, *346*, 35-40.
10. Commercially available (Aldrich) PMN precursors include: $Pb(OAc)_2 \bullet 3H_2O$, $Pb(OAc)_2 \bullet 2Pb(OH)_2$, $Mg(OAc)_2$, $Mg(OAc)_2 \bullet 4H_2O$, $Mg(OEt)_2$, $Mg(OH)_2$, $Nb(OEt)_5$.
11. Bradley, D. C.; Mehrotra, R. C.; Gaur, D. P. *Metal Alkoxides* (Academic Press, New York, 1978).

Electrophoretic Deposition of Mullite In a Continuous Fashion Utilizing Non–Aqueous Polymeric Sols

Peter W. Brown
National Research Council Associate
Wright Laboratory, Materials Directorate
Wright–Patterson AFB, OH, 45433–7817

Abstract

The feasibility of depositing mullite *via* electrophoretic deposition (EPD) onto a 2–*d* substrate, such as Nicalon™ cloth, was determined in this preliminary study. Mullite coatings were deposited onto a C–coated Nicalon™ cloth using a polymeric precursor. XRD data suggest that phase separation of the sol does not occur upon drying and subsequent thermal treatment of the sol. XRD also indicates an amorphous → crystalline transition between 800 and 900 °C. The coatings appeared to be fairly uniform throughout the cloth cross–section, on the order of 250*nm* thick, and displayed a minimal amount of bridging. A large amount of 5 nanometer–sized porosity was uniformly distributed throughout the coating.

Introduction

When Nicalon™ fibers are exposed to high temperatures in an oxidizing environment, a layer of silica (SiO_2) forms on the outer surface.[1,2] This layer of SiO_2 may subsequently bond with the neighboring matrix material (*e.g.*, Al_2O_3), thereby reducing the mechanical properties of the composite.[3,4] Additionally, the reaction producing SiO_2 scale on the fiber surface may cause an inherent reduction in the mechanical properties of the fiber.[5]

The Al_2O_3–SiO_2 binary phase diagram indicates that mullite ($3Al_2O_3{\cdot}2SiO_2$) is the only thermodynamically stable phase between alumina and silica. Hence, upon exposure to elevated temperatures in an oxidizing environment, the resultant formation of a SiO_2 film on Nicalon™ fibers should be stable in the presence of a mullite coating and should therefore reduce the amount of bonding between fibers and matrix. Accordingly, mullite may be a suitable interphase treatment for such composites.[6–8]

It is intended that the Nicalon™ fiber coatings be composed of alternating carbon and mullite layers, as in Figure 1. Figure 1a indicates the desired coating scheme, with an oxide coating layer adjacent to the Nicalon™ fiber. Figure 1b is a schematic of the type of multi–layered coatings which are achievable via EPD. The primary difference between the two coating schemes is in the type of coating directly adjacent to the fiber. Coating fibers *via* EPD requires that the fibers be electrically conductive; hence the presence of the carbon layer as in Figure 1b.

Exposure to high temperatures in an oxidizing environment of a composite reinforced with fibers coated as such should lead to the oxidation and removal of the carbon layer(s), thereby leaving behind an annular gap (ideally) around the fiber. The presence of the mullite layer could retard the outward growth of the SiO_2 layer by slowing the transport of O_2 to the surface of the fiber, thereby reducing the amount of bonding between fibers and matrix.

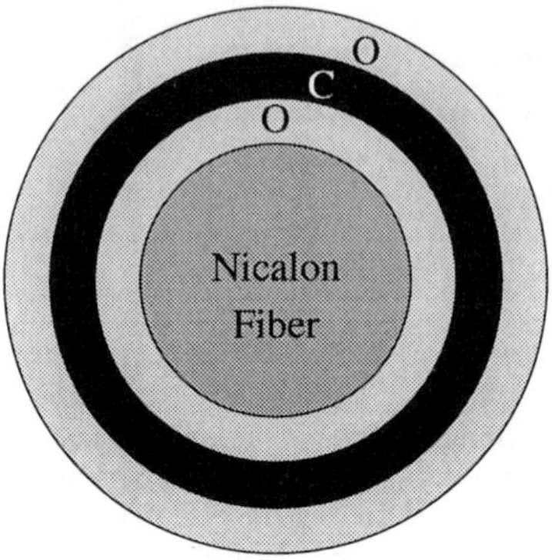

(a) Desired coating scheme.

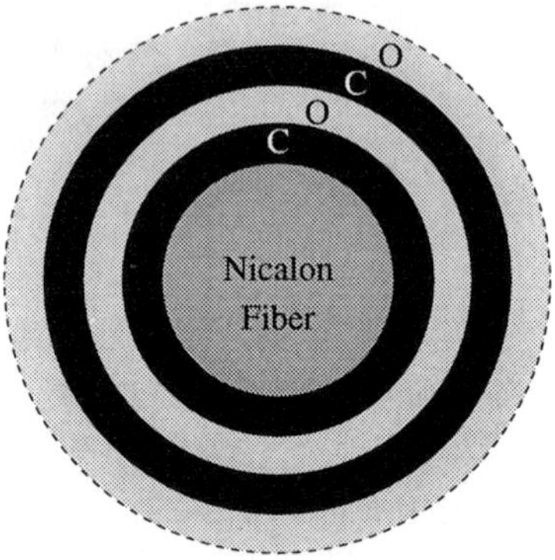

(b) Obtainable coating scheme.

Figure 1: Comparison of the desired and obtainable coating schemes for C–$3Al_2O_3{\cdot}2SiO_2$ coatings on Nicalon™ cloth.

Electrophoresis is the movement of charged particles in an electric field. Usually, EPD is used to coat substrates with submicron sized particles of the coating material suspended in some solvent. The method discussed in the present work uses an electric field to transport charged polymeric molecules, in contrast to the transport of charged particles. EPD has been used in the past to deposit films of several other multicomponent oxides, such as YBCO,[9–13] BSCCO,[14] BPSCCO,[15] (Pb, La)(Zr, Ti)O_3 films,[16] SiO_2,[17,18] YSZ/Al_2O_3 micro–laminates,[19,20] and $BaTiO_3$.[21] The polymeric mullite sol synthesis route discussed in the present work has been investigated previously.[22,23] Other investigators have prepared polymeric mullite sols using a mixture of organic and inorganic precursors[24–29] or other methods based only on organic precursors.[30–32] Other investigators have synthesized mullite using *particulate* precursors[33–39] instead of the *polymeric* (also called *molecular*) route.

Experimental Procedure

The resistivity of the as–received Nicalon™ cloth was prohibitively high ($> 300\frac{M\Omega}{mm}$) to be used as an electrode in an EPD process. Hence, the Nicalon™ substrate required a pre–EPD treatment to reduce the resistivity of the cloth. The method selected for this was to dip–coat the cloth into a pitch/toluene solution, and subsequently heat–treat the dipped cloth at an elevated temperature ($\approx$ 800 °C) in flowing Ar. The resultant resistivity ($\approx 0.1\frac{k\Omega}{mm}$) was sufficiently low for the cloth to be used as an electrode in the EPD process.

The desired thickness of the $3Al_2O_3{\cdot}2SiO_2$ coatings was $\approx 0.25\ \mu m$, which was too thin for a $0.2\ \mu m$ particle suspension. Hence, an alternate precursor was needed which possessed a significantly smaller mullite particle size. Accordingly, a *polymeric* mullite sol was synthesized as follows. *T*etra*eth*o*xy*s*ilane (TEOS) was partially hydrolyzed with a mixture of ethanol (EtOH) and water using a TEOS:EtOH:H_2O volumetric ratio of 90:5:5. Aluminum *sec*–butoxide was subsequently added to the hydrolyzed TEOS in the stoichiometric amount consistent with mullite. The Al *sec*–butoxide reacted with the hydroxyl group of the partially hydrolyzed TEOS to form the Al—O—Si bonded mullite polymer. A simplified reaction sequence may be expressed according to the following series of reactions:

$$Si(OC_2H_5)_4 + H_2O \rightarrow (HO)Si(OC_2H_5)_3 + C_2H_5OH$$
$$(H_9C_4O)_2Al(OC_4H_9) + (HO)Si(OC_2H_5)_3 \rightarrow (H_9C_4O)_2Al(O)Si(OC_2H_5)_3 + C_4H_9OH$$

The concentration of the as–prepared polymeric sol was $170\frac{g}{l}$. The as–prepared sol was diluted with dried EtOH to a final concentration of $6\frac{g}{l}$ for use in the EPD apparatus. Concentrated HNO_3 was then added until the pH of the solution was approximately 4. The pH adjustment step establishes a non–zero surface charge on the polymeric molecules, allowing their use in an EPD coating scheme.

Initially, coating experiments were carried out in a batch mode of operation. However, the resultant coatings displayed extreme variations in thickness, presumably from variations in the electric field. These electric field variations were caused primarily by the voltage drop across the substrate. In order to obtain more uniform coatings, a continuous mode of operation was developed. Figure 2 is a schematic of the apparatus used in a continuous fashion. A strip of the Nicalon™ cloth was used as the anode in the EPD circuit, whilst a Pt/Rh wire mesh was used as the cathode. Typical EPD processing conditions are summarized as follows: V$\approx 100V$, I$\approx 20mA$, residence time of cloth in solution $\approx 1min$.

Experimental Results

The as–prepared polymeric sol was dried and calcined between 200 and 1400 °C, in 100 °C increments, and subsequently analyzed *via* *e*nergy *d*ispersive *s*pectroscopy (EDS) and *x*–*r*ay *d*iffraction (XRD). EDS analysis indicated an Al:Si ratio consistent with that of mullite. XRD results show the sol to be amorphous $\leq$ 800 °C. At temperatures $>$ 800 °C, the XRD data are consistent with the presence of mullite, with no other XRD peaks present.

The mullite–coated cloth samples were subsequently heat–treated at 800–1000 °C in flowing Ar for $1h$. *S*canning *e*lectron *m*icroscopy (SEM), EDS, XRD, and *t*ransmission *e*lectron *m*icroscopy (TEM) analysis indicate the coatings to be mullite approximately $250nm$ thick. TEM analysis indicates the coating to be comprised of equiaxed mullite

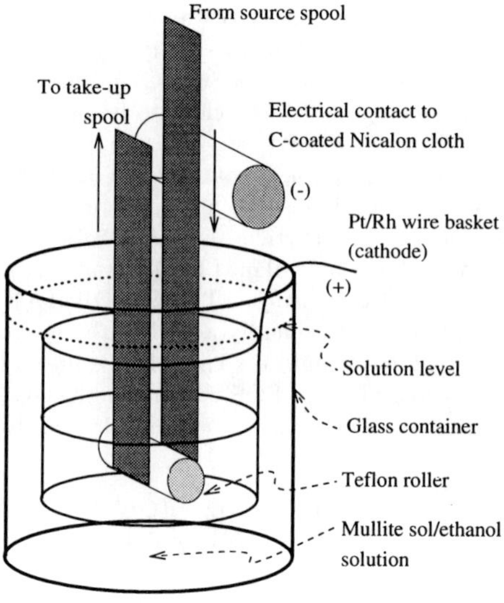

Figure 2: Schematic of the continuous mode EPD coating apparatus.

grains on the order of 20–50nm in diameter. The coating appears to be fairly uniform with many 5nm pores distributed evenly throughout the coating.

The SEM photomicrographs shown in Figures 3 and 4 depict longitudinal and cross–sectional views, respectively, of some $3Al_2O_3{\cdot}2SiO_2$ coatings on Nicalon™ cloth after being subjected to a 900 °C heat treatment in Ar. The coating is on the order of $0.25\mu m$ thick and relatively uniform throughout the cloth.

Conclusions

A method for continuous deposition of multicomponent oxide films by electrophoretic deposition has been developed. Preliminary $3Al_2O_3{\cdot}2SiO_2$ coatings have been deposited successfully using a polymeric mullite sol onto a previously C–coated 2–d Nicalon™ cloth. The sol was found to be homogeneously mixed on the molecular level, due to the lack of any phase separation as seen by XRD during thermal treatment of the sol. The coatings were observed to be fairly uniform throughout the cross–section and along the length of the cloth, and exhibited 5nm–sized pores distributed evenly throughout a 250nm thick coating. After exposure of the coated cloth specimens to 1000 °C in flowing Ar for 1h, the coatings consisted of equiaxed mullite grains on the order of 20–50nm in diameter. Further experiments are planned to resolve the effects of deposition/processing conditions on the resultant coating characteristics.

Figure 3: SEM photomicrograph of a longitudinal view of some representative $3Al_2O_3 \cdot 2SiO_2$ coatings (micron bar represents $50\mu m$).

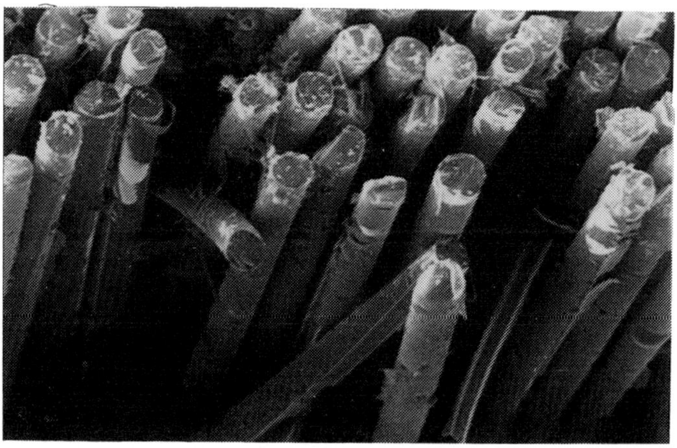

Figure 4: SEM photomicrograph of a cross-sectional view of some representative $3Al_2O_3 \cdot 2SiO_2$ coatings (micron bar represents $50\mu m$).

References

[1] M. Thouless, O. Sbaizero, L. Sigl, and A. Evans, "Effect of interface mechanical properties on pullout in a SiC-fiber-reinforced lithium aluminum silicate glass-ceramic," *Journal of the American Ceramic Society*, vol. 72, pp. 525–532, Apr. 1989.

[2] E. Bischoff, M. Rühle, O. Sbaizero, and A. Evans, "Microstructural studies of the interfacial zone of a SiC-fiber-reinforced lithium aluminum silicate glass-ceramic," *Journal of the American Ceramic Society*, vol. 72, pp. 741–745, May 1989.

[3] W. Kowbel, H. Liu, and H. Tsou, "Fiber-matrix interactions in brittle matrix composites," *Metallurgical Transactions B*, vol. 23A, pp. 1051–1061, 1992.

[4] G. Grathwohl, M. Kuntz, E. Pippel, and J. Woltersdorf, "The real structure of the interlayer between fibre and matrix and its influence on the properties of ceramic composites," *Physica Status Solidi. A: Applied Research*, vol. A146, pp. 393–414, 1994.

[5] T. Mah, N. Hecht, D. McCullum, J. Hoenigman, H. Kim, A. Katz, and A. Lipsitt, "Thermal stability of SiC fibres (Nicalon)," *Journal of Materials Science*, vol. 19, pp. 1191–1201, 1984.

[6] R. Kerans, "Issues in the control of fiber–matrix interface properties in ceramic composites," *Scripta Metallurgica et Materialia*, vol. 31, no. 8, pp. 1079–1084, 1994.

[7] K. Keller, T. Mah, and T. Parthasarathy, "Fugitive interfacial carbon coatings for oxide/oxide composites," in *Ceramic Engineering & Science Proceedings* (J. Wachtman, ed.), vol. 14, (735 Ceramic Place, Westerville, OH, USA, 43081–6136), p. 878, The American Ceramic Society, 1993. Conference held in Cocoa Beach, FL, USA, January 9–14, 1993.

[8] T. Mah, K. Keller, and T. Parthasarathy, "Fugitive interface coating in oxide-oxide composites: A viability study," in *Ceramic Engineering & Science Proceedings* (J. Wachtman, ed.), vol. 12, (735 Ceramic Place, Westerville, OH, USA, 43081-6136), p. 1802, The American Ceramic Society, 1991. Conference held in Cocoa Beach, FL, USA, January 13–16, 1991.

[9] C. Chu and B. Dunn, "Fabrication of $YBa_2Cu_3O_{7-y}$ superconducting coatings by electrophoretic deposition," *Applied Physics Letters*, vol. 55, pp. 492–494, July 1989.

[10] M. Hein, G. M. H. Piel, L. Ponto, M. Becks, U. Klein, and M. Peiniger, "Electrophoretic deposition of textured $YBa_2Cu_3O_{7-x}$ films on silver substrates," *Journal of Applied Physics*, vol. 66, pp. 5940–5943, Dec. 1989.

[11] H. Nojima, H. Shintaku, and M. Nagata, "Fabrication of Ag–doped $Y_1Ba_2Cu_3O_{7-x}$ superconducting films on Cu substrates by electrophoretic deposition," *Japanese Journal of Applied Physics*, vol. 30, pp. L1166–L1168, July 1991.

[12] N. Kouira and H. Shoji, "Preparation of high–t_c superconductors by electrophoretic deposition method," *Physica C*, vol. 190, pp. 163–164, 1991.

[13] N. Minami, N. Koura, and H. Shoji, "Preparation of $YBa_2Cu_3O_{7-\delta}$ superconductor coating on austenitic steel by electrophoretic deposition method—study on buffering layers," *Japanese Journal of Applied Physics*, vol. 31, pp. L784–L786, June 1992.

[14] N. Koura and H. Shoji, "Bscco superconductor coating by the electrophoretic deposition method," *Physica C*, vol. 200, pp. 50–54, 1992.

[15] C. Kao, H. Tang, Y. Shiue, S. Sheen, and M. Wu, "Fabrication of $Bi_{1.7}Pb_{0.4}Sr_{1.6}Ca_{2.4}Cu_{3.6}O_y$ superconducting films on silver substrates by the electrophoretic deposition technique," *Journal of Materials Science Letters*, vol. 12, pp. 1793–1794, 1993.

[16] S. Sugiyama, A. Takagi, and K. Tsuzuki, "(Pb, La)(Zr, Ti)O_3 films by multiple electrophoretic deposition/sintering process," *Japanese Journal of Applied Physics*, vol. 30, pp. 2170–2173, Sept. 1991.

[17] S. Haber and L. Gal-Or, "Deep electrophoretic penetration and deposition of ceramic particles inside porous substrates. i. analytical model," *Journal of the Electrochemical Society*, vol. 139, pp. 1071–1077, Apr. 1992.

[18] L. Gal-Or, S. Liubovich, and S. Haber, "Deep electrophoretic penetration and deposition of ceramic particles inside porous substrates. ii. experimental model," *Journal of the Electrochemical Society*, vol. 139, pp. 1078–1081, Apr. 1992.

[19] P. Sarkar, X. Huang, and P. Nicholson, "Structural ceramic microlaminates by electrophoretic deposition," *Journal of the American Ceramic Society*, vol. 75, pp. 2907–2909, Oct. 1992.

[20] P. Nicholson, P. Sarkar, and X. Huang, "Electrophoretic deposition and its use to synthesize ZrO_2/Al_2O_3 micro–laminate ceramic/ceramic composites," *Journal of Materials Science*, vol. 29, pp. 6274–6278, 1993.

[21] M. Nagai, K. Yamashita, T. Umegaki, and Y. Takuma, "Electrophoretic deposition of ferroelectric barium titanate thick films and their dielectric properties," *Journal of the American Ceramic Society*, vol. 76, pp. 253–255, Jan. 1993.

[22] H. Schneider, L. Merwin, and A. Sebald, "Mullite formation from non–crystalline precursors," *Journal of Materials Science*, vol. 27, pp. 805–812, 1992.

[23] D. Jeng and M. Rahaman, "Sintering and crystallization of mullite powder prepared by sol–gel processing," *Journal of Materials Science*, vol. 28, pp. 4904–4909, 1993.

[24] S. Pradhan, A. Chatterjee, A. Datta, M. De, and D. Chakravorty, "Low temperature mullite formation from sol–gel precursors by hot pressing," *Journal of Materials Research*, vol. 9, pp. 2474–2475, Oct. 1994.

[25] Y. Zhou, J. Vleugels, T. Laoui, and O. van der Biest, "Toughening of sol–gel derived mullite matrix by Al_2O_3 platelets," *Journal of Materials Science Letters*, vol. 13, pp. 1089–1091, 1994.

[26] M. Ge, H. Yang, Y. Wang, and F. Zhang, "Ultrafine pure mullite powder prepared by sol–gel method," *Journal of Non–Crystalline Solids*, vol. 147&148, pp. 565–568, 1992.

[27] J. Huling and G. Messing, "Chemistry–crystallization relations in molecular mullite sols," *Journal of Non–Crystalline Solids*, vol. 147&148, pp. 213–221, 1992.

[28] D. Li and W. Thomson, "Kinetic mechanisms for mullite formation from sol–gel precursors," *Journal of Materials Research*, vol. 5, pp. 1963–1969, Sept. 1990.

[29] Y. Wang, D. Li, and W. Thomson, "The influence of steam on mullite formation from sol–gel precursors," *Journal of Materials Research*, vol. 8, pp. 195–205, Jan. 1993.

[30] T. Heinrich, F. Raether, W. Tappert, and J. Fricke, "Synthesis and characterization of mullite aerogels," *Journal of Non–Crystalline Solids*, vol. 145, pp. 55–59, 1992.

[31] T. Heinrich, F. Raether, and H. Marsmann, "Growth and structure of single phase mullite gels from chelated aluminum alkoxides and alkoxysilanes," *Journal of Non–Crystalline Solids*, vol. 168, pp. 14–22, 1994.

[32] A. Taylor and D. Holland, "The chemical synthesis and crystallisation sequence of mullite," *Journal of Non–Crystalline Solids*, vol. 152, pp. 1–17, 1993.

[33] I. Low, R. Skala, and D. Perera, "Fracture properties of layered mullite/zirconia–toughened alumina composites," *Journal of Materials Science Letters*, vol. 13, pp. 1334–1336, 1994.

[34] J. Lee and S. Yu, "Characteristics of mullite prepared from co–precipitated $3Al_2O_3{\cdot}2SiO_2$ powders," *Journal of Materials Science*, vol. 27, pp. 5203–5208, 1992.

[35] J. Lee and S. Yu, "Mullite formation kinetics of coprecipitated Al_2O_3–SiO_2 gels," *Materials Research Bulletin*, vol. 27, no. 4, pp. 405–416, 1992.

[36] S. Hyun and S. Song, "Low–temperature synthesis of mullite powders by the emulsion route," *Journal of Materials Science Letters*, vol. 13, pp. 177–179, 1994.

[37] B. Kanka and H. Schneider, "Sintering mechanisms and microstructural development of coprecipitated mullite," *Journal of Materials Science*, vol. 29, pp. 1239–1249, 1994.

[38] C. Hong, P. Ravindranathan, D. Agrawal, and R. Roy, "Synthesis and sintering of mullite powders by the decomposition/combustion of aluminum nitrate–amorphous fumed silica–urea mixtures," *Journal of Materials Science Letters*, vol. 13, pp. 1072–1075, 1994.

[39] J. Wu, M. Chen, F. Jones, and P. James, "Mullite and alumina–silica matrices for composites by modified sol–gel processing," *Journal of Non–Crystalline Solids*, vol. 162, pp. 197–200, 1993.

Keyword and Author Index